NUCLEAR MATERIALS AND DISASTER RESEARCH

UNITED STATES NUCLEAR WASTE

CRITICAL MANAGEMENT STRATEGIES AND FUTURE NEEDS

Nuclear Materials and Disaster Research

Additional books in this series can be found on Nova's website under the Series tab.

Additional E-books in this series can be found on Nova's website under the E-book tab.

Energy Science, Engineering and Technology

Additional books in this series can be found on Nova's website under the Series tab.

Additional E-books in this series can be found on Nova's website under the E-book tab.

NUCLEAR MATERIALS AND DISASTER RESEARCH

UNITED STATES NUCLEAR WASTE

CRITICAL MANAGEMENT STRATEGIES AND FUTURE NEEDS

GREGORY HALL

AND

STEVEN L. GREENE

EDITORS

Nova Science Publishers, Inc.

New York

Library of Congress Cataloging-in-Publication Data

ISBN: 978-1-62081-312-6

Published by Nova Science Publishers, Inc. + New York

CONTENTS

PREFACE

The Blue Ribbon Commission on America's Nuclear Future was chartered to recommend a new strategy for managing the back end of the nuclear fuel cycle. The Nation's failure to come to grips with the nuclear waste issue has already proved damaging and costly and it will be more damaging and more costly the longer it continues: damaging to prospects for maintaining a potentially important energy supply option for the future, damaging to state-federal relations and public confidence in the federal government's competence, and damaging to America's standing in the world - not only as a source of nuclear technology and policy expertise but as a leader on global issues of nuclear safety, non-proliferation, and security. This book examines the use of nuclear energy as a low-carbon energy resource with a focus on the management of the nuclear fuel cycle, based on emerging technologies and developments.

Chapter 1 - The Obama Administration's decision to halt work on a repository at Yucca Mountain in Nevada is but the latest indicator of a policy that has been troubled for decades and has now all but completely broken down. The approach laid out under the 1987 Amendments to the Nuclear Waste Policy Act (NWPA)—which tied the entire U.S. high-level waste management program to the fate of the Yucca Mountain site—has not worked to produce a timely solution for dealing with the nation's most hazardous radioactive materials. The United States has traveled nearly 25 years down the current path only to come to a point where continuing to rely on the same approach seems destined to bring further controversy, litigation, and protracted delay.

Chapter 2 - Chairman Shimkus, Ranking Member Green, members of the Subcommittee, it is a pleasure to appear before you today to discuss the final recommendations of the Blue Ribbon Commission on America's Nuclear Future. We appreciate the leadership this Subcommittee has shown in confronting some of our nation's biggest challenges, which certainly include the focus of this hearing - managing spent nuclear fuel and high level nuclear waste in the United States. Thank you for allowing us the opportunity to testify before you today.

Before we begin, we would also like to thank the 13 other members of the Commission who worked so hard in creating our final report. As the Co-Chairmen of the Commission, we were delighted to work with such a talented and dedicated group of fellow Commissioners. We are thankful for the expertise and insights they brought to our endeavors. Their professionalism led to our final report having unanimous approval; all of the Commissioners have agreed to our final report, a fact which we believe speaks to the strength of our recommendations.

As you aware, the Blue Ribbon Commission was formed by the Secretary of Energy at the direction of the President. Our charge was to conduct a comprehensive review of policies for managing the back end of the nuclear fuel cycle and to recommend a new strategy. We came away from our review frustrated by decades of unmet commitments to the American people, yet confident that we can turn this record around.

Chapter 3 - Thank you, Chairman Shimkus and Ranking Member Green, for the opportunity to provide my personal views on the Blue Ribbon Commission (BRC) Report on America's Nuclear Future recommendations. I have spent several decades of my professional life working to implement the Nuclear Waste Policy Act that established, by law; a scientific, regulatory, and administrative process to safely dispose of our Nation's spent nuclear fuel and high level radioactive wastes in an environmentally protective manner. Unfortunately this Administration has succumbed to politics and has reversed much of the progress that has been made over the past 25 years. Although, this Commission was created partially as a cover for the Administration's actions, it has produced a very thoughtful report with excellent useful recommendations that should be incorporated into our statutes and programs.

It is not the fault of the BRC that it was not allowed to examine the obvious and lawful Yucca Mountain high level radioactive waste deep geologic repository disposal solution concluded under the Nuclear Waste Policy Act process. Despite such a politically imposed shortcoming, this highly qualified and dedicated group produced a thoughtful and well-reasoned report with excellent recommendations. I strongly urge the Administration and Congress to act promptly to incorporate the BRC recommendations into our remaining national program to get our nation's high level nuclear waste program moving forward again.

Chapter 4 - I welcome the opportunity to provide my views in this Hearing regarding the Final Report to the Secretary of Energy by the Blue Ribbon Commission on America's Nuclear Future (BRC). Thank you, Chairman Shimkus and Ranking Member Green.

I provide these comments as a private citizen concerned about America's nuclear future, based on my experience. I served from 1989 to 1994 on the Nuclear Waste Technical Review Board (NWTRB), and then for the NAS as a member of the Board on Radioactive Waste Management and as chair of the NAS committee that produced the 2001 report referenced in Endnotes 48 and 125 in the BRC Report, and as a member of the NAS committee that produced the 2008 report referenced in Endnote 136. I have a background in physics, but my career has been in decision analysis and risk analysis. I have served as president of the Society for Risk Analysis, and I am a national associate of the National Research Council. (This honor enables me to use the Members' Library in the National Academy of Sciences Building.)[1]

I strongly endorse the BRC Final Report and its recommendations. The Commission and its staff have produced an excellent document, well researched and well-written, at a level appropriate for a broad audience.

Chapter 5 - Mr. Chairman and members of the Subcommittee, I appreciate the opportunity to provide testimony today regarding the recommendations of the Blue Ribbon Commission on America's Nuclear Future (BRC), released late last week. My name is Martin G. Malsch. I am a partner in the law firm Egan, Fitzpatrick, Malsch & Lawrence, PLLC. I have practiced law in the nuclear energy and nuclear waste fields for over forty years, including many years as the Deputy General Counsel or Acting General Counsel for the U.S.

Nuclear Regulatory Commission. In private practice I have advised both public and private entities regarding nuclear issues.

I believe I am one of the few practitioners in the nuclear energy field who has represented both proponents and opponents of nuclear facilities. I currently represent the State of Nevada on matters related to the Yucca Mountain repository, but I am not testifying today in a representative capacity. My testimony today presents my personal opinions regarding the BRC's recommendations based on my experience and expertise in the nuclear energy and nuclear waste fields; it does not necessarily represent the views of Nevada or anyone else.

Chapter 6 - Good morning. On behalf of the Union of Concerned Scientists, I would like to thank Chairman Shimkus, Ranking Member Green, and the other distinguished members of the Subcommittee for the opportunity to provide our views on the recommendations of the Blue Ribbon Commission on America's Nuclear Future (BRC).

The Union of Concerned Scientists (UCS) is neither pro nor anti-nuclear power, but has served as a nuclear power safety and security watchdog for over 40 years. UCS is also deeply concerned about global climate change and has not ruled out an expansion of nuclear power as an option to help reduce greenhouse gas emissions—provided that it is affordable relative to other low-carbon options and that it meets high standards of safety and security. However, the Fukushima Daiichi crisis has revealed significant vulnerabilities in nuclear safety and has shaken public confidence in nuclear power. Regulators around the world must seriously address these vulnerabilities in order to reduce the risk of another Fukushima in the future. Otherwise, the viability of nuclear power as a reliable electricity option will be in doubt.

Chapter 7 My name is Thomas A. Schatz and I am president of Citizens Against Government Waste (CAGW). CAGW was founded in 1984 by the late industrialist J. Peter Grace and nationally-syndicated columnist Jack Anderson to build support for implementation of President Ronald Reagan's Grace Commission recommendations and other waste cutting proposals. Since its inception, CAGW has been at the forefront of the fight for efficiency, economy, and accountability in government. CAGW has more than one million members and supporters nationwide, and, over the past 28 years, has helped save taxpayers $1.2 trillion through the implementation of Grace Commission findings and other recommendations.

CAGW's mission reflects the interests of taxpayers. All citizens benefit when government programs work cost-effectively, when deficit spending is reduced and government is held accountable. Not only will representative government benefit from the pursuit of these interests, but the country will prosper economically because government mismanagement, fiscal profligacy, and chronic deficits soak up private savings and crowd out the private investment necessary for long-term growth.

Chapter 8 - Good Morning, Chairman Shimkus, Ranking Member Green, and Subcommittee Members. Thank you for the opportunity to appear before you today.

My name is David Wright. I am a commissioner with the South Carolina Public Service Commission and I serve as president of the National Association of Regulatory Utility Commissioners (NARUC), on whose behalf I am speaking this morning. I appreciate the opportunity to present NARUC's views on the subject of disposition of spent or used nuclear fuel from commercial nuclear power plants.

NARUC is a quasi-governmental, non-profit organization founded in 1889. Our membership includes the public utility commissions serving all States and territories. NARUC's mission is to serve the public interest by improving the quality and effectiveness of public utility regulation. Our members regulate the retail rates and services of electric, gas, water, and telephone utilities. We are obligated under the laws of our respective States to assure the establishment and maintenance of such utility services as may be required by the public convenience and necessity and to assure that such services are provided under rates and subject to terms and conditions of service that are just, reasonable, and non-discriminatory.

In: United States Nuclear Waste
Editors: Gregory Hall and Steven L. Greene
ISBN: 978-1-62081-312-6

Chapter 1

BLUE RIBBON COMMISSION ON AMERICA'S NUCLEAR FUTURE: REPORT TO THE SECRETARY OF ENERGY*

Blue Ribbon Commission

PREAMBLE

The Blue Ribbon Commission on America's Nuclear Future (BRC) was formed by the Secretary of Energy at the request of the President to conduct a comprehensive review of policies for managing the back end of the nuclear fuel cycle and recommend a new strategy. It was co-chaired by Rep. Lee H. Hamilton and Gen. Brent Scowcroft.

Other Commissioners are Mr. Mark H. Ayers, the Hon. Vicky A. Bailey, Dr. Albert Carnesale, Sen. Pete Domenici, Ms. Susan Eisenhower, Sen. Chuck Hagel, Mr. Jonathan Lash, Dr. Allison M. Macfarlane, Dr. Richard A. Meserve, Dr. Ernest J. Moniz, Dr. Per Peterson, Mr. John Rowe, and Rep. Phil Sharp.

The Commission and its subcommittees met more than two dozen times between March 2010 and January 2012 to hear testimony from experts and stakeholders, to visit nuclear waste management facilities in the United States and abroad, and to discuss the issues identified in its Charter.

Additionally, in September and October 2011, the Commission held five public meetings, in different regions of the country, to hear feedback on its draft report. A wide variety of organizations, interest groups, and individuals provided input to the Commission at these meetings and through the submission of written materials.

Copies of all of these submissions, along with records and transcripts of past meetings, are available at the BRC website (www.brc.gov).

This report highlights the Commission's findings and conclusions and presents recommendations for consideration by the Administration and Congress, as well as interested state, tribal and local governments, other stakeholders, and the public.

---

* This is an edited, reformatted and augmented version of a Blue Ribbon Commission publication, dated January 2012.

The Honorable Dr. Steven Chu Secretary of Energy
U.S. Department of Energy
1000 Independence Ave., SW Washington, DC 20585

Dear Secretary Chu:

At the direction of the President, you charged the Blue Ribbon Commission on America's Nuclear Future with reviewing policies for managing the back end of the nuclear fuel cycle and recommending a new plan. We thank you for choosing us to serve on the Commission.

We approached our task from different perspectives but with a shared sense of urgency. Put simply, this nation's failure to come to grips with the nuclear waste issue has already proved damaging and costly. It will be even more damaging and more costly the longer it continues: damaging to prospects for maintaining a potentially important energy supply option for the future, damaging to state–federal relations and public confidence in the federal government's competence, and damaging to America's standing in the world—not only as a source of nuclear technology and policy expertise but as a leader on global issues of nuclear safety, non-proliferation, and security.

We have sought to ensure that our review is comprehensive, open and inclusive.

Our Commission has heard from thousands of individuals and organizations on a wide range of issues through formal hearings, site visits, and written letters and comments submitted through the Commission web site. We have visited several communities across the country that have a keen interest in the matters before the Commission and have also visited a number of other countries to gain insights as to how the United States might proceed. We are indebted to the many people who have offered us their expertise, advice and guidance.

Attached for your consideration is the final report of our Commission. Our report includes recommendations covering topics such as the approach to siting future nuclear waste management facilities, the transport and storage of spent fuel and high-level waste, options for waste disposal, institutional arrangements for managing spent nuclear fuel and high-level wastes, reactor and fuel cycle technologies, and international considerations. We also make recommendations regarding critical changes needed in the handling of nuclear waste fees and of the Nuclear Waste Fund. The majority of these recommendations require action to be taken by the Administration and Congress, and offer what we believe is the best chance of success going forward, based on previous nuclear waste management experience in the U.S. and abroad.

We urge that you promptly designate a senior official with sufficient authority to coordinate all of the DOE elements involved in the implementation of the Commission's recommendations.

You directed that the Commission was not to serve as a siting body. Accordingly, we have not evaluated Yucca Mountain or any other location as a potential site for the storage of spent nuclear fuel or disposal of high level waste, nor have we taken a position on the Administration's request to withdraw the Yucca Mountain license application. What we have endeavored to do is recommend a sound waste management approach that can lead to the resolution of the current impasse; an approach that neither includes nor excludes Yucca Mountain as an option for a repository and can and should be applied regardless of what site or sites are ultimately chosen to serve as the permanent disposal facility for America's spent nuclear fuel and other high-level nuclear wastes.

We are committed to seeing action taken on our recommendations because we believe it is long past time for the government to make good on its commitments to the American people to provide for the safe disposal of nuclear waste.

This generation has an obligation to avoid burdening future generations with finding a safe permanent solution for nuclear wastes they had no part in creating, while also preserving their energy options. To that end we commit ourselves to provide whatever assistance you deem necessary as you consider how to act on the final recommendations of our Commission. Please do not hesitate to call on us at any time.

Respectfully submitted,

Lee H. Hamilton, Co-Chairman
Brent Scowcroft, Co-Chairman
Mark H. Ayers
Vicky A. Bailey
Albert Carnesale
Pete V. Domenici
Susan Eisenhower
Chuck Hagel
Jonathan Lash
Allison M. Macfarlane
Richard A. Meserve
Ernest J. Moniz
Per F. Peterson
John W. Rowe
Phil Sharp

Executive Summary

America's Nuclear Waste Management Program an Impasse

The Obama Administration's decision to halt work on a repository at Yucca Mountain in Nevada is but the latest indicator of a policy that has been troubled for decades and has now all but completely broken down. The approach laid out under the 1987 Amendments to the Nuclear Waste Policy Act (NWPA)—which tied the entire U.S. high-level waste management program to the fate of the Yucca Mountain site—has not worked to produce a timely solution for dealing with the nation's most hazardous radioactive materials. The United States has traveled nearly 25 years down the current path only to come to a point where continuing to rely on the same approach seems destined to bring further controversy, litigation, and protracted delay.

The Blue Ribbon Commission on America's Nuclear Future (the Commission) was chartered to recommend a new strategy for managing the back end of the nuclear fuel cycle. We approached this task from different perspectives but with a shared sense of urgency. Put simply, this nation's failure to come to grips with the nuclear waste issue has already proved damaging and costly and it will be more damaging and more costly the longer it continues: damaging to prospects for maintaining a potentially important energy supply option for the future, damaging to state–federal relations and public confidence in the federal government's competence, and damaging to America's standing in the world—not only as a source of nuclear technology and policy expertise but as a leader on global issues of nuclear safety, non-proliferation, and security. Continued stalemate is also costly—to utility ratepayers, to communities that have become unwilling hosts of long-term nuclear waste storage facilities, and to U.S. taxpayers who face mounting liabilities, already running into billions of dollars, as a result of the failure by both the executive and legislative branches to meet federal waste management commitments.

The need for a new strategy is urgent, not just to address these damages and costs but because this generation has a fundamental ethical obligation to avoid burdening future generations with the entire task of finding a safe permanent solution for managing hazardous nuclear materials they had no part in creating.

At the same time, we owe it to future generations to avoid foreclosing options wherever possible so that they can make choices—about the use of nuclear energy as a low-carbon energy resource and about the management of the nuclear fuel cycle—based on emerging technologies and developments and their own best interests.

Almost exactly one year after the Commission was chartered and less than five months before our initial draft report was due, an unforeseen event added yet more urgency to our charge and brought the problem of nuclear waste into the public eye as never before.

A massive earthquake off the northeastern coast of Japan and the devastating tsunami that followed set off a chain of problems at the Fukushima Daiichi nuclear power station that eventually led to the worst nuclear accident since Chernobyl.

In the weeks of intense media coverage that followed, many Americans became newly aware of the presence of tens of thousands of tons of spent fuel at more than 70 nuclear power plant sites around this country—and of the fact that the United States currently has no physical capacity to do anything with this spent fuel other than to continue to leave it at the sites where it was first generated.[1]

The strategy we recommend in this report has eight key elements:

1. A new, consent-based approach to siting future nuclear waste management facilities.
2. A new organization dedicated solely to implementing the waste management program and empowered with the authority and resources to succeed.
3. Access to the funds nuclear utility ratepayers are providing for the purpose of nuclear waste management.
4. Prompt efforts to develop one or more geologic disposal facilities.
5. Prompt efforts to develop one or more consolidated storage facilities.[2]
6. Prompt efforts to prepare for the eventual large-scale transport of spent nuclear fuel and high-level waste to consolidated storage and disposal facilities when such facilities become available.
7. Support for continued U.S. innovation in nuclear energy technology and for workforce development.
8. Active U.S. leadership in international efforts to address safety, waste management, non-proliferation, and security concerns.

The elements of this strategy will not be new to those who have followed the U.S. nuclear waste program over the years. All of them are necessary to establish a truly integrated national nuclear waste management system, to create the institutional leadership and wherewithal to get the job done, and to ensure that the United States remains at the forefront of technology developments and international responses to evolving nuclear safety, non-proliferation, and security concerns.

A few general points about the Commission's proposed strategy are worth emphasizing before we discuss each of the above elements in greater detail. First is the issue of cost.

In this time of acute concern about the federal budget deficit and high energy prices, we have been sensitive to the concern that our recommendations—particularly those that involve launching a new approach and a new organization for nuclear waste management—could add to the financial burden on the U.S. Treasury and on American taxpayers and utility ratepayers.[3]

Certainly it will cost something to implement a successful U.S. waste management program; however, trying to implement a deeply flawed program is even more costly, for all the reasons already mentioned. In fact, U.S. ratepayers are *already* paying for waste disposal (through a fee collected on each kilowatt-hour of nuclear-generated electricity)—but the program they're paying for isn't working.

Taxpayers are paying too—in the form of damage payments from the taxpayer-funded Judgment Fund to compensate utilities for the federal government's failure to meet its contractual waste acceptance commitments.

Overall, we are confident that our waste management recommendations can be implemented using revenue streams *already dedicated for this purpose* (in particular the Nuclear Waste Fund and fee). Other Commission recommendations—particularly those concerning nuclear technology programs and international policies—are broadly consistent with the program plans of the relevant agencies.

Another overarching point concerns timing and implementation. All of our recommendations are interconnected and will take time to implement fully, particularly since many elements of the strategy we propose require legislative action to amend the NWPA and other relevant laws (see text box).

Nevertheless, prompt action can and should be taken in several areas, without waiting for legislative action, to get the waste management program back on track.

The last section of this report (section 13) identifies a number of concrete next steps; in addition, the text box on page ix of this Executive Summary lists several ways to get started on the specific task of siting new waste disposal and consolidated storage facilities.

Finally, there are several questions the Commission was not chartered to address. We have not:

- Rendered an opinion on the suitability of the Yucca Mountain site or on the request to withdraw the license application for Yucca Mountain. Instead, we focused on developing a sound strategy for future storage and disposal facilities and operations that we believe *can and should be implemented regardless of what happens with Yucca Mountain*.
- Proposed any specific site (or sites) for any component of the waste management system.

- Offered a judgment about the appropriate role of nuclear power in the nation's (or the world's) future energy supply mix.
- These are all important questions that will engage policy-makers and the public in the years ahead. However, none of them alters the urgent need to change and improve our strategy for managing the high-level wastes and spent fuel that already exist and will continue to accumulate so long as nuclear reactors operate in this country. That is the focus of the Commission's work and of the specific recommendations that follow.

Proposed Legislative Changes

Fully implementing the Commission's recommendations will require several changes to the Nuclear Waste Policy Act or other legislation:

Establishing a new facility siting process – The NWPA, as amended in 1987, now provides only for the evaluation and licensing of a single repository site at Yucca Mountain, Nevada. The Act should be amended to authorize a new consent-based process to be used for selecting and evaluating sites and licensing consolidated storage and disposal facilities in the future, similar to the process established in the expired Nuclear Waste Negotiator provisions of the Act (but under new organizational leadership, as described below).

Authorizing consolidated interim storage facilities – The NWPA allows the government to construct one consolidated storage facility with limited capacity, but only after construction of a nuclear waste repository has been licensed. One or more consolidated storage facilities should be established, independent of the schedule for opening a repository.

The Act should be modified to allow for a consent-based process to site, license, and construct multiple storage facilities with adequate capacity when needed and to clarify that nuclear waste fee payments can be used for this purpose.

Broadening support to jurisdictions affected by transportation – The NWPA provides funding and technical assistance for training public safety officials to states and tribes whose jurisdictions would be traversed by shipments of spent fuel to a storage or disposal facility.

The Act should be amended to give the waste management organization the broader authorities given to DOE in the WIPP Land Withdrawal Act that supported the successful large-scale transport of transuranic waste to WIPP (including a public information program, support for the acquisition of equipment to respond to transportation incidents, and broad assistance for other waste-related transportation safety programs).

Establishing a new waste management organization – Responsibility for implementing the nation's program for managing spent nuclear fuel and high-level radioactive wastes is currently assigned to the U.S. Department of Energy.

Legislation will be needed to (1) move this responsibility to a new, independent, government-chartered corporation focused solely on carrying out that program and (2) establish the appropriate oversight mechanisms.

Ensuring access to dedicated funding – Current federal budget rules and laws make it impossible for the nuclear waste program to have assured access to the fees being collected from nuclear utilities and ratepayers to finance the commercial share of the waste program's expenses. We have recommended a partial remedy that should be implemented promptly by the Administration, working with the relevant congressional committees and the Congressional Budget Office. A long-term remedy requires legislation to provide access to the Nuclear Waste Fund and fees independent of the annual appropriations process but subject to rigorous independent financial and managerial oversight.

Promoting international engagement to support safe and secure waste management – Congress may need to provide policy direction and new legislation to implement some measures aimed at helping other countries manage radioactive wastes in a safe, secure, and proliferation-resistant manner, similar to the expired NWPA provisions for technical assistance to non-nuclear weapons states in the area of spent nuclear fuel storage and disposal.

1. A New Consent-Based Approach to Siting

Siting storage or disposal facilities has been the most consistent and most intractable challenge for the U.S. nuclear waste management program. Of course, the first requirement in siting any facility centers on the ability to demonstrate adequate protection of public health and safety and the environment. Beyond this threshold criterion, finding sites where all affected units of government, including the host state or tribe, regional and local authorities, and the host community, are willing to support or at least accept a facility has proved exceptionally difficult. The erosion of trust in the federal government's nuclear waste management program has only made this challenge more difficult. And whenever one or more units of government are opposed, the odds of success drop greatly. The crux of the challenge derives from a federal/state/tribal/local rights dilemma that is far from unique to the nuclear waste issue—no simple formula exists for resolving it. Experience in the United States and in other nations suggests that any attempt to force a top-down, federally mandated solution over the objections of a state or community—far from being more efficient—will take longer, cost more, and have lower odds of ultimate success.

By contrast, the approach we recommend is explicitly adaptive, staged, and consent-based. Based on a review of successful siting processes in the United States and abroad—including most notably the siting of a disposal facility for transuranic radioactive waste, the Waste Isolation Pilot Plant (WIPP) in New Mexico, and recent positive outcomes in Finland, France, Spain and Sweden—we believe this type of approach can provide the flexibility and sustain the public trust and confidence needed to see controversial facilities through to completion.

In practical terms, this means encouraging communities to volunteer to be considered to host a new nuclear waste management facility while also allowing for the waste management organization to approach communities that it believes can meet the siting requirements. Siting processes for waste management facilities should include a flexible and substantial incentive program.

The approach we recommend also recognizes that successful siting decisions are most likely to result from a complex and perhaps extended set of negotiations between the implementing organization and potentially affected state, tribal, and local governments, and other entities. It would be desirable for these negotiations to result in a partnership agreement or some other form of legally enforceable agreement with the organization to ensure that commitments to and by host states, tribes, and communities are upheld. All affected levels of government must have, at a minimum, a meaningful consultative role in important decisions; additionally, both host states and tribes should retain—or where appropriate, be delegated—direct authority over aspects of regulation, permitting, and operations where oversight below the federal level can be exercised effectively and in a way that is helpful in protecting the interests and gaining the confidence of affected communities and citizens. At the same time, host state, tribal and local governments have responsibilities to work productively with the federal government to help advance the national interest.

In this context, any process that is prescribed in detail up front is unlikely to work. Transparency, flexibility, patience, responsiveness, and a heavy emphasis on consultation and cooperation will all be necessary—indeed, these are attributes that should apply not just to siting but to every aspect of program implementation.

This discussion raises another issue highlighted in numerous comments to the BRC: the question of how to define "consent." The Commission takes the view that this question ultimately has to be answered by a potential host jurisdiction, using whatever means and timing it sees fit.

Siting New Nuclear Waste Management Facilities – Getting Started

First, the Environmental Protection Agency and the Nuclear Regulatory Commission should develop a generic disposal standard and supporting regulatory requirements early in the siting process. Generally-applicable regulations are more likely to earn public confidence than site-specific standards. In addition, having a generic standard will support the efficient consideration and examination of multiple sites.

Once the new waste management organization is established it should:

- Develop a set of basic initial siting criteria – These criteria will ensure that time is not wasted investigating sites that are clearly unsuitable or inappropriate.
- Encourage expressions of interest from a large variety of communities that have potentially suitable sites – As these communities become engaged in the process, the implementing organization must be flexible enough not to force the issue of consent while also being fully prepared to take advantage of promising opportunities when they arise.
- Establish initial program milestones – Milestones should be laid out in a mission plan to allow for review by Congress, the Administration, and stakeholders, and to provide verifiable indicators for oversight of the organization's performance.

We believe a good gauge of consent would be the willingness of affected units of government – the host states, tribes, and local communities – to enter into legally binding agreements with the facility operator, where these agreements enable states, tribes, and communities to have confidence that they can protect the interests of their citizens.

All siting processes take time; however, an adaptive, staged approach may seem particularly slow and open-ended. This will be frustrating to stakeholders and to members of the public who are understandably anxious to know when they can expect to see results. The Commission shares this frustration—greater certainty and a quicker resolution would have been our preference also. Experience, however, leads us to conclude that there is no short-cut, and that any attempt to short-circuit the process will most likely lead to more delay. That said, we also believe that attention to process must not come at the expense of progress and we are sympathetic to the numerous comments we received asking us to include a more detailed and specific set of milestones in our final report. Obviously there is an inherent tension between recommending an adaptive, consent-based process and setting out deadlines or progress requirements in advance. But we agree that it will be important—without imposing inflexible deadlines—to set reasonable performance goals and milestones for major phases of program development and implementation so that Congress can hold the waste management organization accountable and so that stakeholders and the public can have confidence the program is moving forward. Other countries have taken this approach, in several cases identifying target timeframes, rather than specific dates for completing stages in their process. For example the implementing organization might consider a range of, say, 15 to 20 years to accomplish site identification and characterization and to conduct the licensing process for a geologic repository. A notional timeframe for siting and developing a consolidated storage facility would presumably be shorter, perhaps on the order of 5 to 10 years.

2. A New Organization to Implement the Waste Management Program

The U.S. Department of Energy (DOE) and its predecessor agencies have had primary responsibility for implementing U.S. nuclear waste policy for more than 50 years. In that time, DOE has achieved some notable successes, as shown by the WIPP experience and recent improvements in waste cleanup performance at several DOE sites. The overall record of DOE and of the federal government as a whole, however, has not inspired widespread confidence or trust in our nation's nuclear waste management program. For this and other reasons, the Commission concludes that a new, single-purpose organization is needed to provide the stability, focus, and credibility that are essential to get the waste program back on track. We believe a congressionally chartered federal corporation offers the best model, but whatever the specific form of the new organization it must possess the attributes, independence, and resources to effectively carry out its mission.

The central task of the new organization would be to site, license, build, and operate facilities for the safe consolidated storage and final disposal of spent fuel and high-level nuclear waste at a reasonable cost and within a reasonable timeframe.

In addition, the new organization would be responsible for arranging for the safe transport of waste and spent fuel to or between storage and disposal facilities, and for undertaking applied research, development, and demonstration (RD&D) activities directly relevant to its waste management mission (e.g., testing the long-term performance of fuel in dry casks and during subsequent transportation).

For the new organization to succeed, a substantial degree of implementing authority and assured access to funds must be paired with rigorous financial, technical, and regulatory oversight by Congress and the appropriate government agencies.

We recommend that the organization be directed by a board nominated by the President, confirmed by the Senate, and selected to represent a range of expertise and perspectives. Independent scientific and technical oversight of the nuclear waste management program is essential and should continue to be provided for out of nuclear waste fee payments.

In addition, the presence of clearly independent, competent regulators is essential; we recommend the existing roles of the U.S. Environmental Protection Agency in establishing standards and the Nuclear Regulatory Commission (NRC) in licensing and regulating waste management facilities be preserved but that steps be taken to ensure ongoing cooperation and coordination between these agencies.

Late in our review we heard from several states that host DOE defense waste that they agree with the proposal to establish a new organization to manage civilian wastes, but believe the government can more effectively meet its commitments if responsibility for defense waste disposal remains with DOE.

Others argued strongly that the current U.S. policy of commingling defense and civilian wastes should be retained. We are not in a position to comprehensively assess the implications of any actions that might affect DOE's compliance with its cleanup agreements, and we did not have the time or the resources necessary to thoroughly evaluate the many factors that must be considered by the Administration and Congress in making such a determination.[4]

The Commission therefore urges the Administration to launch an immediate review of the implications of leaving responsibility for disposal of defense waste and other DOE-owned waste with DOE versus moving it to a new waste management organization.

The implementation of other Commission recommendations, however, should not wait for the commingling issue to be resolved. Congressional and Administration efforts to implement our recommendations can and should proceed as expeditiously as possible.

3. Access to Utility Waste Disposal Fees for Their Intended Purpose

The 1982 NWPA created a "polluter pays" funding mechanism to ensure that the full costs of disposing of commercial spent fuel would be paid by utilities (and their ratepayers), with no impact on taxpayers or the federal budget. Nuclear utilities are assessed a fee on every kilowatt-hour of nuclear-generated electricity as a *quid pro quo* payment in exchange for the federal government's contractual commitment to begin accepting commercial spent fuel by January 31, 1998. Fee revenues go to the government's Nuclear Waste Fund, which was established for the sole purpose of covering the cost of disposing of civilian nuclear waste and ensuring that the waste program would not have to compete with other funding priorities. In contrast, costs for disposing of defense nuclear wastes are paid by taxpayers through appropriations from the Treasury.

The Fund does not work as intended. A series of executive branch and congressional actions has made annual fee revenues (approximately $750 million per year) and the unspent $27 billion balance in the Fund effectively inaccessible to the waste program. Instead, the waste program must compete for federal funding each year and is therefore subject to exactly the budget constraints and uncertainties that the Fund was created to avoid. This situation must be remedied to allow the program to succeed.

In the near term, the Administration should offer to amend DOE's standard contract with nuclear utilities so that utilities remit only the portion of the annual fee that is appropriated for waste management each year and place the rest in a trust account, held by a qualified third-party institution, to be available when needed. At the same time, the Office of Management and Budget should work with the congressional budget committees and the Congressional Budget Office to change the budgetary treatment of annual fee receipts so that these receipts can directly offset appropriations for the waste program. These actions are urgent because they enable key subsequent actions the Commission recommends.

Therefore, we urge the Administration to act promptly to implement these changes (preferably in fiscal year 2013). For the longer term, legislation is needed to transfer the unspent balance in the Fund to the new waste management organization so that it can carry out its civilian nuclear waste obligations independent of annual appropriations (but with congressional oversight)—similar to the budgeting authority now given to the Tennessee Valley Authority and Bonneville Power Administration.

We recognize that these actions mean no longer counting nuclear waste fee receipts against the federal budget deficit and that the result will be a modest negative impact on annual budget calculations. The point here is that the federal government is contractually bound to use these funds to manage spent fuel. The bill will come due at some point. Meanwhile, failure to correct the funding problem does the federal budget no favors in a context where taxpayers remain liable for mounting damages, compensated through the Judgment Fund, for the federal government's continued inability to deliver on its waste management obligations. These liabilities are already in the billions of dollars and could increase by hundreds of millions of dollars annually for each additional year of delay.

4. Prompt Efforts to Develop a New Geologic Disposal Facility

Deep geologic disposal capacity is an essential component of a comprehensive nuclear waste management system for the simple reason that very long-term isolation from the environment is the *only* responsible way to manage nuclear materials with a low probability of re-use, including defense and commercial reprocessing wastes and many forms of spent fuel currently in government hands.

The conclusion that disposal is needed and that deep geologic disposal is the scientifically preferred approach has been reached by every expert panel that has looked at the issue and by every other country that is pursuing a nuclear waste management program.

Some commenters have urged the prompt adoption of recycling of spent fuel as a response to the waste disposal challenge, as well as a means to extend fuel supply. *It is the Commission's view that it would be premature for the United States to commit, as a matter of policy, to "closing" the nuclear fuel cycle given the large uncertainties that exist about the merits and commercial viability of different fuel cycles and technology options.* Future evaluations of potential alternative fuel cycles must account for linkages among all elements of the fuel cycle (including waste transportation, storage, and disposal) and for broader safety, security, and non-proliferation concerns. Moreover, all spent fuel reprocessing or recycle options generate waste streams that require a permanent disposal solution. In any event, we believe permanent disposal will very likely also be needed to safely manage at least some portion of the commercial spent fuel inventory even if a closed fuel cycle were adopted.

The Difference between "Storage" and "Disposal"

Disposal, intended as the final stage of waste management, is isolation that relies in the long term only on the passive operation of natural environmental and man-made barriers, does not permit easy human access to the waste after final emplacement, and does not require continued human control and maintenance. Storage, intended as an intermediate step in waste management, is isolation that permits managed access to the waste after its emplacement, with active human control and maintenance to assure isolation. After a period in storage, waste is subject to disposal. As used in this report, the term "disposal" is understood to mean permanent disposal; the term "storage" is understood to mean storage for an interim period prior to disposal or other disposition.

We recognize that current law establishes Yucca Mountain in Nevada as the site for the first U.S. repository for spent fuel and high-level waste, provided the license application submitted by DOE meets relevant requirements.

The Blue Ribbon Commission was not chartered as a siting commission. Accordingly we have not evaluated Yucca Mountain or any other location as a potential site for the storage or disposal of spent nuclear fuel and high-level waste, nor have we taken a position on the Administration's request to withdraw the license application.[5] We simply note that regardless what happens with Yucca Mountain, the U.S. inventory of spent nuclear fuel will soon exceed the amount that can be legally emplaced at this site until a second repository is in operation. So under current law, the United States will need to find a new disposal site even if Yucca Mountain goes forward. We believe the approach set forth here provides the best strategy for assuring continued progress, regardless of the fate of Yucca Mountain.

5. Prompt Efforts to Develop One or More Consolidated Storage Facilities

Safe and secure storage is another critical element of an integrated and flexible national waste management system. Fortunately, experience shows that storage—either at or away from the sites where the waste was generated—can be implemented safely and cost-effectively. Indeed, *a longer period of time in storage offers a number of benefits because it allows the spent fuel to cool while keeping options for future actions open*.

Developing consolidated storage capacity would allow the federal government to begin the orderly transfer of spent fuel from reactor sites to safe and secure centralized facilities independent of the schedule for operating a permanent repository. The arguments in favor of consolidated storage are strongest for "stranded" spent fuel from shutdown plant sites. Stranded fuel should be first in line for transfer to a consolidated facility so that these plant sites can be completely decommissioned and put to other beneficial uses. Looking beyond the issue of today's stranded fuel, the availability of consolidated storage will provide valuable flexibility in the nuclear waste management system that could achieve meaningful cost savings for both ratepayers and taxpayers when a significant number of plants are shut down in the future, can provide back-up storage in the event that spent fuel needs to be moved quickly from a reactor site, and would provide an excellent platform for ongoing R&D to better understand how the storage systems currently in use at both commercial and DOE sites perform over time.

For consolidated storage to be of greatest value to the waste management system, the current rigid legislative restriction that prevents a storage facility developed under the NWPA from operating significantly earlier than a repository should be eliminated. At the same time, efforts to develop consolidated storage must not hamper efforts to move forward with the development of disposal capacity. To allay the concerns of states and communities that a consolidated storage facility might become a *de facto* disposal site, a program to establish consolidated storage must be accompanied by a parallel disposal program that is effective, focused, and making discernible progress in the eyes of key stakeholders and the public. Progress on both fronts is needed and must be sought without further delay.

Even with timely development of consolidated storage facilities, a large quantity of spent fuel will remain at reactor sites for many decades before it can be accepted by the federal waste management program. Current at-reactor storage practices and safeguards are being scrutinized in light of the lessons that are emerging from Fukushima. In addition, the Commission recommends that the National Academy of Sciences (NAS) conduct a thorough assessment of lessons learned from Fukushima and their implications for conclusions reached in earlier NAS studies on the safety and security of current storage arrangements for spent nuclear fuel and high-level waste in the United States. This effort would complement investigations already underway by the NRC and other organizations. More broadly, it will also be vital to continue vigorous public and private research and regulatory oversight efforts in areas such as spent fuel and storage system degradation phenomena, vulnerability to sabotage and terrorism, full-scale cask testing, and others. As part of this process, it is appropriate for the NRC to examine the advantages and disadvantages of options such as "hardened" onsite storage that have been proposed to enhance security at storage sites.

6. Early Preparation for the Eventual Large-Scale Transport of Spent Nuclear Fuel and High-Level Waste to Consolidated Storage and Disposal Facilities

The current system of standards and regulations governing the transport of spent fuel and other nuclear materials appears to have functioned well, and the safety record for past shipments of these types of materials is excellent. But the current set of transport-related regulations will need to be updated to accommodate changes in fueling practices. Moreover, past performance does not guarantee that future transport operations will match the record to date, particularly as the logistics involved expand to accommodate a much larger number of shipments. Experiences in the United States and abroad, and extensive comments to the Commission, indicate that many people fear the transportation of nuclear materials. Thus greater transport demands are likely to raise new public concerns.

As with siting fixed facilities, planning for associated transportation needs has historically drawn intense interest. Transport operations typically also have the potential to affect a far larger number of communities.

The Commission believes that state, tribal and local officials should be extensively involved in transportation planning and should be given the resources necessary to discharge their roles and obligations in this arena. Accordingly, DOE should (1) finalize procedures and regulations for providing technical assistance and funds for training to local governments and tribes pursuant to Section 180(c) of the NWPA and (2) begin to provide such funding, independent from progress on facility siting. While it would be premature to fully fund a technical assistance program before knowing with some certainty where the destination sites for spent fuel are going to be, substantial benefits can be gained from a modest early investment in planning for the transport of spent fuel from shutdown reactor sites.

Planning and providing for adequate transportation capacity while simultaneously addressing related stakeholder concerns will take time and present logistical and technical challenges. Given that transportation represents a crucial link in the overall storage and disposal system, it will be important to allow substantial lead-time to assess and resolve transportation issues well in advance of when materials would be expected to actually begin shipping to a new facility.

For many years, states have been working cooperatively with DOE to plan for shipments, often through agreements with regional groupings of states and in ways that involve radiological health, law enforcement, and emergency response personnel.

As has been shown with the WIPP program and other significant waste shipping campaigns, planning, training and execution involves many different parties and takes time. In addition, specialized equipment may be required that will need to be designed, fabricated and tested before being placed into service. Historically, some programs have treated transportation planning as an afterthought. No successful programs have done so.

7. Support for Advances in Nuclear Energy Technology and for Workforce Development

Advances in nuclear energy technology have the potential to deliver an array of benefits across a wide range of energy policy goals. The Commission believes these benefits—in light of the environmental and energy security challenges the United States and the world will confront this century— justify sustained public- and private-sector support for RD&D on advanced reactor and fuel cycle technologies.

In the near term, opportunities exist to improve the safety and performance of existing light-water reactors and spent fuel and high-level waste storage, transport, and disposal systems. Longer term, the possibility exists to advance "game-changing" innovations that offer potentially large advantages over current technologies and systems.

The Commission believes the general direction of the current DOE research and development (R&D) program is appropriate, although we also urge DOE to take advantage of the Quadrennial Energy Review[6] process to refine its nuclear R&D "roadmap."

We are not making a specific recommendation concerning future DOE funding for nuclear energy RD&D; in light of the extraordinary fiscal pressures the federal government will confront in coming years, we believe that budget decisions must be made in the context of a broader discussion about priorities and funding for energy RD&D more generally.

One area where the Commission recommends increased effort involves ongoing work by the NRC to develop a regulatory framework for advanced nuclear energy systems. Such a framework can help guide the design of new systems and lower barriers to commercial investment by increasing confidence that new systems can be successfully licensed. Specifically, the Commission recommends that adequate federal funding be provided to the NRC to support a robust effort in this area.

We also support the NRC's risk-informed, performance based approach to developing regulations for advanced nuclear energy systems, including NRC's ongoing review of the current waste classification system. Changes to the existing system may eventually require a change in law.

Another area where further investment is needed is nuclear workforce development. Specifically, the Commission recommends expanded federal, joint labor-management and university-based support for advanced science, technology, engineering, and mathematics training to develop the skilled workforce needed to support an effective waste management program as well as a viable domestic nuclear industry.

At the same time, DOE and the nuclear energy industry should work to ensure that valuable existing capabilities and assets, including critical infrastructure and human expertise, are maintained. Finally, the jurisdictions of safety and health agencies should be clarified and aligned. New site-independent safety standards should be developed by the safety and health agencies responsible for protecting nuclear workers through a coordinated joint process that actively engages and solicits input from all relevant constituencies.

Efforts to support uniform levels of safety and health in the nuclear industry should be undertaken with federal, industry, and joint labor–management leadership. Safety and health practices in the nuclear construction industry should provide a model for other activities in the nuclear industry.

8. Active U.S. Leadership in International Efforts to Address Safety, Non-Proliferation and Security Concerns

As more nations consider pursuing nuclear energy or expanding their nuclear programs, U.S. leadership is urgently needed on issues of safety, non-proliferation, and security/ counter-terrorism. Many countries, especially those just embarking on commercial nuclear power development, have relatively small programs and may lack the regulatory and oversight resources available to countries with more established programs. International assistance may be required to ensure they do not create disproportionate safety, physical security, and proliferation risks. In many cases, mitigating these risks will depend less on technological interventions than on the ability to strengthen international institutions and safeguards while promoting multilateral cooperation and coordination. From the U.S. perspective, two further points are particularly important: First, with so many players in the international nuclear technology and policy arena, the United States will increasingly have to lead by engagement and by example. Second, the United States cannot exercise effective leadership on issues related to the back end of the nuclear fuel cycle so long as its own program is in disarray; effective domestic policies are needed to support America's international agenda.

The Fukushima accident has focused new attention on nuclear safety worldwide. Globally, some 60 new reactors are under construction and more than 60 countries that do not have nuclear power plants have expressed interest in acquiring them. These nations will have to operate their facilities safely and plan for safe storage and disposition of spent nuclear fuel. The United States should help launch a concerted international safety initiative—encompassing organizations like the International Atomic Energy Agency (IAEA) as well as regulators, vendors, operators, and technical support organizations—to assure the safe use of nuclear energy and the safe management of nuclear waste in all countries that pursue nuclear technology.

Nuclear weapons proliferation has been a central concern of U.S. nuclear policy from the earliest days of the nuclear era. These concerns are still prominent, especially where the deployment of uranium enrichment, reprocessing, and recycled fuel fabrication technology is being contemplated. As countries with relatively less nuclear experience acquire nuclear energy systems, the United States should work with the IAEA, nuclear power states, private industry, and others in the international community to ensure that all spent fuel remains under effective and transparent control and does not become "orphaned" anywhere in the world with inadequate safeguards and security.

Longer term, the United States should support the use of multi-national fuel-cycle facilities,[7] under comprehensive IAEA safeguards, as a way to give more countries reliable access to the benefits of nuclear power while simultaneously reducing proliferation risks. U.S. sponsorship of the recently-created IAEA global nuclear fuel bank is an important step toward establishing such access while reducing a driver for some states to engage in uranium enrichment. But more is needed. The U.S. government should propose that the IAEA lead a new initiative, with active U.S. participation, to explore the creation of one or more multi-national spent fuel storage or disposal facilities.

In addition, the United States should support the evolution of spent fuel "take-away" arrangements as a way to allow some countries, particularly those with relatively small national programs, to avoid the costly and politically difficult step of providing for spent fuel disposal on their soil and to reduce associated safety and security risks. An existing program to accept highly-enriched uranium fuel from research reactors abroad for storage in the United States has provided a demonstration—albeit a limited one—of the national security value of such arrangements.

The capability to accept limited quantities of spent fuel from foreign commercial reactors could be similarly valuable from a national security perspective. As the United States moves forward with developing its own consolidated storage and disposal capacity, it should work with the IAEA and with existing and emerging nuclear nations to establish conditions under which one or more nations, including the United States, can offer to take foreign spent fuel for ultimate disposition.

The susceptibility of nuclear materials or facilities to intentional acts of theft or sabotage for terrorist purposes is a relatively newer concern but one that has received considerable attention since 9/11. The United States should continue to work with countries of the former Soviet Union and other nations through initiatives such as the Nunn-Lugar Cooperative Threat Reduction Program and the Global Initiative to Combat Nuclear Terrorism to prevent, detect, and respond to nuclear terrorism threats. Domestically, evolving terrorism threats and security risks must be closely monitored by the NRC, the Department of Homeland Security, and other responsible agencies to ensure that any additional security measures needed to counter those threats are identified and promptly implemented.

The recent events at Fukushima have – as they should – prompted the NRC and the industry to re-examine the adequacy of "mitigative strategies" for coping with large-scale events (like an explosion or fire) or catastrophic system failures (like a sudden loss of power or cooling); as noted previously, we also recommend that Congress charter the National Academy of Sciences to assess lessons learned from Fukushima with respect to the storage of spent fuel.

Tying It Together

The overall record of the U.S. nuclear waste program has been one of broken promises and unmet commitments. And yet the Commission finds reasons for confidence that we can turn this record around. To be sure, decades of failed efforts to develop a repository for spent fuel and high-level waste have produced frustration and a deep erosion of trust in the federal government. But they have also produced important insights, a clearer understanding of the technical and social issues to be resolved, and at least one significant success story – the WIPP facility in New Mexico. Moreover, many people have looked at aspects of this record and come to similar conclusions.

The problem of nuclear waste may be unique in the sense that there is wide agreement about the outlines of the solution. Simply put, we know what we have to do, we know we have to do it, and we even know how to do it. Experience in the United States and abroad has shown that suitable sites for deep geologic repositories for nuclear waste can be identified and developed. The knowledge and experience we need are in hand and the necessary funds have been and are being collected.

Rather the core difficulty remains what it has always been: finding a way to site these inherently controversial facilities and to conduct the waste management program in a manner that allows all stakeholders, but most especially host states, tribes and communities, to conclude that their interests have been adequately protected and their well-being enhanced—not merely sacrificed or overridden by the interests of the country as a whole.

This is by no means a small difficulty—in fact, many other countries have not resolved this problem either. However, we have seen other countries make significant progress with a flexible approach to siting that puts a high degree of emphasis on transparency, accountability, and meaningful consultation. We have had more than a decade of successful operation of WIPP. And most recently, we have witnessed an accident that has reminded Americans that we have little physical capacity at present to do anything with spent nuclear fuel other than to leave it where it is. Against this backdrop, the conditions for progress are arguably more promising than they have been in some time. But we will only know if we start, which is what we urge the Administration and Congress to do, without further delay.

LIST OF ACRONYMS AND ABBREVIATIONS

AEA	Atomic Energy Act
AEC	Atomic Energy Commission
AMFM	Alternative Means of Financing and Managing
BEA	Budget Enforcement Act of 1990
BRC	Blue Ribbon Commission on America's Nuclear Future
C&C	consultation and cooperation
CFR	Code of Federal Regulations
CRA	Congressional Review Act
CSA	comprehensive safeguards agreement CUTGO cut-as-you-go
DHS	Department of Homeland Security
DOE	U.S. Department of Energy
DOJ	Department of Justice
DOT	U.S. Department of Transportation
DRR	Domestic Research Reactor
EDRAM	Environmentally Safe Disposal of Radioactive Materials
EM	DOE Office of Environmental Management
EPA	Environmental Protection Agency
EPRI	Electric Power Research Institute
ERDA	Energy Research and Development Administration
FERC	Federal Energy Regulatory Commission
FRA	Federal Railroad Administration
FRR	Foreign Research Reactor
FY	fiscal year
GAO	Government Accountability Office
GRH	Gramm-Rudman-Hollings
GTCC	Greater than Class C
HAZMAT	hazardous material
HEU	high-enriched uranium
HLW	high-level waste
HOSS	hardened on-site storage
IAEA	International Atomic Energy Agency
INL	Idaho National Laboratory

ISFSI	Independent Spent Fuel Storage Installation
LEU	low-enriched uranium
LLW	low-level waste
LWR	light water reactor
LWT	legal-weight truck
MC&A	material control and accountability
MOX	mixed oxide
MRS	monitored retrievable storage
MTHM	metric tons heavy metal
MTU	metric tons of uranium
NAS	National Academy of Sciences
NEA	Nuclear Energy Agency
NEI	Nuclear Energy Institute
NGO	non-governmental organizations
NNPP	Naval Nuclear Propulsion Program
NNWS	non-nuclear weapon states
NPT	Nuclear Non-Proliferation Treaty
NRC	Nuclear Regulatory Commission
NRF	Naval Reactors Facility
NWF	Nuclear Waste Fund
NWMO	Nuclear Waste Management Organization
NWPA	Nuclear Waste Policy Act
NWS	nuclear weapon states
NWTRB	Nuclear Waste Technical Review Board
O&M	operations and maintenance OCRWM Office of Civilian Radioactive Waste Management (DOE)
OFF	oldest fuel first
OMB	Office of Management and Budget
OSHA	Occupational Safety and Health Administration
PAYGO	pay-as-you-go
PFSPrivate	Fuel Storage, LLC
PWR	Pressurized Water Reactor
R&D	research and development
RCRA	Resource Conservation and Recovery Act
RD&D	research, development, and demonstration
SNF	spent nuclear fuel
SRG	state regional group
STGWG	State and Tribal Government Working Group
TEC	Transportation External Coordination
TRU	transuranic
TVA	Tennessee Valley Authority
WANO	World Association of Nuclear Operators
WINS	World Institute for Nuclear Security
WIPP	Waste Isolation Pilot Plant

1. INTRODUCTION

> The Blue Ribbon Commission on America's Nuclear Future (BRC) was formed by the Secretary of Energy, at the direction of the President, to conduct a comprehensive review of policies for managing the back end of the nuclear fuel cycle and recommend a new plan.

The Commission charter and a list of Commissioners may be found in appendix A.

As required by our charter, a draft of this final report was released for public comment on July 29, 2011. This report reflects comments received on the draft report, which generated thousands of written submissions to the Commission, as well as input gathered at five public meetings that were held in Denver, Boston, Washington D.C., Atlanta, and Minneapolis in September and October 2011 to solicit feedback on the draft report. A comprehensive record of comments and feedback received on the Commission's draft report, along with other materials and work products generated in the course of the Commission's work are available at the Commission web site at www.brc.gov. Fulfilling our charter has required the Commission to investigate a wide range of issues. To aid our investigations, we sought and obtained the approval of the Secretary of Energy to form three subcommittees: one to examine disposal issues, a second to address issues of transportation and storage, and a third focused on reactor and fuel cycle technology. We also requested and received DOE approval in late 2011 to establish an *ad hoc* subcommittee to investigate the issue of commingling of defense and civilian wastes. The *ad hoc* subcommittee reported its findings at a public meeting held on December 2, 2011, but did not make a recommendation to the full Commission. The disposal, transportation and storage, and reactor and fuel cycle subcommittees issued draft reports for public comment in June 2011. Updated versions of these reports will be issued shortly after the submission of this final report and will likewise be available—along with a white paper developed by the *ad hoc* committee on commingling— on the Commission web site.

Throughout, we have sought to ensure that our review is comprehensive, open and inclusive. To that end, the Commission and its subcommittees have heard from thousands of individuals and organizations through formal hearings, site visits, written letters and comments submitted to the Commission web site. We have visited several communities across the country that have a keen interest in the matters before the Commission. We have also visited a number of other countries to gain insights as to how the United States might proceed. A list of Commission meetings is included in appendix B. We are indebted to the many people who have given us the benefit of their expertise, advice, and guidance.

As the Commission prepared its draft report, an earthquake and tsunami of historic proportions struck the eastern coast of Japan triggering the worst accident at a nuclear facility since the 1986 Chernobyl disaster.[1] Various parts of four reactors at the Fukushima Daiichi nuclear power station suffered significant to massive damage when cooling systems failed due to the loss of primary and backup power. Substantial amounts of radiation were released, contamination has occurred offsite, and people were evacuated from a large area around the plant.

Commission members and staff were deeply saddened by these events. We are also acutely aware that the Fukushima disaster has altered the technical, social, and political context into which our findings and recommendations are being released. To the extent possible we have tried to reflect in this document, if not the lessons of Fukushima (since those lessons are only beginning to be elucidated and understood) then at least the recognition that U.S. policy going forward will have to be responsive to the new knowledge and changed circumstances brought about by the accident. This report also reflects an awareness of the changing and far from certain global outlook for nuclear power and the effects of America's diminishing ability to influence where and how nuclear energy is used.

As more information about the Fukushima accident has emerged, we have sought to ensure that it is reflected in this report, recognizing that many of the expert inquiries now underway have not been completed as this report goes to press. That said, we are confident the strategy we are proposing—with its strong emphasis on flexibility, adaptation, responsiveness, accountability, and continuous learning—is the right one for a post-Fukushima world and can help the United States recapture some of its lost influence over international nuclear developments.

While the scope of our review has been broad, it has not been without limits:

- The Commission is not a siting body. We have not made any findings about the Yucca Mountain repository site or about any alternative sites; in fact, we have not recommended specific locations for any component or facility of the U.S. nuclear waste management system.
- The Commission was not asked to make recommendations regarding the advisability or the appropriate level of future U.S. reliance on nuclear power. Some witnesses urged the Commission to recommend that nuclear power plants be shut down until a disposal solution for spent nuclear fuel (SNF) is found, while others urged the Commission to encourage a widespread expansion in the domestic use of nuclear energy. These questions fall outside our charter and we have declined to address them. We have, however, considered multiple scenarios for the future of nuclear energy in the United States to ensure that our recommendations can accommodate a full range of possibilities.

As we have listened to testimony and public comment, we have been constantly reminded of the lack of trust that exists today in the federal government's ability to meet its waste cleanup and management obligations. Past decisions—first to truncate the siting process for two repositories that was established in 1982; then to limit all efforts to a single site at Yucca Mountain, Nevada; and then, after more than 20 years of work on the site, to request to withdraw the license application for that site—have only increased this deficit of trust, particularly among nuclear utility ratepayers and in communities that host nuclear waste storage facilities. These people and others believe they have been let down repeatedly by a government that has yet to make good on its commitment to provide a disposal solution for the most hazardous nuclear wastes.

By contrast, we are convinced by our investigations that such a disposal solution *can* be found.

While there is no reasonably foreseeable technology that could eliminate the need for a high-level nuclear waste disposal facility, progress on deep, mined geologic repositories—particularly in Sweden and Finland, but in other nations as well —has dramatically increased confidence in the ability to identify and license acceptable sites. Here in the United States, more than 10 years of operating experience at the Waste Isolation Pilot Plant (WIPP) in New Mexico, which is successfully accepting and disposing of certain radioactive wastes from our nation's nuclear weapons program, show that nuclear wastes can be transported safely over long distances and placed securely in a deep, mined repository.

In this report, we focused on developing a sound strategy for future storage and disposal facilities and operations that we believe can and should be implemented regardless of what happens with Yucca Mountain. We are confident this strategy can dramatically increase the U.S. waste management program's chances for success regardless of what site or sites are chosen to provide for the ultimate disposal of America's SNF and other high level nuclear wastes.

This report is organized as follows: sections 2 and 3 provide policy context and background information on the nuclear fuel cycle, the history of U.S. waste management efforts, and existing waste and spent fuel inventories. sections 4 and 5 then describe the underlying rationale for expeditious action to establish geologic disposal and consolidated storage capacity for SNF and high-level waste (HLW) in the United States. The next three sections (sections 6 through 8) describe the key institutional and policy changes that we are recommending in pursuit of those objectives, including changes in the approach to siting new facilities, the need for new institutional leadership of the nation's waste management program, and the need for fundamental reforms to the way the waste program is being financed. sections 9 and 10 discuss transportation and regulatory issues. sections 11 and 12 take up the subjects of advanced reactor and fuel cycle technologies and international issues, respectively. The last section discusses next steps for Congress and the Administration to implement the new strategy the Commission is recommending.

2. Foundations of a New Strategy

Our charter directs the Commission to focus its attention on the back end of the nuclear fuel cycle.

Based on the information we gathered and the input we heard, we are optimistic that a new strategy can (1) better meet the challenge of managing nuclear waste and providing for its long-term disposition in a way that meets this generation's ethical obligations to current and future generations; (2) help address the safety, weapons proliferation, and security concerns that could otherwise accompany the international spread of nuclear technology; and (3) allow future generations to rely on nuclear power if they so choose. Implementing the strategy we propose will not be quick or easy, but we believe it is doable. This section describes the important program features, policy objectives, and guiding principles that we believe will be central to success.

Our charter also recognizes that the nuclear power industry is not the only source of spent fuel[2] and HLW in need of management and disposal; indeed, it gives equal attention to the need to consider alternatives for disposing of wastes from the nation's defense programs. The first HLW and spent fuel were produced more than 60 years ago as part of the U.S. defense program and a large quantity of these materials is now being stored at U.S. Department of Energy (DOE) sites with the expectation that they will be sent to a repository for disposal. While the activities that generated most of those materials ceased decades ago, the nuclear-powered vessels of the U.S. Navy continue to be a small but important source of spent fuel. Safe disposal of these materials is a national obligation that will exist regardless of the future use of civilian nuclear power.

2.1. Elements of a Successful Strategy

Effectively managing the back end of the nuclear fuel cycle requires a vision and a strategy. Both have been lacking in the U.S. waste management program to date. The vision must be stable, comprehensive, clear, and compelling. The strategy must combine durability with flexibility so that it can endure over the years and decades needed for policies and programs to unfold while being continually responsive to new experience and information and changes in values and circumstances, such as the future use of nuclear power. Multiple views and interests will need to be balanced and decision-making processes will need to be designed so as to not only facilitate, but actually benefit from the participation of a wide range of stakeholders.

A comprehensive strategy must be attentive to the scientific, technical, political, and societal dimensions of nuclear fuel cycle choices and it must account for impacts and risks from "cradle to grave" (i.e., from the mining of uranium ore to the disposal of wastes). It must accommodate a range of perspectives and interests and advance broadly held policy goals with respect to safety, security, the environment, economics, non-proliferation, equity, and public and political acceptance. Importantly, it must respect the sovereignty, aspirations, and realities of other nations while preserving America's own options and her interest in retaining a position of international leadership in technology and global efforts to promote safety, security, and environmental protection.

Finally, the U.S. nuclear waste management program must consistently honor promises and commitments in order to regain the trust and confidence of important constituencies.

Obviously the full ramifications of recent events in Japan have yet to be felt but they warrant a thorough reexamination of the safety performance and other operational features, both of the current fleet of nuclear plants and of new designs that are being constructed or proposed. Indeed, Fukushima will prompt re-assessments, not only of reactor design and performance, but of different management strategies for storing, transporting, and ultimately disposing of spent fuel.

Prudence would dictate that the United States continue to insist on rigorous efforts by the industry and its regulators to improve the existing fleet, while also promoting the development of new plant designs that demonstrably improve safety, security, economics, and performance.

Agencies with regulatory oversight authority will likewise be scrutinized for demonstrating independence from short-term political considerations and an unwavering focus on safety and security.

2.2. Core Interests and Objectives for U.S. Waste Management Policy

Success in the complex, controversial, and long-term endeavor of implementing a sound strategy for the management and disposition of SNF and high-level radioactive waste will require careful and continuous attention to a number of core interests and objectives. These are not interests or objectives to be traded off in a zero sum sense. Rather they are synergistic interests that can all be served through an approach that consistently strives to meet high standards of organization, implementation, governance, and leadership.

2.2.1. Public and Occupational Health and Safety

The first objective in all decision-making regarding nuclear materials or activities must be to protect public health and safety and to protect the health and safety of the nuclear workforce. This must also be the U.S. government's priority when engaging the international nuclear community. It will not be possible to gain public trust at home or exercise leadership internationally if the U.S. program is seen as trading off or compromising public health and safety for other objectives. A commitment to continual safety improvement is essential to further reduce existing risks and to address new ones as they arise.

2.2.2. Environmental Protection

Understanding and awareness of environmental issues generally—and of the environmental impacts of different energy technologies in particular—has only grown over the more than 60-year history of the nuclear power industry. Since some materials generated by the back end of the nuclear fuel cycle will be radioactive over many millennia, they must be properly isolated from the biosphere to avoid posing a long-term hazard to other living organisms and ecosystems, as well as to human populations.

The Commission's view and that of many experts is that these risks can be managed, but the nature and longevity of the environmental hazard clearly demand an extra measure of care, rigorous planning, and continued vigilance. Environmental concerns and trade-offs must also be viewed in a broader context. All energy supply options have significant advantages and disadvantages.

Nuclear power today provides two-thirds of the nation's low-carbon electricity production. If this generation or future generations see an imperative to meet rising energy demand while substantially reducing carbon dioxide emissions, continued access to nuclear power as an established low-carbon energy option may have significant environmental, as well as economic and social, value.

2.2.3. Cost-Effectiveness

The purpose of civilian nuclear energy systems is to provide safe, reliable, and affordable energy. The nation's strategy for managing the back end of the nuclear fuel cycle must be consistent with that purpose.

It must also reflect a recognition that the money to implement waste storage, transport, and disposal solutions will ultimately come from U.S. citizens—primarily from nuclear utility ratepayers in the case of commercial spent fuel and from U.S. taxpayers in the case of defense wastes. The federal government therefore has an obligation to ensure that all funds being collected from ratepayers (or appropriated from the federal budget in the case of defense wastes) are being used wisely and efficiently to achieve the nuclear waste program's objectives.

2.2.4. Non-Proliferation and National Security

The growth and (more importantly) the diffusion of nuclear energy technology and expertise require careful attention to weapons proliferation and security considerations. While the vast majority of conventional (predominantly light-water) nuclear power plants in operation worldwide and the fresh low-enriched uranium fuel they use do not present significant proliferation concerns, the uranium enrichment facilities used to produce this fuel and the spent fuel that results do pose such risks. Enrichment and reprocessing facilities, in particular, have the potential to be misused to develop materials for nuclear weapons. Spent fuel contains plutonium which, if separated, could be used to make weapons. The United States has an important stake in ensuring that strong international norms emerge for safety, physical security, and non-proliferation. U.S. policies for managing the back end of the nuclear fuel cycle must support these goals and must strengthen key elements of the international non-proliferation and security regimes, including the Nuclear Non-Proliferation Treaty (NPT) and the work of the International Atomic Energy Agency (IAEA) and the World Institute for Nuclear Security (WINS) in this domain. The U.S. nuclear industry also has an important role to play in meeting nonproliferation goals—indeed, this role is recognized in principles of conduct that were recently announced by nuclear power plant exporters.[3]

2.3. Core Values and Principles for a Successful Waste Management Program

2.3.1. Ethical Responsibility

Ethical considerations have been at the heart of many of the comments and presentations the Commission has heard. From this standpoint, the case for developing disposal capacity for the high-level radioactive wastes that have accumulated over decades of weapons program activity and commercial nuclear power production is clear: the generations who created these wastes and benefited from the activities that produced them have an obligation to ensure that the entire burden of providing for their disposal does not fall to future generations.[4] That means mustering, without further delay, the financial, programmatic, institutional, and political wherewithal to implement a functional system to manage these materials that provides for their safe transportation, consolidated storage, and disposal. While the process should not be rushed, the capability to provide for disposal must exist and the process of emplacing long-lived radioactive wastes, including particularly those materials with no realistic possibility of being re-used, must be underway within a reasonable timeframe.

Finally, this generation's responsibility to future generations includes taking care not to foreclose options that future generations may see as being in their best interest. In this context, with the benefit of advances in technology, future generations may want to use spent fuel as an energy resource. A well-constructed waste management program, with the flexibility we recommend, can do both—provide a solution and leave choices.

2.3.2. Fairness

The ethical argument made in the foregoing section is grounded in the principle of intergenerational equity. But it will also be critical to provide a demonstrably fair process to those who are immediately engaged in and affected by the waste management program. The program must be—and must be viewed as being—both fair and inclusive.

This is a significant challenge. Different and sometimes competing interests are at play. Communities with current and accumulating inventories of waste may see issues of fairness quite differently than communities being considered as potential host sites for storage or repository facilities. Communities near existing DOE sites where spent fuel and HLW are being stored and utilities that have entered into legal commitments with the federal government concerning the timing of spent fuel acceptance and disposition may have been promised actions that cannot be delivered. While there will be different perspectives, future decisions must be reached in a way that makes these fairness and equity considerations explicit.

2.3.3. Transparency

Transparency is an important feature and one that deserves careful attention in designing a successful program. The aim should be openness and inclusiveness with respect to program plans and decisions, the handling of input from affected parties, and the application of different mechanisms for demonstrating accountability.

Useful guidance for achieving transparency in practice can be found in an "Open Government Directive" developed by the Office of Management and Budget in late 2009.[5] According to the directive: "Transparent decisions are decisions in which the decision maker clearly presents to others the normative and factual premises behind the conclusions and explains the reasoning leading from these premises to the conclusion. Transparency thus involves uncovering, describing, documenting and communicating all the argumentative steps in the line of reasoning. It also involves acknowledging the weighting of any evidence drawn upon in reaching the final decision. It is recommended that each decision should be accompanied by an audit trail describing the premises justifying it. Uncertainties should be presented in connection with each possible adverse effect to indicate alternative scenarios to the most likely risk characterization together with an evaluation of the reliability of each of the alternative scenarios."

2.3.4. Values

U.S. programs and policies for managing the back end of the fuel cycle must continually be informed by and reflect the values of those directly affected by the program and the values of the broader citizenry. These priorities and values will change with time, making it essential to design an adaptable and flexible program. In a context where conditions, interests, perceptions, and values are constantly shifting and where different parties hold different views and values, perfect consensus and solutions that satisfy all constituencies will be rare. In most cases, decision-makers will need to balance competing interests, make trade-offs in the face of uncertainty, and be willing to move forward without full consensus. In these cases, stakeholders and the public are entitled to a clear understanding of how decisions were reached and how different values and interests were considered and resolved in the process.

2.3.5. Informed Participation

In a democracy, informed participation is at the heart of durable solutions to significant policy challenges. Managing the back end of the nuclear fuel cycle is a technically complex, institutionally demanding, and inherently long-term task. This task is made both more challenging and more important by the fact that many Americans view the risks associated with radiation and nuclear energy as fundamentally different in nature from other kinds of risks.

Radiation, in particular, has a number of properties that tend to heighten people's fear of being exposed to it: radiation is invisible, it can be penetrating, its long-term health effects (which can include cancer and birth defects) can be severe but may not be immediately detectable, and some materials that are radioactive can remain so for extremely long periods of time. Broad public support for a new strategy will depend on some shared understanding of the nature and extent of the problem, available options for resolving the problem, and the consequences and risks associated with different actions—including the consequences and risks of further inaction.

The job of better communicating information and effectively engaging different constituencies must be seen as one of the core missions of a revitalized waste management program. Likewise, the commitment of technical and financial support to enable informed participation by a wide range of stakeholders in key decision-making processes must be viewed as an appropriate and indeed necessary use of resources for successful program implementation. This point features prominently in the sections that follow because we believe it is central to our recommendations.

2.3.6. Governance and Leadership

The key insight that permeates this report is that the best hope for greater success lies not in changing the objectives of the waste management program, but in changing the approach taken to reach those objectives. Central to this approach is establishing the institutional leadership and a governance framework matched to the challenges at hand. Both must endure over the very long timeframes involved in managing and planning for the disposition of nuclear materials. As discussed in more detail later in this report, the Commission has heard and considered many options and has concluded that the situation calls for a new waste management organization with a clear mission and the independent authority and access to resources needed to carry out that mission. At the same time, we recognize that no institutional change or policy reform can substitute for outstanding, inspired leadership—both at the level of the new organization itself and within Congress and the Administration. Whatever new strategy is adopted, it must encourage such leaders and give them ample opportunity to succeed, while also holding key policy-makers, oversight agencies, and the new organization accountable for results.

3. Technical and Historical Background

We begin with a review of the nuclear fuel cycle and a brief history of nuclear waste management policy in the United States.

3.1. Overview of the Nuclear Fuel Cycle

The nuclear fuel cycle is the series of industrial processes used to produce electricity from uranium in a nuclear reactor. The nuclear fuel cycle can be described as having three major parts (see figure 1): the "front end" where uranium is mined and processed into fuel for use in a nuclear reactor; the use of that fuel in a reactor; and the "back end" where the spent fuel is first stored and ultimately sent for disposal or reprocessing (if the spent fuel is reprocessed, remaining wastes would still require disposal).

- Uranium enrichment – The nuclear fuel cycle begins with the extraction of uranium from ores or other natural sources. Uranium provides the basic fissile material or "fuel" for nearly all nuclear reactors. Extracted uranium consists almost entirely of two isotopes or types of uranium atoms,[6] mostly uranium-238 (99.3 percent) together with a much smaller fraction (0.7 percent) of the fissionable isotope uranium-235 or "U-235." In its natural state, mined uranium is only weakly radioactive—meaning that it can be handled without the need for radiation shielding. Before it can be used in a commercial reactor, natural uranium must be purified and enriched to boost the amount of fissionable U-235 present in the fuel. Most of the commercial nuclear power plants in operation today are light-water reactors that require fuel enriched to a U-235 concentration of anywhere from 3 to 5 percent[7]—a typical figure for fuel used in commercial U.S. reactors is 4 percent. Techniques for enriching uranium are well developed, with the most prominent methods involving gaseous diffusion or centrifuge technology.

- Use as reactor fuel – Enriched uranium oxide is cast into hard pellets and stacked inside long metal tubes or "cladding" to form fuel rods (figure 2). The fuel rods are bundled into fuel assemblies (each assembly is about 12 to 14 feet long). The core of a typical light-water commercial nuclear power reactor in the United States contains anywhere from 200 to 500 fuel assemblies, totaling approximately 100 metric tons of uranium oxide. Inside the reactor, the enriched uranium sustains a series of controlled nuclear reactions that collectively liberate substantial quantities of energy. The energy is converted to steam and used to drive turbines that generate electricity. Meanwhile, the fission process inside the reactor creates new elements or "fission products," and gives rise to some heavier elements, collectively known as "transuranics," which may take part in further reactions (among the most important is plutonium-239).
- Wet (pool) storage – Nuclear fuel will remain in a commercial power reactor for about four to six years, after which it can no longer efficiently produce energy and is considered used or spent. The spent fuel that has been removed from a reactor is thermally hot and emits a great deal of radiation; upon removal from the reactor, each spent fuel assembly emits enough to deliver a fatal radiation dose in minutes to someone in the immediate vicinity who is not adequately shielded. To keep the fuel cool and to protect workers from the radiation, the spent fuel is transferred to a deep, water-filled pool where it is placed in a metal rack. Typically, spent fuel is kept in the pool for at least five years, although spent fuel at many U.S. reactor sites has been in pool storage for several decades (figure 3). Approximately 50,000 metric tons of commercial spent fuel are currently stored in pools in the United States.
- Dry (cask) storage – After the fuel has cooled sufficiently in wet storage, it may be transferred to dry storage. Dry storage systems take many forms but generally consist of a fuel storage grid placed within a steel inner container and a concrete and steel outer container (figure 4). The amount of commercial spent fuel stored in dry casks in the United States totals about 15,000 metric tons.
- Transportation – Because of the residual hazard it poses, spent fuel must be shipped in containers or casks (figure 5) that shield and contain the radioactivity and dissipate the heat. In the United States, spent fuel has typically been transported via truck or rail; other nations also use ships for spent fuel transport.[8]
- Reprocessing or recycling – Even after commercial fuel is considered "spent," it still contains unused uranium along with other re-usable elements (primarily plutonium which is generated within the fuel while it is in the reactor) and fission products (elements produced by the fissioning of uranium and plutonium in the reactor core). Current reprocessing technologies separate the spent fuel into three components: uranium; plutonium (or a plutonium-uranium mix); and waste, which contains fission products and so-called transuranic elements that are produced within the fuel. The plutonium is mixed with uranium and fabricated into new fuel while the fission products and other waste elements are packaged into a new form for disposal (the uranium can also be re-used to make new fuel but because recovered uranium is more difficult to use than freshly mined uranium, this has only been done to a limited extent). Coupled with new reactor types, future reprocessing technologies could, if

they can be successfully developed and deployed, allow for a greater fraction of the material in spent fuel to be recovered and re-used.

- Disposal – Regardless of whether spent fuel is reprocessed or directly disposed of, every foreseeable approach to the nuclear fuel cycle still requires a means of disposal that assures the very long-term isolation of radioactive wastes from the environment.[9] Many nations, including those engaged in reprocessing, are working to develop disposal facilities for spent fuel and/or HLW, but no such facility has yet been put into operation. Every nation that is developing disposal capacity plans to use a deep, mined geologic repository for this purpose. Other disposal options (e.g., deep boreholes) have been considered and may hold promise in the long-term but are at a much earlier stage of development.

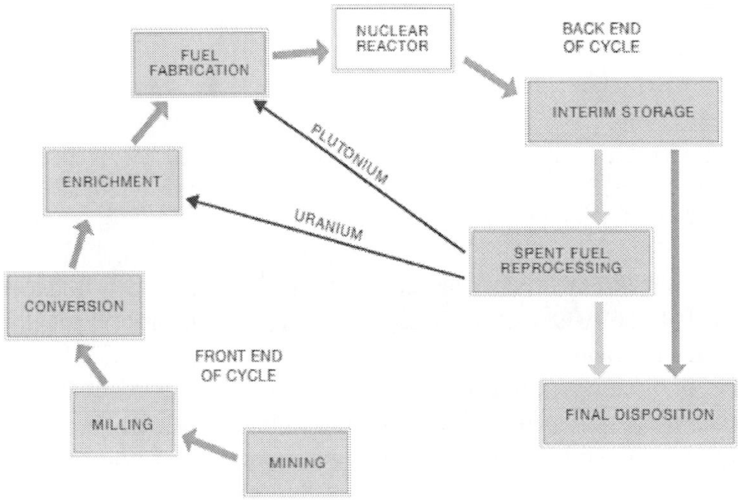

Figure 1. The nuclear fuel cycle.

Source: http://www.nrc.gov/images 20071114-045.jpg.

Figure 2. Fuel assembly.

Figure 3. Wet pool storage.

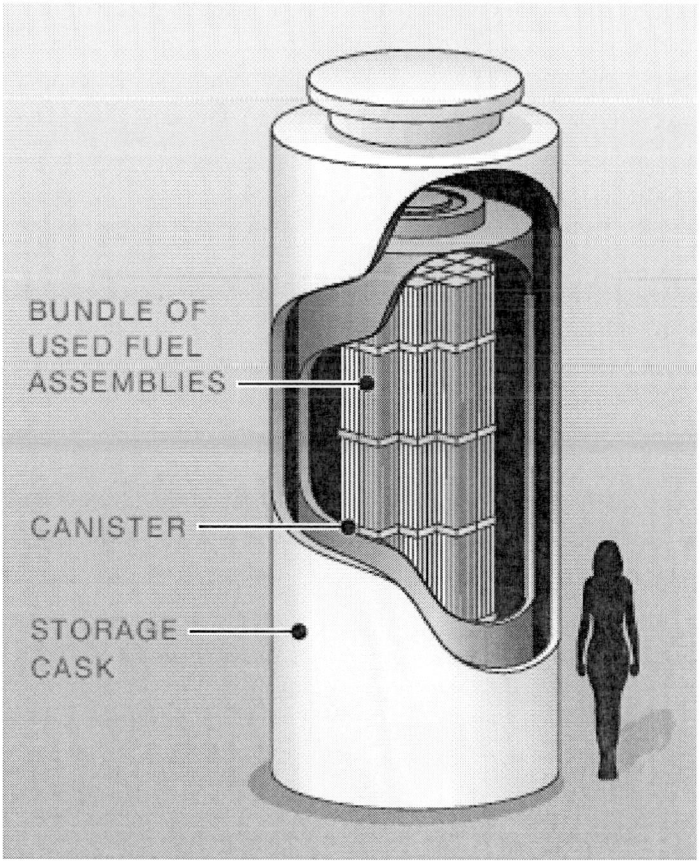

Figure 4. Dry storage system.

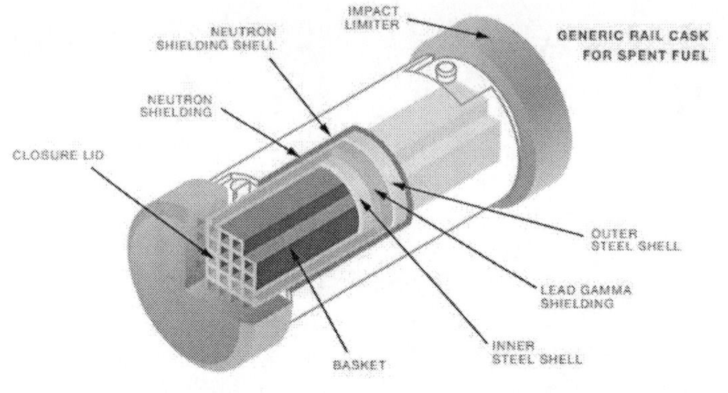

Figure 5. Shipping container for spent fuel.

3.2. The Nature and Longevity of Hazard Posed by Different Types of Nuclear Waste

Spent nuclear fuel and HLW are hazardous if not properly managed and controlled, primarily as a result of the radiation emitted by the radioactive decay of unstable elements in the fuel. Spent fuel emits high levels of radiation and thus requires shielding to be handled safely. In wet storage, shielding is provided by a large volume of water in a storage pool. In dry storage configurations, shielding is primarily provided by thick layers of steel and concrete.

The other major hazard from spent fuel arises if its radioactive constituents (see figure 6) are mobilized into air or water. There is no risk of this occurring as long as fuel assemblies are intact: the fuel is encased in metal tubes or cladding; the tubes in turn are configured in bundles that are designed to withstand four to six years of exposure to very high temperatures and high levels of radiation in a reactor core. But during the initial period after fuel is removed from a reactor core, the rapid decay of short-lived fission products generates sufficient heat that overheating has the potential to damage the fuel cladding and release radioactive material if sufficient cooling is not provided. Over the very long time periods associated with geologic disposal, by contrast, the concern is that gradual corrosion processes or disruptive events (such as seismic or volcanic activity) may allow for radioactive material to be mobilized in ground water and migrate out of an engineered disposal facility.

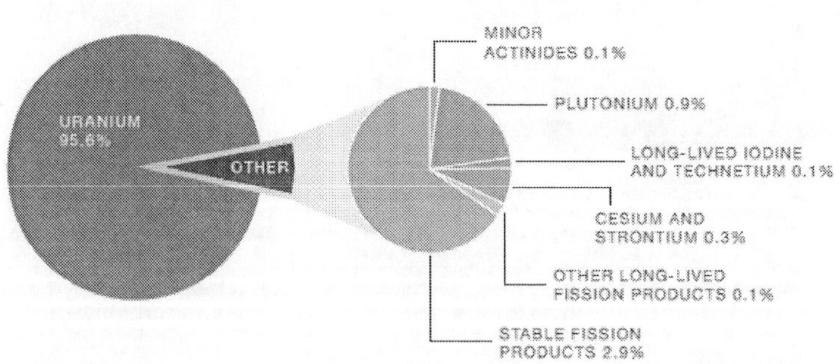

Figure 6. Composition of spent nuclear fuel after 10 years of cooling.[10]

High-level radioactive wastes arise from the chemical reprocessing of spent fuel. Modern reprocessing facilities convert waste streams into solid glass, ceramic, cement, or metal waste forms that are typically contained in stainless steel canisters (like SNF). Like spent fuel, HLW emits high levels of radiation and thus requires similar shielding and handling methods. Likewise, the concern from a disposal standpoint centers on the possibility that corrosion processes or disruptive events may, over a very long period of time, mobilize radioactive material into groundwater. Spent fuel and HLW may also contain materials that are chemically hazardous; uranium is an example. While these chemical hazards may have to be considered in developing some regulations and undertaking safety analyses, they are generally small compared to the radiation hazards associated with high-level nuclear wastes.

Exposure to radioactive materials—whether natural or man-made—can be damaging because many forms of radiation have the ability to change the structure of molecules, including the structure of molecules found in the tissues of living organisms. Human beings are exposed continuously to very low levels of naturally-occurring and man-made radiation (see text box and figure 7).

In most cases the body responds to that exposure[11] by repairing or replacing damaged molecules at the cellular level. In other cases, however, the harm that can result from radiation exposure can be very serious. The exposed individual could develop cancer, for example, or suffer genetic effects (i.e., mutations in the reproductive cells that could be damaging to offspring).

Typical Radiation Exposures for the American Population

In a 2009 report, the National Council on Radiation Protection estimated that the average American is exposed to 6.2 millisieverts of natural and man-made radiation each year. About half of this exposure comes from natural sources (primarily radon gas, a decay product of U-238), while the other half comes from man-made sources (almost entirely from CT scans and other medical procedures that involve nuclear materials).

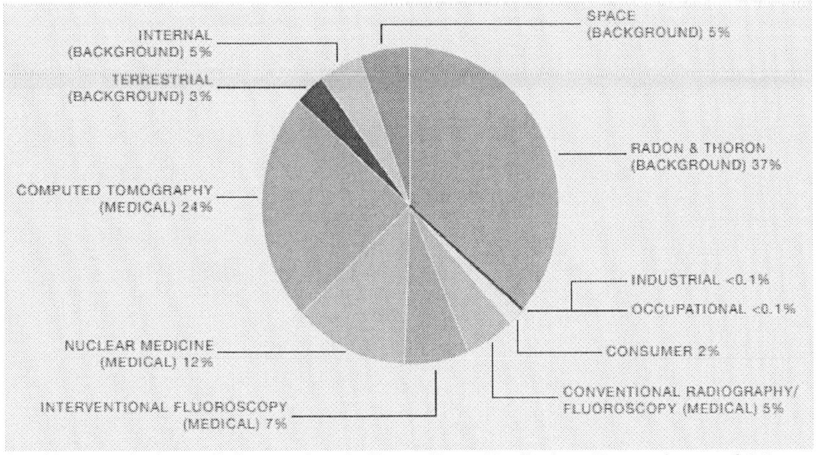

Reprinted with permission of the National Council on Radiation Protection and Measurements, http://NCRPpublications.org.

Extremely high doses of radiation can cause burns or acute radiation damage, which can lead to death in a relatively short period of time (hours to weeks).

For these and other reasons – including the legacy of nuclear weapons use, testing, and high-profile accidents at Three Mile Island, Chernobyl and, now, Fukushima, many people have an understandable fear of radiation and of radioactive materials. Compounding this fear is the fact that radiation cannot be detected by any human senses and effects, if any, often show up long after exposure occurred. These concerns, combined with an oft-stated lack of confidence in government's ability to protect the public from radiation hazards, play a large role in making it difficult to site facilities that handle nuclear materials.[12]

The materials associated with the back end of the nuclear fuel cycle (including both spent fuel and HLW), if not managed properly, can deliver much higher levels of radiation than humans are normally exposed to from natural background, medical procedures, and the like. Safe management is required to ensure that these materials don't deliver elevated radiation doses to humans and other organisms.

Some categories of nuclear waste (generally including all HLW and virtually all spent fuel) remain radioactive for thousands of years because of the long half-lives[13] of some of the radioisotopes they contain. The radioactive decay of a typical spent fuel assembly over time is shown in figure 8.

It is worth mentioning, however, that (1) radiation levels in HLW and spent fuel drop considerably over time and (2) very long-lived isotopes also tend to pose less of an external radiation hazard. By comparison the most dangerous isotopes tend to be those that decay more quickly (the more rapid the decay, the higher the initial level of radioactivity).

3.3. Scale of the Waste Management Challenge in the United States

3.3.1. Current Inventory of Spent Nuclear Fuel Being Managed by the U.S. Commercial Nuclear Power Industry

There are 104 commercial nuclear power reactors operating in the United States today; together they supply approximately 20 percent of the nation's electricity needs. Given that each reactor uses about 20 metric tons of uranium fuel per year, the industry as a whole generates 2,000 to 2,400 metric tons of spent fuel on an annual basis (1 metric ton equals about 2,200 pounds).[14] At present, nearly all of the nation's existing inventory of SNF is being stored at the reactor sites where it was generated—about three-quarters of it in shielded concrete pools and the remainder in dry casks above ground. The quantity of commercially-generated spent reactor fuel currently being stored in this manner totals close to 65,000 metric tons—roughly speaking, it would cover one football field to a depth of approximately 20 feet. This inventory includes approximately 3,000 metric tons of spent fuel in storage at nine sites where commercial reactors have been shut down and are no longer operating.

Figure 9 shows the location of operating commercial nuclear power reactors in the United States today, along with years of operation for each facility.

How much spent fuel will be added to the existing inventory in the future, and at what rate, depends on a number of factors. Market conditions, climate policy, government support, the evolution of reactor technology, and nuclear-related regulatory and policy developments will all influence the nuclear power industry's prospects going forward and will play a role in determining what type and quantity of nuclear waste is produced in the future. At present, some uncertainty surrounds all of these factors. Under a no-growth scenario that assumes continued operation of existing reactors to the end of their current licenses only, and no further expansion of the industry, the total inventory of spent fuel that will have accumulated by 2050 can be expected to remain below 150,000 metric tons. Under a high-growth scenario that assumes substantial numbers of new reactors coming on line in the next few decades, the nation's accumulated spent fuel inventory would be predicted to substantially exceed 200,000 metric tons by mid-century. Even if all commercial reactors in the United States were shut down tomorrow, about 75,000 metric tons—equal to the current spent fuel inventory plus the fuel currently in commercial reactor cores —would require disposal.

These figures illustrate the uncertainty inherent in making predictions about the future. Obviously changing any of the input assumptions—including not only assumptions about future nuclear-based electricity production, but also assumptions about future reactor technology and fuel cycle characteristics—would produce very different results.

3.3.2. Current Inventory of Spent Nuclear Fuel and High-Level Waste Being Managed by DOE

In addition to the spent fuel currently being stored at commercial nuclear power plant sites around the country, there are substantial quantities of spent fuel and HLW at a number of government-owned facilities managed by DOE. Most but not all of this material derives from national defense nuclear activities and is therefore often referred to as "defense waste."

It is important to be clear, however, that these materials were produced and have always been managed by DOE and its predecessor agencies, which had responsibility for nuclear weapons production—not by the Department of Defense.

DOE's spent fuel was mainly produced at Hanford, the Idaho National Laboratory (INL), and the Savannah River Site and most of it is still being stored there. Smaller quantities of spent fuel have also been or are being produced at other facilities, including at Oak Ridge National Laboratory, Brookhaven

National Laboratory, and at various university and commercial research reactors, but after a short period of storage this spent fuel is transferred to one of two sites—INL or the Savannah River Site.

The current inventory of DOE-managed spent fuel represents a relatively small fraction of the nation's total spent-fuel inventory: approximately 2,500 metric tons. In general, DOE has not taken commercial spent fuel for storage at its facilities except in special cases. For example, the damaged reactor core from the 1979 Three Mile Island (TMI) accident was moved to INL for study; in addition, DOE has assumed responsibility for spent fuel in a graphite matrix from the unique, gas-cooled Fort Saint Vrain reactor in Colorado (some of that spent fuel has been shipped to INL for storage, while the rest is currently being stored on site) as well as damaged fuel removed from the TMI Unit 2 reactor. The federal inventory also includes a small quantity of spent fuel from nuclear reactors that power the nation's submarines and other U.S. Navy ships. Spent naval reactor fuel is shipped to INL for evaluation and subsequent storage.

Figure 10 shows the quantity and location of SNF at DOE sites. Both wet and dry methods of storage are in use by DOE, although at the Hanford site in Washington State—where by far the largest portion of DOE's current SNF inventory is being stored—all spent fuel has been moved to dry storage.

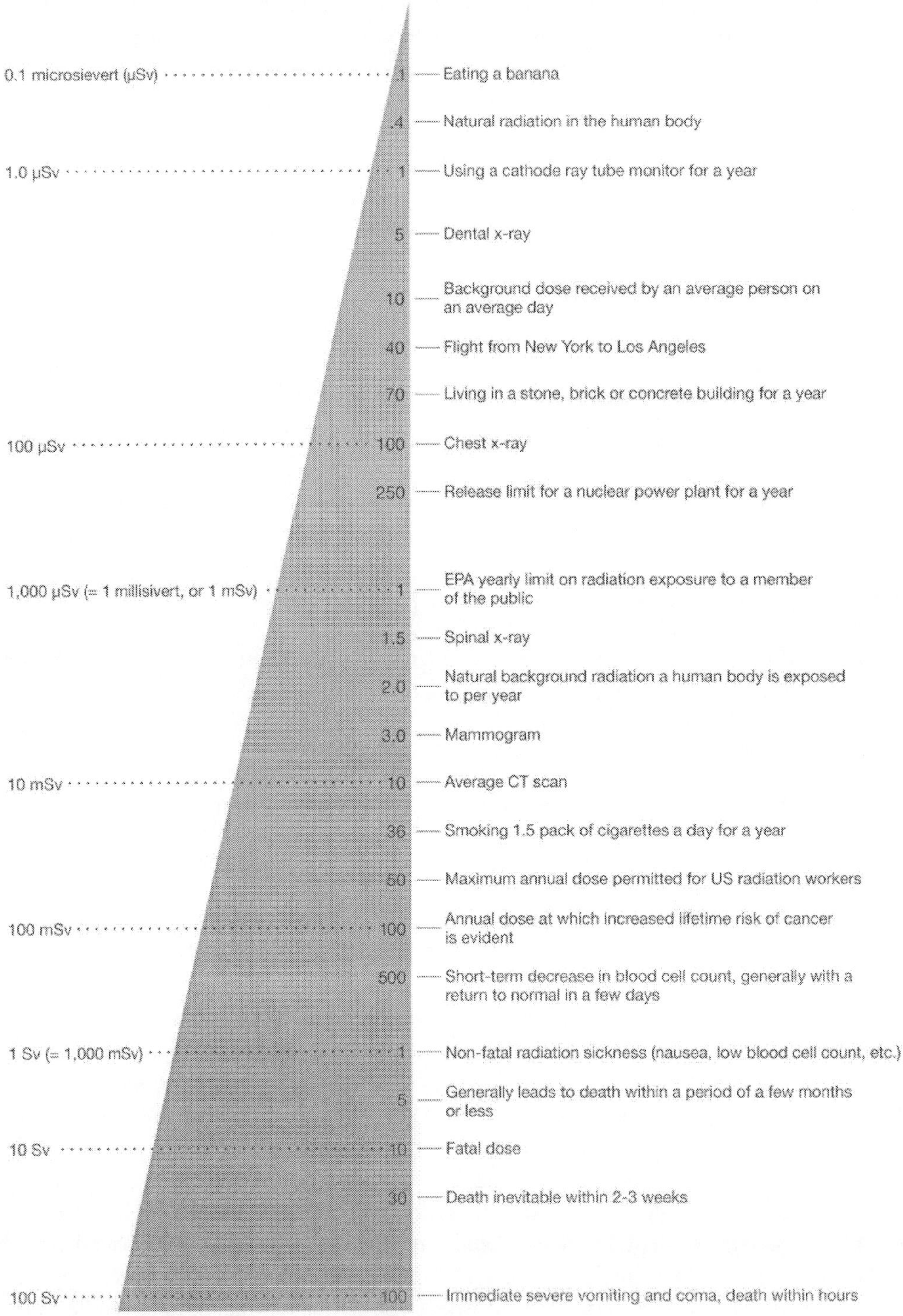

Source: BRC staff using information from various sources[15]

Note: Risk depends on both the total dose and how quickly it was received.

Rem is an older term for radiation exposure, now replaced by the sievert, with 1 Sv = 100 rem.

Figure 7. Comparison of radiation doses.

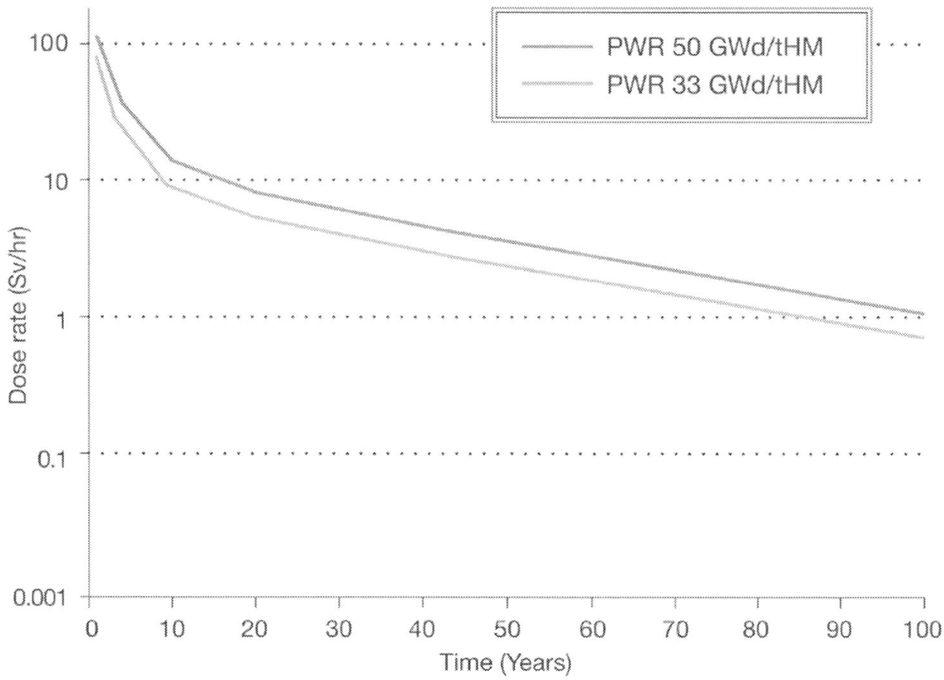

Figure 8. Radioactive decay of typical spent fuel.

In addition, DOE accepts relatively small quantities of spent fuel under the Foreign Research Reactor (FRR) and Domestic Research Reactor (DRR) programs. The DRR program accepts spent fuel from U.S. universities and other government research reactors.[16] The FRR program was established to support U.S. nuclear security and non-proliferation goals. It accepts spent fuel from research reactors in other countries that operated on highly enriched uranium (HEU), which poses particular concerns because it could potentially be used without further enrichment in nuclear explosive devices. So far, more than 9,000 spent fuel assemblies (about 6 metric tons) have been accepted from 29 countries under the FRR program, which is currently slated to run until 2019.

Until recently, DOE had been processing much of this HEU fuel through the H Canyon facility at its Savannah River Site, where the HEU was recovered and blended down for use as fuel in Tennessee Valley Authority nuclear power reactors.

Along with SNF, DOE's HLW inventory includes some 90 million gallons of HLW liquids, sludges and solids from past fuel reprocessing operations for weapons production. Most of this waste is being stored at DOE's Hanford, INL, and Savannah River sites—for the most part in large underground tanks made of stainless or carbon steel.

More recently, DOE has begun converting most of its inventory of liquid HLW into glass forms suitable for on-site storage in canisters. In addition, DOE manages a small quantity of HLW from the short-lived operation of a commercial reprocessing facility at West Valley, New York in the late 1960s and early 1970s.

This waste is now stored dry in the chemical process cell of the main plant and is slated for dry cask storage pending the availability of a repository. Figure 11 shows the geographic distribution of DOE's HLW inventory.

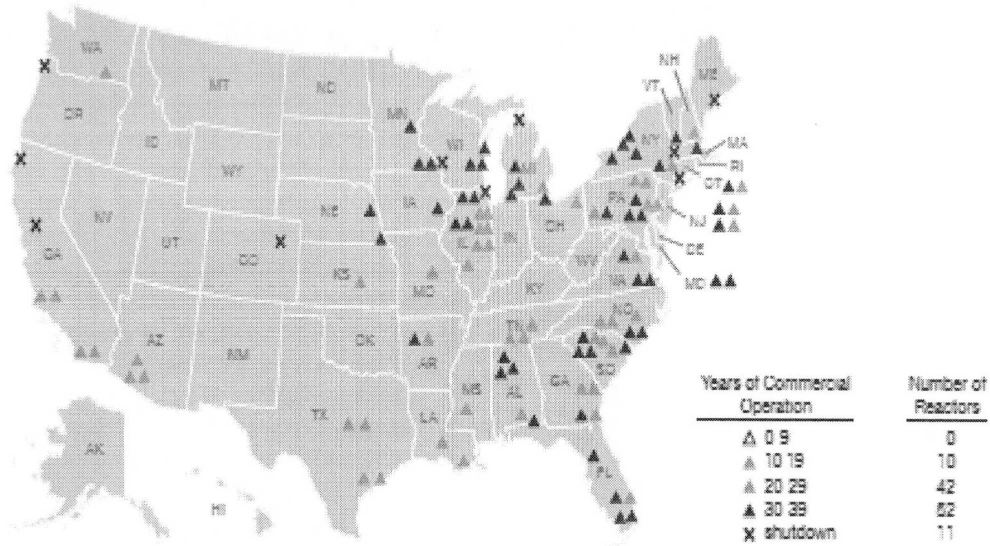

Source: U.S. Nuclear Regulatory Commission.

Figure 9. Operating and shutdown commercial nuclear power reactors in the united states.

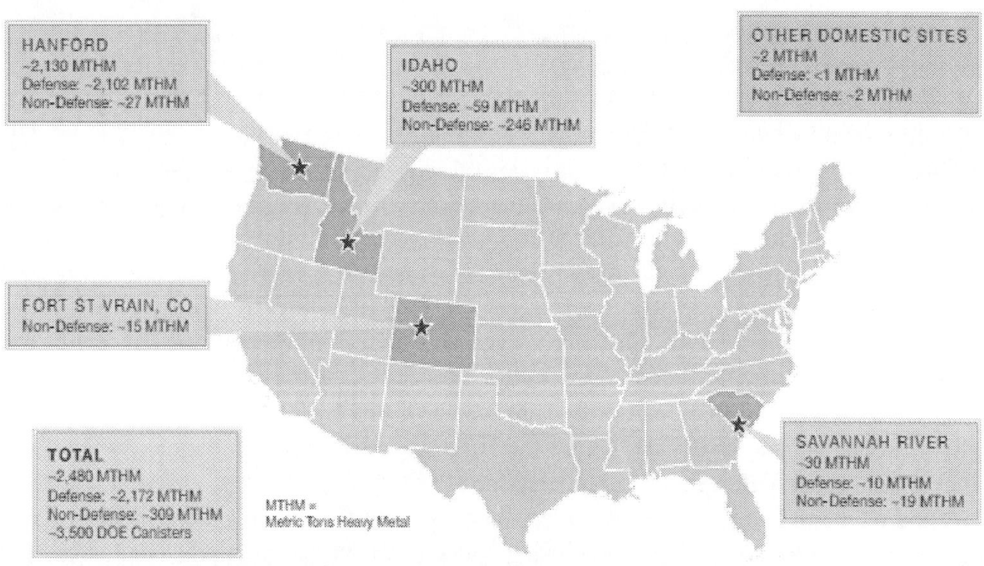

Source: BRC staff using information from DOE and other sources.

Figure 10. U.S. DOE spent nuclear fuel inventory in 2010.

To address state concerns about the indefinite storage of spent fuel and HLW at existing federal facilities, DOE has entered into agreements with Idaho and Colorado to remove all spent fuel and other HLW by 2035.

Failure to meet this deadline will trigger monetary penalties and restrictions on further shipments of waste material into these states, including the shipment of Navy spent fuel into Idaho (see further discussion of Navy spent fuel in section 3.3.3 below).

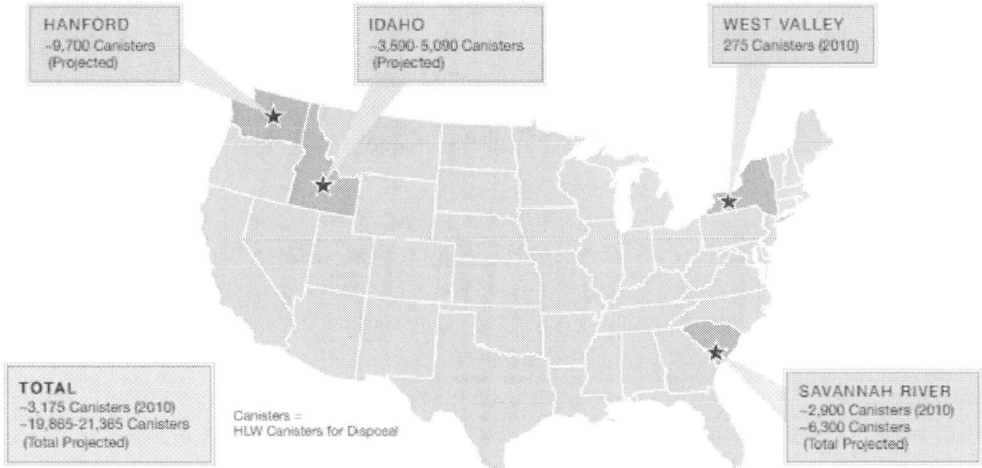

Source: BRC staff using information from DOE and other sources.

Figure 11. U.S. DOE high-level waste inventory in 2010.

Finally, DOE has statutory responsibility for disposing of greater than Class C (GTCC) low-level radioactive waste. This category of waste includes activated metals from decommissioned power plants, some sealed sources, and nondefense-related transuranic (TRU) waste. The current volume of GTCC waste totals approximately 1,100 cubic meters; future decommissioning of existing nuclear power plants is expected to generate an additional 4,200 cubic meters. GTCC waste may require deep geologic disposal. A path for the ultimate disposal of this class of waste has yet to be identified, although DOE has developed a draft environmental impact statement that evaluates GTCC disposal alternatives and is working toward a final environmental impact statement and record of decision.[17] The alternatives being considered include disposal in a deep geologic repository and disposal in boreholes at depths up to 1,000 ft.

3.3.3. Navy Spent Fuel

The federal inventory currently includes a relatively small quantity of spent fuel—approximately 27 metric tons—from naval reactors that power the nation's fleet of 83 nuclear-powered submarines and aircraft carriers. The inventory of naval SNF is growing slowly, at a rate of 1 to 2 metric tons per year, due to the continued operation and necessary re-fueling of reactors on these ships. The Navy's current projection is that a total of 65 metric tons will be generated by 2035, all of which would be destined for disposal in a repository (the Navy does not consider reprocessing as an option for its SNF).[18]

The Naval Nuclear Propulsion Program (NNPP), an integrated program carried out jointly by the Navy and DOE, manages spent naval reactor fuel, which for many years has been shipped to INL for technical studies and storage pending final disposal. Current practice is to transport the Navy's SNF from the shipyards where refueling occurs by rail, in specially-designed casks, to the Naval Reactors Facility (NRF) on the INL. At NRF, the spent fuel is placed in a water pool similar to those used for commercial and other DOE spent fuel, examined to confirm that its actual condition is consistent with expectations, and evaluated for other technical studies (e.g., to improve the efficiency of future nuclear fuel). After an appropriate cooling period, the SNF is transferred to specifically-designed multipurpose canisters suitable for dry storage at INL as well as subsequent transportation and disposal; the naval SNF will, under current plans, never be removed from these canisters.[19] At present, the Navy has about 50 loaded canisters in dry storage at INL; by 2035, it estimates there will be just over 350 canisters ready for disposal. For perspective, the Yucca Mountain license application allocated space for 400 canisters of naval SNF in the total of 11,000 canisters it was designed to hold.

In 1995, DOE and the Navy entered into a formal agreement with the State of Idaho (known as the Batt Agreement). Among numerous other provisions, the Batt Agreement covers the storage, treatment, and disposal of DOE and Navy SNF stored at INL.[20] It allows limited quantities of naval SNF to continue to be shipped to INL (at an average rate of about 20 casks per year).

It also sets two deadlines: first, that all SNF then at INL be placed in dry storage by December 31, 2023 and second, that spent fuel be removed from Idaho by January 1, 2035. If this last milestone is not met, the Navy will face a significant financial penalty of $60,000 for each day the waste remains in Idaho after January 1, 2035.[21] (A 2008 addendum to the Agreement modified its terms to allow for continued management and technical evaluation of a modest in-process inventory of naval SNF at NRF beyond 2035.) The Agreement also allows the State of Idaho to stop further shipments of Navy fuel to INL at any time if any key parts of the Agreement are not upheld.

In a recent review of how the suspension of work on Yucca Mountain could impact SNF storage at DOE sites, the Government Accountability Office (GAO) reported[22] that the Navy's "greater concern" was not the financial penalties in the Idaho agreement if the 2035 deadline is not met, but instead the possibility that Idaho would bar further Navy shipments of SNF to the state. This would dramatically affect the Navy's ability to refuel its nuclear fleet.

The Batt Agreement also requires that naval SNF be included "among the early shipments to a permanent geologic repository or interim storage site." However, at the BRC's September 13, 2011 meeting in Denver, a representative of the State of Idaho stated that "It may not make sense to send DOE spent nuclear fuel to interim storage as most of that waste is already in dry storage and some of it (Navy fuel) is ready for final disposal."[23] The BRC Transportation and Storage Subcommittee reached a similar but more general conclusion, stating in its draft report that "[t]here appear to be no technical or safety-related reasons to move defense high-level waste and spent fuel from temporary storage at the DOE sites where these materials are now located, before final disposal capacity becomes available." The Commission concurs with these conclusions.

Furthermore, in comments on the draft BRC report,[24] the Navy has stated that the focus should be on disposing of naval SNF in a geologic repository when one becomes available. The Navy's comments point out that "naval SNF as a waste form is well suited for geologic disposal" and that "the NNPP has invested significant resources in a packaging and transportation infrastructure based on geologic disposal." That said, it is important to stress that under current law, DOE (not the Navy) is responsible for final disposal of this federally-owned spent fuel.

The importance of providing a path forward for the disposition of Navy spent fuel is yet another reason why the Commission recommends that the United States promptly resume a program leading to the development of one or more deep geologic repositories.

3.4. History of Nuclear Waste Management Policy in the United States

Spent fuel and HLW have been produced in the United States since the 1940s, first as a byproduct of nuclear weapons research and production and later also as a byproduct of the civilian nuclear power industry. The record of past efforts to manage and dispose of these materials is long and complicated, so the overview presented here is necessarily condensed. A more complete history of nuclear waste policy in the United States is available from many sources (links to some of these sources are available at www.brc.gov).

3.4.1. Early U.S. Policy on Nuclear Waste Management (1940s–1982)

In the 1940s, during the early days of nuclear weapons development in the United States, national security considerations took precedence over concerns about the safe disposal of nuclear waste. With the emphasis on rapid production of plutonium for use in weapons, storage in large, underground steel tanks was deemed adequate as an interim means of isolating the highly radioactive liquid waste that remained after acid was used to dissolve irradiated nuclear fuel as part of the plutonium separation process. Even at the time, however, the underground tanks were not considered a long-term solution; in a 1949 report the Atomic Energy Commission (AEC)[25] emphasized that "better means of isolating, concentrating, immobilizing, and controlling wastes will ultimately be required."

In 1957, the National Academy of Sciences (NAS)[26] issued a report (titled *The Disposal of Radioactive Waste on Land*[27]) that looked specifically at the question of nuclear waste disposal. That report reached several conclusions, among them that "radioactive waste can be disposed of safely in a variety of ways and at a large number of sites in the United States" and that geologic disposal in salt deposits represents "the most promising method of disposal." The NAS further concluded that solidification of liquid waste for transport and disposal would be "advantageous" and that transportation issues would need to be considered in the location of waste disposal facilities.

Prompted by these recommendations, the AEC began investigating mined geologic disposal options and potential salt bed repository sites in the late 1950s. Its early efforts included experiments with solids and liquids in salt mines and exploratory work on methods for solidifying liquid wastes. In June 1970, the AEC announced plans to investigate an abandoned salt mine in Lyons, Kansas as a potential demonstration site for the disposal of HLW and low-level waste (LLW). At the time, the AEC anticipated that the Lyons site could begin accepting LLW as early as 1974 and HLW by 1975. By 1971, however, state opposition to the project was growing and in 1972, after a number of technical problems had emerged that called into question the geological integrity of the Lyons site, the AEC announced that it would seek alternative sites and also pursue the development of long-term surface storage facilities for the waste.

During the same time period (i.e., in the early 1970s), the AEC—at the invitation of the local community—began exploring an area of deep salt beds near Carlsbad, New Mexico as a potential repository site for high-level radioactive waste. Disposal at the site, which became known as the Waste Isolation Pilot Plant or WIPP, was subsequently limited to defense-related TRU waste. Congress authorized WIPP to begin receiving waste as early as 1979 but it took until 1999, 20 years later, before the first shipments began arriving at the facility (see text box).

The search for a suitable site for long-term geologic disposal of spent fuel and HLW continued throughout the 1970s, first under the AEC and later under its successor agency, the Energy Research and Development Administration (ERDA).[28] Among the sites considered during this period were bedded salt formations in Michigan, Texas, and Utah; salt domes in Louisiana and Mississippi; basalt formations at Hanford; and welded volcanic tuff at Yucca Mountain in Nevada. Meanwhile, the focus of future waste management efforts had begun to shift as a result of policy changes prompted by weapons proliferation concerns.

Responding to these concerns, President Ford in 1976 issued a presidential directive deferring commercial reprocessing and recycling of plutonium in the United States. In 1977, President Carter extended this deferral indefinitely and directed the relevant federal agencies to focus on alternative fuel cycles and re-assess future spent fuel storage needs. The Carter policy was later reversed by President Reagan but for a variety of reasons, including cost, commercial reprocessing was never resumed.

Recognizing that the commitment to an open fuel cycle with no spent fuel reprocessing would have an impact on the quantity and type of waste produced by the commercial nuclear power industry in the future, a DOE-led Interagency Review Group in 1979 recommended that a number of potential repository sites for spent fuel and HLW be identified in different geologic environments and in different parts of the country. Specifically, the Interagency Review Group recommended "several repositories sited on a regional basis insofar as technical considerations permit." The Group saw multiple regional repositories as a way to respond to several concerns, including: (1) accommodating uncertainties inherent in future nuclear waste inventory projections; (2) reducing system-wide transportation requirements; (3) promoting regional equity in the siting of high-level radioactive waste facilities; and (4) providing "redundancy that would hedge against the possibility of operational difficulties causing unexpected repository shutdown." At the same time, the Interagency Review Group was aware that with a regional approach "there is a risk that organizational and political commitments might develop to particular regions or locations to such an extent that less than full attention would be given to safety, environmental and security considerations." For this reason the Group urged DOE to "be certain that technical adequacy is a prerequisite for site selection" and to "provide adequate assurance to the public in this regard."

3.4.2. U.S. Policy under the Nuclear Waste Policy Act (1982–Present)

Passage of the Nuclear Waste Policy Act (NWPA) in 1982 marked the beginning of a new chapter in U.S. efforts to deal with the nuclear waste issue. The legislation was the product of four years of Congressional debate marked, on the one hand, by growing concern about an imminent shortage of spent-fuel storage pool capacity at operating reactors and, on the other hand, by an equally urgent concern on the part of individual states that they not be selected to host a repository site.

Recognizing the need for a Congressional mandate to overcome opposition to the selection of any given site, Congress sought through the NWPA to establish a fair and technically sound process for selecting repository locations. In fact, to avoid the perception that any one state or locale would be asked to bear the entire burden of the nation's nuclear waste management obligations, the Act provided for the selection of two repository sites (though not stipulated in the legislation itself, it was widely assumed that one of these sites would be located in the West, the other in the East).

The Waste Isolation Pilot Plant (WIPP)

WIPP is the world's only operating deep geological repository for long-lived nuclear waste.

It is located in an ancient 2000-foot deep salt bed, 26 miles southeast of Carlsbad in Eddy County, New Mexico. WIPP is a DOE facility and accepts only defense TRU waste—that is, nuclear waste from past weapons programs that is not considered high-level waste, but that contains long-lived radioactive transuranic elements such as plutonium.

The Atomic Energy Commission first began looking at salt beds in southeastern New Mexico for the disposal of defense wastes in the early 1970s.

The current WIPP site was selected for exploratory work in 1974 after local officials expressed interest in being considered; five years later Congress authorized an R&D facility at the site. By this time, tensions had begun to emerge between the federal government and New Mexico, which was concerned about the inclusion of high-level waste and commercial spent nuclear fuel in some of the early plans for WIPP. Authorizing legislation adopted by Congress in 1979 stipulated that WIPP could not be used for the permanent disposal of spent fuel and high-level waste but it also heightened tensions by denying the state veto power and removing the project from the licensing authority of the NRC. Two years later, when DOE attempted to move forward with construction, New Mexico filed suit against both DOE and the U.S. Department of the Interior (which had jurisdiction over the land at the site).

That suit was eventually settled out of court, but over the next decade difficulties arose in a number of areas, from problems with the design of transport casks to concerns about funding for road improvements, controversies over health and environmental standards, and plans for an early test phase during which waste could be stored at the facility without meeting final disposal standards. In 1987, DOE began withdrawing land around WIPP from general use and announced that the facility would open in 1988. This proved unrealistic, as efforts to complete the land withdrawal failed over the next few years. In 1991, the state again filed suit—this time to prevent the transfer of land from public uses to use for a WIPP testing phase.

In response, the courts issued an injunction against proceeding with the facility according to DOE's plans.

Progress on WIPP resumed when Congress passed the Land Withdrawal Act in 1992. This legislation required EPA (not DOE) to certify that WIPP met applicable standards and gave the state authority to regulate mixed waste at WIPP under the Resource Conservation and Recovery Act (RCRA), including issuing a hazardous waste permit for the facility. Other provisions prohibited high-level waste at WIPP, even for experiments; provided additional funding for highways and emergency preparedness; and directed DOE to prepare plans for retrievability and eventual decommissioning. DOE later announced that it would move radioactive waste experiments out of WIPP and into the national laboratories.

In 1998, EPA certified that WIPP met all applicable federal regulations for the disposal of TRU waste. Soon after, the 1992 court injunction was lifted and in 1999 WIPP received its first shipment of waste.

As of mid-November 2011, WIPP had received 10,181 shipments for a total waste volume of approximate 68,200 cubic meters. DOE currently estimates that work to begin closing WIPP could commence as early as 2030.

In contrast to the years of controversy and delay that surrounded the development of the facility, WIPP now enjoys considerable support at the state and local level.

And to further ensure that the end result would not be a single, national repository, Congress included provisions explicitly limiting the capacity of the first repository to 70,000 metric tons until a second repository was opened.

Beyond establishing a process for the selection of two permanent geologic spent fuel and HLW repositories, the NWPA included a number of other provisions:

1. Established a new Office of Civilian Radioactive Waste Management (OCRWM) within DOE, with a director appointed by the President and confirmed by the Senate.
2. Authorized DOE to enter into contracts with utilities for federal removal of spent fuel from reactor sites beginning by 1998 in return for a fee on utilities' sales of nuclear-generated electricity.
3. Directed DOE propose a site and design for "monitored retrievable storage" of nuclear waste prior to its being shipped to a disposal site.
4. Provided for federal storage of civilian spent fuel/HLW on an interim basis in cases of need.
5. Granted states certain rights with respect to oversight over waste storage or disposal sites within their borders and the ability to veto DOE siting decisions, subject to override by both houses of Congress.

6. Gave the Nuclear Regulatory Commission (NRC) responsibility for licensing the construction and operation of waste facilities, subject to public health and environmental standards established by the Environmental Protection Agency (EPA).

In May 1986, Energy Secretary John Herrington recommended the Hanford site in Washington State, Deaf Smith County in Texas, and Nevada's Yucca Mountain for detailed site characterization as leading candidates for the nation's first permanent high-level geologic waste repository. By that time, however, DOE's efforts to identify promising sites—not only for the two permanent repositories but also for a monitored retrievable storage (MRS) facility—were drawing strong opposition from the elected officials of all potentially affected states. (As an aside, we note that while the federal government's performance on nuclear waste management has left a lot to be desired, state opposition has played a significant role in the federal government's failures. As we discuss at length in later sections, it is clear that the cooperation of affected state governments will be vital to the success of the nuclear waste program going forward.)

Citing rising costs and lower projections for nuclear waste production in the future, Secretary Herrington announced that DOE was suspending efforts to identify and develop a second permanent geologic repository. This announcement also came in May 1986—not surprisingly, it served to intensify the opposition of the three states that had been selected as potential hosts for the first repository.

Faced with a deteriorating political situation[29] and growing recognition that the NWPA's original timelines and cost assumptions were unrealistic, Congress revisited the issue of nuclear waste management in 1987. The resulting NWPA Amendments Act of 1987 halted then ongoing research in crystalline rock of the type found in the Midwest and along the Atlantic coast, cancelled the second repository program, nullified the selection of Oak Ridge, Tennessee as a potential MRS site, and designated Yucca Mountain as the sole site to be considered for a permanent geologic repository. The decision was widely viewed as political and it provoked strong opposition in Nevada, where the 1987 legislation came to be known as the "Screw Nevada" bill.

To address concerns about the technical integrity of DOE's assessments, Congress established a new federal agency—the U.S. Nuclear Waste Technical Review Board (NWTRB)—for the sole purpose of providing independent scientific and technical oversight of DOE's waste management and disposal program. Congress also tried a new approach to overcoming state and local opposition: under the 1987 amendments, states could receive up to $20 million per year for hosting a repository and up to $10 million per year for hosting an MRS site. The amendments also created the Office of the United States Nuclear Waste Negotiator with a presidentially appointed head authorized to reach agreements with states or Indian tribes to host nuclear waste facilities under any "reasonable and appropriate terms."

At the time, a negotiated, voluntary agreement seemed the best hope for siting a MRS facility that would enable DOE to meet its obligation to begin accepting waste from commercial reactors by 1998.[30] The hope was that a voluntary process that offered economic incentives might succeed where other siting efforts had failed.

This hope proved short-lived. The Office of the Nuclear Waste Negotiator closed in 1995, after just a few years in operation; the first head of the agency had not been appointed by President George H.W. Bush until 1990. And neither he nor his successor (who was appointed by President Clinton in 1993) succeeded in reaching an agreement despite reaching out to hundreds of potential host communities and Indian tribes and identifying a number of potentially promising candidate sites.

At one point in 1992, seven communities (including five Indian tribes) had formally notified the government of their interest in being considered.[31] Each of these communities was entitled to receive $100,000 in DOE grants, while those that agreed to participate in a second phase of study could potentially have been eligible for several million dollars in grants. In no case, however, was a host state supportive of having the process go forward.

3.4.3. Experience with the Yucca Mountain Repository Program

Following the dictates of the 1987 NWPA Amendments, DOE continued detailed site characterization studies at Yucca Mountain through the 1990s and issued a formal finding on the suitability of the site in 2002—four years past the 1998 deadline by which the federal government was obliged to begin accepting commercial nuclear waste for disposal under the NWPA. The President's subsequent recommendation of the site to Congress prompted Nevada, which had remained staunchly opposed to the project throughout, to file an official "Notice of Disapproval." A Congressional resolution to override the state's veto, however, was signed by the President, clearing the way for DOE to apply to the NRC for a license to commence construction. The latter step was supposed to follow fairly quickly (within 90 days), but due to litigation over the repository safety standards, persistent funding shortfalls, and other problems it took another six years before the application for construction authorization was filed with the NRC.

In the end, DOE succeeded in completing the world's first license application for a HLW repository. Submitted to the NRC in June 2008, the license application was deemed suitable for review three months later. Within a year, however, the new Administration declared its intent to suspend further work on Yucca Mountain and later moved to withdraw the application for a construction license to the NRC. At this point, with key decisions by the courts and the NRC still pending, the future of the Yucca Mountain project remains uncertain.

Several attributes of the nation's approach to nuclear waste management generally, and to the selection and characterization of the Yucca Mountain site in particular, are widely viewed as having contributed to the long delays and significant difficulties encountered in implementing the NWPA Amendments. First, DOE's termination of the siting process for the second repository, combined with Congress's subsequent action to short-circuit the site selection process established under the original NWPA and single out Yucca Mountain as the sole site for consideration, created a widespread perception that the repository location was being determined on the basis of primarily political, rather than technical and scientific, considerations.[32] Second, neither the original site selection process established by the Act nor the subsequent legislative designation of Yucca Mountain as the sole site for consideration could be viewed as consent-based since the State of Nevada was not asked for, and did not provide, consent for the site to be selected for investigation. On the contrary, the state and a majority of its citizens strongly opposed the selection of Yucca Mountain as a potential repository site, although the project did have some support from local constituencies. (In comments submitted to the Blue Ribbon Commission during the course of its deliberations, several counties in Nevada—including Nye, Mineral, and Lincoln counties—have expressed support for the Yucca Mountain project or for at least allowing the license approval process for Yucca Mountain to go forward.)

A third issue, and one that pre-dated the decision to focus only on Yucca Mountain, was the practice of setting unrealistic and rigid deadlines. As DOE failed time and again to meet various deadlines, confidence in the federal government's competence to manage either the Yucca Mountain project or its broader obligations concerning the management of civilian and defense nuclear waste eroded among all parties involved. Key stakeholders, including not only citizens of the communities where these materials were being stored, but also nuclear utilities and their customers, who continued to pay into the Nuclear Waste Fund even as the repository program fell further and further behind, became increasingly frustrated. All the while, the federal government was also exposing itself (and U.S. taxpayers) to liability and large financial damages arising from its failure to comply with its obligations under the Act and DOE contracts with utilities (discussed in section 3.6) in a timely manner.

Another fundamental flaw of the repository development process established under the 1982 Act, and one that carried over to Yucca Mountain after it was designated, was its relative inflexibility and prescriptiveness. This made it difficult to adapt or respond to new developments, whether in the form of new scientific information, technological advances, or (just as important) the expressed concerns of potentially affected publics and their representatives. The 1987 NWPA Amendments made no provision for an alternative path forward if Yucca Mountain proved untenable. This lack of adaptability further undermined confidence in the analysis and planning conducted by DOE and other federal agencies, making it easy to view these efforts as mere paper exercises, rigged to justify a preordained conclusion. Similarly, by directing EPA to develop safety standards specific to the Yucca Mountain site in the Energy Policy Act of 1992, Congress undermined confidence that those standards represented an independent scientific judgment about what was necessary to protect human health and the environment.

These attributes of the Yucca Mountain siting process led to a serious erosion of trust, especially among the people of the state of Nevada. The recent decision by the Administration to attempt to withdraw the Yucca Mountain license application has further diminished confidence in the government's ability to provide a safe and timely solution for the disposal of spent fuel and HLW. This is not a comment on the merits of the Administration's decision; the Commission was not asked to examine that issue and offers no opinion. However, it is clear to the Commission that waste cleanup commitments were made to states and communities across the United States, and to the nuclear utility industry and its ratepayers and shareholders, that have not been upheld. The decision to suspend work on the repository has left all of these parties wondering, not for the first time, if the federal government will ever deliver on its promises.

3.5. Utility Initiatives

Following the federal government's abandonment of efforts to site an MRS facility through the Office of the Nuclear Waste Negotiator, a group of eight nuclear utilities formed a private consortium, called Private Fuel Storage, LLC (or PFS), with the objective of finding a community willing to host such a facility. In 1996, PFS signed an agreement with the leadership of the Skull Valley Goshute Indian Tribe to open an MRS facility on the Tribe's reservation in Utah. Details about the amount of compensation being offered have not been disclosed, but reportedly include millions of dollars in promised payments. The effort has generated controversy within the Tribe, however, and is strongly opposed by the state of Utah and a majority of Utah citizens, according to media reports.[33]

PFS subsequently applied for and received a license to construct the proposed facility from the NRC. In a comment letter from the Governor of Utah on the BRC draft report, one of the many reasons cited for state-level opposition to the PFS project was that the BRC-recommended "consent-based, transparent and standards- and science-based approach to nuclear waste management...was totally lacking in the NRC proceeding to license a private SNF storage facility."[34] The PFS project was later halted when the Department of the Interior's Bureau of Indian Affairs did not approve the Tribe's lease of land for the storage facility (citing the risk that it would become a permanent repository by default) and the Bureau of Land Management denied needed railroad rights of way over federal land. These decisions were recently found by a federal court to be arbitrary and capricious and were remanded for reconsideration, leaving the future of the facility, according to a recent (2010) article that appeared in the *Environmental Law and Policy Review*, "uncertain."[35]

3.6. Current Waste Acceptance Commitments and Litigation

The NWPA established the Nuclear Waste Fund (NWF) and authorized DOE to enter into Standard Contracts with commercial reactor licensees. During the 1980s, DOE entered into 76 such contracts. Under the Standard Contract, DOE agreed to take title to spent fuel or HLW and, in return for a payment of fees to the NWF, dispose of the materials beginning not later than January 31, 1998 (the fee amount was initially set at 1 mill or one-tenth of one cent per kilowatt-hour; it is reviewed annually to ensure that it is adequate to cover program costs and has never been changed). The NWPA also stipulated that the NRC may not issue or renew a commercial reactor license without a Standard Contract in place. In 2008, DOE amended the Standard Contract for new reactors. Under the amended Standard Contract, DOE is not required to accept spent fuel until 20 years after the expiration of the reactor's operating license and any extensions thereto.

Despite the NWPA mandate to begin accepting spent fuel and HLW for delivery to and disposal at a permanent repository no later than January 31, 1998, no permanent repository has yet been licensed by the NRC.

This has led numerous utilities to file suit to recover damages associated with the government's failure to meet the 1998 waste acceptance deadline. The status of this litigation and of associated taxpayer liabilities, which are already running into the billions of dollars, is discussed in greater detail in section 8.5 of this report.

3.7. Linkages between the Back-End of the Fuel Cycle and the Future of Nuclear Power

All forms of energy production have impacts; in many cases, these impacts include generating wastes or by-products. The spent fuel from nuclear power reactors gets (and deserves) special attention because of the hazard it poses and because it contains certain elements (primarily plutonium) that can be extracted and re-used either for power production or in nuclear weapons.

For these reasons, the successful management of SNF has long been viewed as necessary if nuclear power is going to remain a viable energy option. As discussed earlier, the assumption in the early days of the industry was that uranium was scarce and that the back end of the nuclear fuel cycle should be managed in a way that provided plutonium and other elements to power future nuclear reactors.

Several nations continue to extract plutonium (and uranium) from spent fuel for planned or ongoing re-use, but as uranium has been found to be more naturally abundant than first expected,[36] many nations are now primarily focused on developing options for the near-term safe storage of HLW and spent fuel and for the long-term isolation of these materials from people and the environment.

The United States may someday find it advantageous to extract useful elements from spent fuel for re-use (later sections of this report discuss the value of research, development, and demonstration (RD&D) to ensure that future generations have a wide range of nuclear fuel cycle options to choose from).

In the nearer term, laws in several states that put a moratorium on new nuclear plant construction until certain waste management conditions have been met, together with the NRC's Waste Confidence rulemaking, which was first initiated in October 1979, create the most direct linkage between progress on nuclear waste disposal and the future prospects of the domestic nuclear power industry.

3.7.1. State Moratoria

Efforts to establish a formal legal link between the use of nuclear power and solutions for the back end of the nuclear fuel cycle began in California in the mid-1970s when it became clear that the prospects for successfully completing either reprocessing capacity or a waste disposal system were increasingly dim.[37]

At that time, the California legislature adopted a law that allows the state to grant permits for new nuclear power plants only if the California Energy Resources Conservation and Development Commission can make a finding that the federal government has identified and approved a demonstrated technology for the disposal of spent fuel/high-level nuclear waste.

The California law was challenged on grounds that federal law preempts state statutes concerning nuclear power, but it was upheld by the Supreme Court, which found that California had acted on the basis of an economic rather than a nuclear regulatory rationale.

Subsequently, eight other states—Connecticut, Illinois, Kentucky, Maine, New Jersey, Oregon, West Virginia, and Wisconsin—adopted statutes that tied approval of new reactors to (at a minimum) progress on the issue of waste disposal.[38] Recent years have seen efforts to repeal those laws in some states, although none have succeeded so far.

3.7.2. NRC Waste Confidence Proceeding

The NRC's Waste Confidence proceeding grew out of an NRC statement that, as a matter of policy, it "would not continue to license reactors if it did not have reasonable confidence that the wastes can and will in due course be disposed of safely."[39] While the Waste Confidence Rule is narrowly applied so that waste management and disposal issues don't have to be re-litigated every time the NRC reviews a license application, the NRC itself has indicated that this proceeding has broader policy implications.

The NRC's first waste confidence decision was issued in 1984. In it, the NRC found reasonable assurance that safe disposal of HLW and spent fuel in a geologic repository is technically feasible, and that repository capacity would become available in the 2007–2009 timeframe.[40]

The NRC also found that HLW and spent fuel will be safely managed until repository capacity is available, that spent fuel generated in any reactor can be stored safely and without significant environmental impacts for extended periods, and that spent fuel storage will be available as needed.

In its initial waste confidence rulemaking, the NRC said it would revisit the issue periodically. Five years later, the NRC took another look and basically reaffirmed and expanded the original finding. Specifically, the NRC made clear that its confidence in the environmental soundness of on-site storage extended for at least 30 years beyond the licensed lifespan of operating reactors, including life-extensions that might occur from license renewals. At the same time, the NRC clarified its thinking about repository timing to say that there was reasonable assurance one or more repositories would be made available within the first quarter of the 21st century.

In 1999, the NRC reviewed the matter again and found that experience and developments in the interim confirmed the confidence it had earlier expressed. The Commission said it would look at the issue again after the ongoing repository process had run its course, or if "significant" and "unexpected" events occurred that warranted a reassessment.

In 2007, the nuclear industry called on the NRC to reaffirm its waste confidence decision, citing concern that uncertainties about waste management were affecting investment decisions about new nuclear power plants.[41] In 2010 the NRC issued revisions to the agency's waste confidence findings and regulations.

The revisions expressed the NRC's confidence that: (1) the nation's SNF can be safely stored for at least 60 years beyond the licensed life of any reactor and (2) that sufficient repository capacity will be available when necessary (though on this occasion the NRC did not specify an anticipated timeframe).[42] *The NRC also made clear, however, that by revising its earlier waste confidence findings it did not intend to signal that it was endorsing the indefinite storage of spent fuel at reactor sites.*

On February 17, 2011, the Natural Resources Defense Council filed a petition for review with the United States Court of Appeals for the DC Circuit challenging the NRC's most recent waste confidence rule. The states of New Jersey, New York, Vermont and Connecticut have also challenged the rule.

3.7.3. Impact of Waste Management Uncertainty on Nuclear Plant Investment Decisions

Beyond the formal linkages discussed in the preceding sections, there is a broader question about the impact of current uncertainty about waste management on decision making about new nuclear plants in the United States. As already noted, the NRC's most recent waste confidence position was prompted by industry concerns, including specifically concerns related to uncertainty about the licensing process for Yucca Mountain (at the time, DOE had not yet submitted the license application).

The decision to attempt to withdraw the license application appears to have heightened these concerns. A witness at a recent Congressional hearing on the subject argued that the current "complete lack of direction on nuclear waste management and...dereliction of responsibility on the part of the federal government...creates substantial government-imposed risk on the nuclear industry, which is the primary obstacle to an expansion of U.S. nuclear power."[43]

3.8. International Context/Comparison

In the course of its deliberations, the Commission sent delegations to Finland, France, Japan, Russia, Sweden and the United Kingdom to learn about these countries' waste management programs.

The Commission also heard presentations about the nuclear waste management programs of Canada and Spain.

We found that the experiences of other countries, some of which are at or near the stage of licensing a deep geologic repository, offer many useful insights for the U.S. program. Some of those insights are discussed in section 6 of this report which provides recommendations on a new approach to siting.

In addition, appendix C summarizes the status of other countries' waste programs drawing primarily from information collected by the NWTRB as part of a 2009 report to Congress and the Secretary of Energy on the status of nuclear waste management efforts around the world.[44]

4. The Need for Geologic Disposal

> The central flaw of the U.S. nuclear waste management program to date has been its failure to develop permanent disposal capability.

This failure has occurred despite decades of efforts and a legislative mandate in the form of the Nuclear Waste Policy Act (NWPA). The conclusion that disposal is needed and that deep geologic disposal is the scientifically preferred approach has been reached by every expert panel that has looked at the issue and by every other country that is pursuing a nuclear waste management program.

Moreover, all spent fuel reprocess or recycle options generate waste streams that require a permanent disposal solution.

Lack of disposal capability is not only at the heart of the U.S. government's inability to honor its waste management obligations to date, it is—especially after Fukushima—a source of renewed concern to the general public, a growing liability to taxpayers, and a burden to nuclear utilities, their ratepayers,[45] and the nuclear energy industry's prospects going forward.

One of our central recommendations, therefore, is that the *United States should undertake an integrated nuclear waste management program that leads to the timely development of one or more permanent deep geological facilities for the safe disposal of spent fuel and high-level nuclear waste.*

This section discusses the ethical, technical, and practical grounding for that recommendation and elaborates on the options available for developing disposal capacity.

4.1. The Rationale for Developing Disposal Capacity

Spent nuclear fuel and other high-level radioactive wastes[46] contain elements that present a potentially significant radiation hazard to exposed populations and ecosystems. These hazards diminish over time, often declining significantly in the first few hundred years and thereafter much more gradually.

As detailed in section 3 of this report, the decay processes for some constituents of spent fuel and HLW take hundreds of thousands of years or more. Therefore, the central challenge for managing these materials is to store and finally dispose of them in a way that provides adequate protection of the public and the environment over very long periods of time.

The need for a disposal solution is quite clear in the case of nuclear materials with a low probability of reuse—a category that includes defense and commercial reprocessing wastes and many forms of spent fuel currently in government hands. From a practical standpoint, the Commission believes it is also very likely that disposal will be needed to safely manage at least some portion of the existing commercial SNF inventory.

This is because there is no cost-effective way using existing technology to separate the most hazardous and long-lived radioactive elements in spent fuel and convert them to short-lived or stable isotopes.[47] In the meantime, the more frequently discussed option is to re-cycle and re-use some of the constituents of spent fuel.

This option involves reprocessing spent fuel to separate and remove the still usable constituents for re-use as reactor fuel. Options for partially or fully "closing" the nuclear fuel cycle are the subject of ongoing research and development in the United States and elsewhere and are discussed in section 11 of this report. *The central point is that all spent fuel reprocessing or recycle options generate waste streams.*

Moreover, some of these waste streams contain sufficient amounts of long-lived radioactive elements that the need for a long-term disposal solution cannot be eliminated with any foreseeable separations technology.[48]

In concluding that disposal capacity will be needed, the Commission is echoing the consensus view, not only of numerous former expert panels[49] that have looked at the situation in the United States but also of all countries with significant nuclear waste inventories (including those that are currently reprocessing spent nuclear fuel) and of major international organizations such as the IAEA and the OECD's Nuclear Energy Agency (NEA).[50,51]

4.2. The Case for Disposal

The ethical case for developing disposal capacity for spent fuel and high-level nuclear wastes from the nation's past weapons programs and civilian nuclear power industry is outlined in section 2.3.1 of this report, which highlights the obligation to avoid placing an undue burden on future generations. From a legal standpoint, the U.S. government's general obligation to provide a timely disposal solution has been established for more than three decades. In fact, under current law the federal government was obliged to begin accepting commercial spent fuel by January 31, 1998.

Apart from commercial spent fuel, the federal government is also liable for the eventual disposition of waste from defense production facilities. Enforceable commitments to remove federally owned waste have been made in cleanup agreements with the host states of Washington, South Carolina, Colorado and Idaho. Direct disposal of both defense HLW and the West Valley HLW at an appropriate site (without interim storage at another location) should be pursued, as this material will never be further processed or re-used.

Finally, although much of the federally-owned HLW and spent fuel was generated to produce materials used in nuclear weapons, a smaller inventory of spent fuel exists and is being generated by the U.S. Navy's nuclear fleet. Continued Navy operations to examine and store this fuel in Idaho depend upon the future availability of disposal capacity at a suitable repository site for this fuel.

As we have already noted, the Commission's central conclusion concerning the need for disposal capability is consistent with decades of expert opinion and policy consensus in the United States and abroad. That the use of nuclear technologies—whether for defense purposes or for energy production—would necessitate a means for permanently disposing of their radioactive by-products has been known from the beginning. In short, because these materials exist, the ethical, legal, and practical obligation to dispose of them also exists. Regardless how one views the nuclear energy industry or its future prospects, all parties should be able to agree that there is little to be gained—and potentially a very high price to be paid—for continued deferral and delay in developing the capability for disposal. Moreover, only by moving forward can some of the key questions and uncertainties about a future disposal path for spent fuel and high-level nuclear waste be identified and resolved.

An underground chamber at the WIPP facility.

4.3. Options for Disposal

While several options for disposing of spent fuel and high-level nuclear waste have been considered in the United States and elsewhere, international scientific consensus clearly endorses the conclusion that *deep geological disposal is the most promising and accepted method currently available for safely isolating spent fuel and high-level radioactive wastes from the environment for very long periods of time.*[52]

In its deliberations, the Commission focused chiefly on two deep geologic disposal options: disposal in a mined geological formation and disposal in deep boreholes. The former has been the front-running disposal strategy in the United States for more than 50 years; it is also the approach being taken in other countries with spent fuel or HLW disposal programs. (An artist's rendering of the mined geologic disposal concept is shown in figure 12.) By contrast, disposal in deep boreholes may hold promise but this option is less well understood.

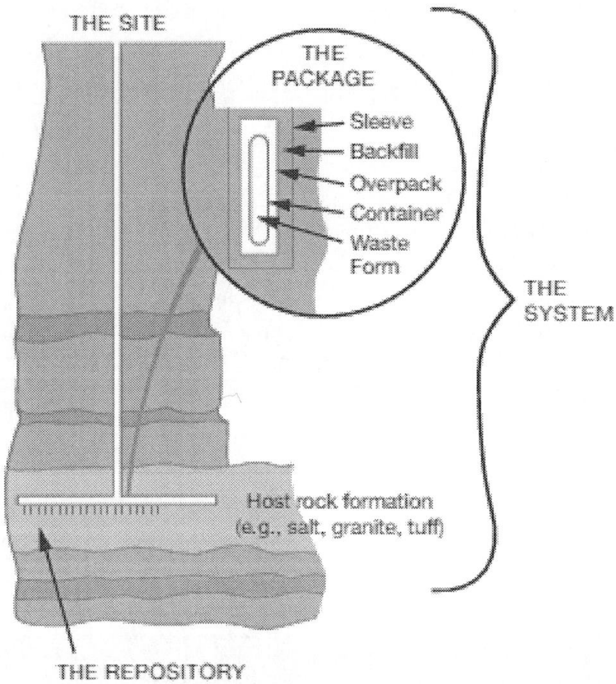

Mined geologic disposal will use a system comprised of engineered barriers (the waste package and the mined repository) and naturally occuring barriers (the host rock formation and the chemical and physical properties of the repository site itself) to provide long-term isolation of waste from the biosphere.

Department of Energy.

Figure 12. Mined geologic disposal concept.

Further RD&D is needed to fully assess its potential advantages and disadvantages and should be performed in parallel with the development of an updated safety standard (consistent with the new disposal safety standard the Commission recommends for mined geologic repositories).

In a mined geologic repository, wastes would be placed in engineered arrays in conventionally mined cavities deep beneath the earth's surface. The waste itself would be contained in canisters or other packages appropriate to its particular form, chemical content, and radiation intensity. As developed and studied around the world, proposals for geologic disposal also employ the concept of multiple barriers.[53] These include both engineered and geologic barriers that improve confidence that radioactive constituents will not return to the biosphere in biologically significant concentrations. Engineered barriers include the waste form itself, canisters, fillers, overpacking, sleeves, shaft and tunnel seals, and backfill materials. Each of these components may be designed to reduce the likelihood that radioactive material would be released and would be selected on the basis of site- and waste-specific considerations. Geologic barriers include the repository host rock and surrounding rock formations. While engineered barriers would be tailored to a specific containment need, geologic barriers would be chosen for their *in-situ* properties with respect to both waste containment and isolation.

The basic objective or standard of performance for a permanent waste repository was articulated by the IAEA in a 2003 report on the scientific and technical basis for geologic disposal of radioactive wastes: "to provide sufficient isolation, both from human activity and from dynamic natural processes, that eventual releases of radionuclides will be in such low concentrations that they do not pose a hazard to human health and the natural environment."[54]

Decades of research and site investigations in the United States and elsewhere suggest that a wide variety of rock types and geologic environments could—in combination with appropriate repository design—be suitable for achieving this objective. The rock types that have been considered for a deep geologic repository have included bedded and domed rock salts, crystalline rocks (i.e., granite or gneiss), clay, shale, volcanic tuffs, basalt, and various other types of sedimentary rocks.[55]

Each of these rock types and their particular geologic environments have advantages and disadvantages from a strictly technical perspective, and different geologic settings and emplacement methods may be better for particular types of waste. However, many or all of them may ultimately be found to demonstrate acceptable performance for a wide range of wastes. The geologic environment into which waste would be emplaced is a related and perhaps more important consideration than the type of rock by itself. The BRC has benefitted from visits to several facilities in different geologic settings in the United States and abroad. This exposure contributes to our collective observation that deep geologic disposal constitutes a vital element of all international waste management programs. It also reinforces our confidence that many geologic formations and sites that would be technically suitable for hosting a permanent repository can be found within the borders of the United States.[56]

Deep boreholes represent another form of deep geologic disposal that may offer benefits, particularly for the disposal of certain forms of waste. As we have already noted, however, this concept is less well understood than disposal in a mined repository and requires further exploration.[57] Basically, a deep borehole is a cased hole on the order of 45 centimeters (approximately 20 inches) in diameter drilled into crystalline basement rock to a depth of 4 to 5 kilometers (2.5 to 3 miles). In most designs, the bottom 1 to 3 kilometers would be filled with either vitrified HLW or spent fuel and some backfill or sealant would be added to fill in the gaps between the wastes and the well casing. Figure 13 illustrates the deep borehole disposal concept.

A number of possible advantages have been cited that support further efforts to investigate the deep borehole option. These include the potential to achieve (compared to mined geologic repositories) reduced mobility of radionuclides and greater isolation of waste, greater tolerance for waste heat generation, modularity and flexibility in terms of expanding disposal capacity, and compatibility with a larger number and variety of possible sites. On the other hand, deep boreholes may also have some disadvantages in terms of the difficulty and cost of retrieving waste (if retrievability is desired) after a borehole is sealed, relatively high costs per volume of waste capacity, and constraints on the form or packaging of the waste to be emplaced.

Other Disposal Concepts

A number of alternative disposal concepts or alternative types of sites for geologic disposal have been advanced over the years. For example, disposal on or beneath unoccupied islands has been considered in the context of options for siting an international repository or monitored storage facility.[58] Another option, sub-seabed disposal in stable clay sediments, was investigated in the 1970s and 1980s and was thought by a number of experts to hold potential advantages over land-based disposal. Other disposal concepts that have been proposed, at least for some forms of waste, include disposal by *in situ* melting (this was suggested in the 1970s as a way to dispose of liquid wastes from reprocessing, perhaps by using already contaminated underground nuclear test cavities) or space disposal—that is, shooting nuclear wastes into solar orbit or even into the sun. For reasons of practicality, public and international acceptance, and/or cost these options have generally not received as much attention as disposal in a deep, land-based, mined geologic repository. In sum, based on the evidence available to date, the Commission sees no reason to change the current focus of the U.S. program on developing mined geologic repositories.

Overall, the Commission recommends further RD&D to help resolve some of the current uncertainties about deep borehole disposal and to allow for a more comprehensive (and conclusive) evaluation of the potential practicality of licensing and deploying this approach, particularly as a disposal alternative for certain forms of waste that have essentially no potential for re-use. Likewise, EPA and NRC should begin work on a regulatory framework for borehole disposal, in parallel with their development of a site-independent safety standard for mined geologic repositories, to support the RD&D effort leading to licensed demonstration of the borehole concept.[59]

4.4. Retrievability and Reversibility

The concepts of retrievability and reversibility have long been part of the discussion about how best to safely dispose of highly radioactive materials. While no standardized definition exists for either term, reversibility means the more generic ability to reconsider and reverse course at any time during the development and implementation of a geologic disposal program. By contrast, retrievability refers more specifically to the ability to retrieve waste after it has been placed underground in a geologic disposal facility.[60,61] This could be considered a desirable or necessary feature of facility design for two main reasons: (1) so that it remains possible to monitor the nuclear waste to confirm the behavior of the repository and remove the waste if necessary and (2) to preserve the option of retrieving spent fuel for future reprocessing and recycling if that proves warranted.

The Commission considers retrievability and reversibility as closely related but distinct issues. Our view is that existing requirements concerning retrievability at mined repository sites (at 40 Code of Federal Regulations [CFR] 191 and 10 CFR 60.111 (b)) are appropriate and should be retained. These requirements are intended to ensure that emplaced waste can be removed if the repository is not behaving as anticipated or if its performance is called into question for any reason prior to permanent closure[62]— they are not intended for the purpose of retaining easy access to emplaced materials for possible later recovery and reuse. Past evaluations have indicated that a wide range of candidate mined repository sites in different geologic media (including granite, salt and volcanic tuff) could meet these existing retrievability requirements.

On the other hand, we recognize that the same level of retrievability may not be practical or necessary in the context of other disposal approaches, such as deep boreholes. In that case, related regulatory requirements and time periods can and should be reassessed as part of a larger evaluation of disposal system performance objectives.

On the subject of reversibility, the Commission views this attribute as an important part of what we believe should be a staged, adaptive approach to waste management and disposal in the United States more generally. The details of such an approach are discussed at length in section 6 of this report. For purposes of this discussion, it suffices to note that for a program to be adaptive there needs to be some capacity to reverse course, at least for a period of time. Flexibility of this kind is needed because implementing a disposal program will take at least several generations, during which technology and values are sure to evolve albeit in unpredictable ways. While there is general consensus that we cannot rely on active management of nuclear waste disposal facilities over the many millennia of safety and environmental concern, an adaptive, staged approach requires this flexibility in the near term, when it is reasonable to have confidence that the institutional oversight and management capacity to implement responsible course corrections will be available.

4.5. Cost of Disposal

The Commission heard many comments regarding the costs of nuclear waste management generally and the costs of nuclear waste disposal in particular. While it is impossible to prepare detailed cost estimates for an integrated U.S. nuclear waste management system without knowing the specific facilities and sites that will be used for waste management and many other pertinent details, some general conclusions about cost can be reached based on work performed for the Yucca Mountain project and other information.

A 2008 DOE life-cycle cost estimate[64] arrived at a figure of $96.2 billion (in 2007 dollars) to license, construct, operate and close a repository at Yucca Mountain of sufficient size to dispose of a total of 122,000 metric tons of commercial and defense-origin spent fuel and high-level waste (note that the legislated capacity of Yucca Mountain is 70,000 metric tons until a second repository is in operation). The cost share assigned to the 109,000 tons of commercially-generated wastes assumed for disposal was about 80 percent of the $96.2 billion total, or approximately $77 billion.

That same year, DOE produced a detailed report "to evaluate whether the collection of the [nuclear waste] fee will provide sufficient revenues to offset the commercial utilities' share of the total life cycle costs of the Civilian Radioactive Waste Management Program."[65] The report concluded that the "the fee is adequate and [DOE] finds no reason to adjust the fee at this time." This conclusion echoes past fee evaluations which, over two-plus decades of the nation's nuclear waste management program, concluded that the current one-tenth of one cent per kilowatt-hour collected for spent fuel management would be sufficient to pay for disposal of the nation's spent commercial reactor fuel.

It is important to stress that judgments about the adequacy of current fee payments to cover anticipated disposal costs are separate from the question of whether the current fee mechanism is working as intended to make fee payments available to fund the waste management program. The latter issue is the subject of section 8 of this report.

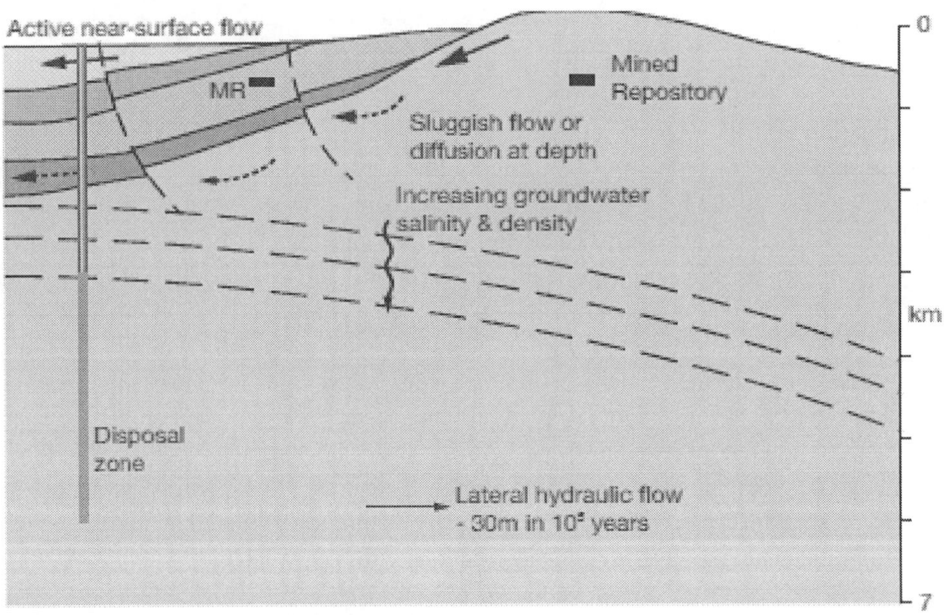

Figure 13. Deep borehole disposal concept. [63]

5. STORAGE AS PART OF AN INTEGRATED WASTE MANAGEMENT STRATEGY

Storage is a necessary and important element of a comprehensive strategy for managing the back end of the nuclear fuel cycle.

Implemented with a strong emphasis on safety and security and designed for compatibility with other steps in the fuel cycle, storage facilities have the potential to increase the flexibility, resiliency, and robustness of the system as a whole. Current arrangements for the storage of SNF in the United States, however, have evolved in an *ad hoc* fashion. Changes to the current approach are needed for several reasons: to support progress toward the development of disposal capability; to address immediate and growing financial and legal liabilities stemming from the federal government's failure to meet its waste acceptance obligations in a timely manner; and to improve confidence in the safety and security of current storage arrangements, including addressing any new concerns that emerge in the wake of the March 2011 accident at Japan's Fukushima Daiichi nuclear power facility.

Having investigated a range of issues related to the storage of spent fuel and HLW, the Commission has two central recommendations.

First, we recommend that *the United States establish a program that leads to the timely development of one or more consolidated storage facilities*. Access to consolidated storage capacity, even on a limited basis, would—when coupled with a viable transportation system—provide valuable flexibility as part of an integrated nuclear waste management system. Without this capacity the federal government will have essentially no physical capability to accept spent fuel for emergency or any other purposes until a permanent repository is in operation.

Second, we urge *vigorous, ongoing efforts by industry and by the appropriate regulatory authorities to ensure that all near-term forms of storage meet high standards of safety and security for the multi-decade-long time periods that they are likely to be in use*. Based on the evidence and safety record to date, the Commission sees no unmanageable safety or security risks with current storage arrangements. That said, active research, monitoring, and continued responsiveness to new information and lessons learned—including lessons learned from a more complete understanding of recent events at Fukushima—are clearly needed to sustain this confidence. Any realistic assessment of the time it can be expected to take to site, construct, license and begin operating consolidated storage and disposal facilities underscores the need for continued vigilance and attention to safety and security concerns at existing storage sites.

This section elaborates on the above points and on other recommendations developed by the Commission's Transportation and Storage Subcommittee. We begin by discussing the role of storage as part of a comprehensive waste management strategy, before developing the rationale for consolidated storage. Subsequent sections of this section discuss safety and security issues at existing dispersed storage sites. Further discussion of these subjects can be found in supporting Commission materials and in the report of the Commission's Transportation and Storage Subcommittee (available at www.brc.gov).

5.1. The Role of Storage

Storage in some form, for some period of time, is an inevitable part of the nuclear fuel cycle. This is simply because spent fuel, upon being removed from the reactor core, needs to be allowed to cool before it can be handled further. In the early days of the nuclear energy industry it was assumed that storage times for spent fuel would be relatively short—on the order of several years to a decade or two at most before spent fuel would be sent either for reprocessing or final disposal. The current reality, of course, is much different. Storage is not only playing a more prominent and protracted role in the nuclear fuel cycle than once expected, it is the *only* element of the back end of the fuel cycle that is currently being deployed on an operational scale in the United States. In fact, much larger quantities of spent fuel are being stored for much longer periods of time than policy-makers envisioned or utility companies planned for when most of the current fleet of reactors were built.

Section 3 of this report describes how the current situation—in which the vast majority of spent fuel is still being stored at the reactor sites where it was generated— arose by default as the U.S. government first decided not to pursue reprocessing and then fell further and further behind in developing a disposal repository. With DOE in breach of its contractual waste acceptance obligations, individual utilities have been left to cope on their own with the problem of growing spent-fuel inventories. Over the years, most of them have responded by packing spent fuel more tightly in cooling pools and, increasingly, by moving the spent fuel from wet storage to on-site dry cask storage when available space in the pools is exhausted. At plants that have implemented this form of storage, canisters containing spent fuel are typically placed on concrete pads in an open air enclosure or in horizontal concrete vaults on site where they are monitored on an ongoing basis. Other storage methods, such as dry vault storage for fuel from the St. Vrain plant and wet storage at the GE Morris facility, have also been licensed.

Existing dry storage systems at nuclear facilities are robust. In the most widely used type of dry storage system, a canister containing used fuel is placed inside a concrete structure. The canister typically consists of 1/2 inch to 5/8 inches thick stainless steel; it serves as the primary boundary to confine radioactive material (see figure 4).

Figure 14. Dry cask storage facility at the decommissioned maine Yankee reactor site.

Depending on the cask system design, the canister may be oriented vertically or horizontally inside a thick reinforced concrete structure. These reinforced concrete structures, which are typically 2.5 feet thick for vertical systems and 3 feet thick for horizontal systems, provide shielding from radiation and protect the canister. The total weight of current dry storage systems (canister and concrete structure) is typically between 160 and 180 tons (320,000–360,000 pounds).

After an initial period of cooling in wet storage (generally at least five years), dry storage (in casks or vaults) is considered to be the safest and hence preferred option available today for extended periods of storage (i.e., multiple decades up to 100 years or possibly more). Unlike wet storage systems, dry systems are cooled by the natural circulation of air and are less vulnerable to system failures. Nevertheless, it is important to emphasize that spent fuel pools are essential to operating a nuclear power plant given the need to be able to cool newly discharged fuel in a water-filled pool close to the reactor core. Pools are also advantageous for the transfer of spent fuel into and out of casks.

In the United States, pools remain the dominant form storage for spent fuel at still-operating reactor sites (pools are currently also used for at-reactor and consolidated storage in other countries, including France, Russia, and Sweden). Currently, less than one-fourth of the nation's commercial spent fuel stockpile is being stored in dry casks, although the Electric Power Research Institute (EPRI) projects this fraction will grow steadily in coming years and that all operating power reactors will have dry storage facilities in operation by 2025.66 Figure 15 shows EPRI's projection for the expected amount and distribution of commercial spent fuel in dry versus wet storage over the next several decades.[67]

While current storage arrangements have been judged adequately safe and secure by the relevant regulatory authorities—in fact, as discussed in section 3, the NRC in2010 updated its "Waste Confidence Decision" to state that at-reactor or away-from-reactor spent fuel could be stored safely for up to 60 years after the termination of an operating reactor's license (with extensions up to 60 more years)[68]—it is clear that today's institutional arrangements and storage technologies were not designed for the lengthy storage timescales that now appear inevitable for at least some portion of the nation's spent fuel inventory.

Assuring safe and secure storage of SNF and HLW over extended periods of time will require continued public and private efforts—including efforts by the NRC, DOE, and industry organizations such as EPRI —to conduct rigorous research and oversight and continuously incorporate lessons learned from new developments or events.

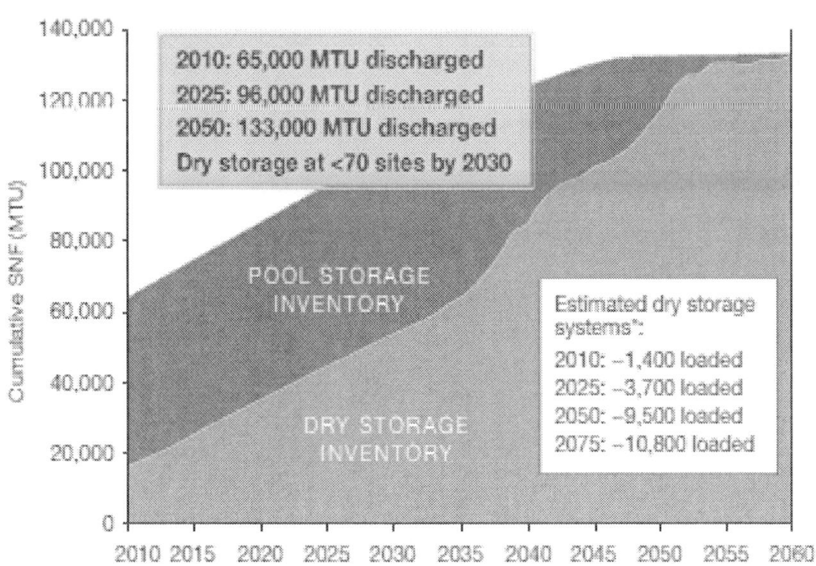

Aug 2010, Energy Resources International, Inc.

Figure 15. Epri projection of cumulative spent nuclear fuel from commercial nuclear power plants in pool storage and dry storage, 2010–2060.

For example, it will be important to continue exploring fuel degradation mechanisms in dry storage, particularly since many current safety assessments are based on an examinations of fuel with lower burnup than is now "standard" and do not account for storage times of the length now being contemplated. Further research may identify unanticipated problems with extended fuel storage (e.g., unexpected corrosion rates) and will help ensure that problems are detected and appropriately mitigated if they emerge.

Given the history of the U.S. waste management program, it is perhaps not surprising that the need for extended storage has come to symbolize the program's larger failure to perform so far. Nonetheless we find that developing extended storage capacity, if approached in a way that maximizes its system benefits, could strengthen the U.S. waste management program as a whole.

First, having extended storage capability preserves options and enhances flexibility while other elements of a comprehensive waste management system—including options for the final disposition of HLW and SNF—are developed and tested.

The United States may ultimately dispose of spent fuel or make use of reprocess and recycle technologies if closing the fuel cycle becomes advantageous in the future. Storage capability preserves the option of going in either direction.

If the ultimate disposition path for spent fuel involves disposal in a geologic repository, allowing the fuel to cool through a period of storage reduces the siting challenge for a disposal facility and/or increases the disposal capacity of a given facility.

These system benefits apply whether storage is provided at consolidated facilities or at dispersed sites, as is currently the case. But the storage arrangements in place today were not designed to maximize operational efficiency at a system level or to respond to unforeseen events or changes in management strategy, much less for indefinite storage at reactor sites after the reactors themselves have been decommissioned. These issues are addressed in later sections of this section; below we turn first to the rationale for developing one or more consolidated storage facilities.

5.2. The Case for Consolidated Storage

The fundamental policy question for spent fuel storage in the United States today is whether the federal government should proceed to develop one or more consolidated storage facilities as a way to begin the orderly transfer of the fuel to federal control pending its ultimate disposition through reuse or disposal. The Commission concludes that there are several compelling reasons to move as quickly as possible to develop safe, consolidated storage capacity on a regional or national basis.

5.2.1. Consolidated Storage Would Allow for the Removal of 'Stranded" Spent Fuel from Shutdown Reactor Sites

There are currently nine shutdown commercial nuclear power plant sites in the United States (see table 1) and one DOE-owned spent fuel facility (at Fort St. Vrain in Colorado) where the reactor itself has been or is being removed and the spent fuel—often referred to as "stranded fuel"—is being stored on site. At seven of the nine commercial sites and at the DOE site, the spent fuel is in dry storage. At all of these sites, which are formally known as "independent spent fuel storage installations" or ISFSIs, the spent fuel is both monitored and well-guarded and hence is not thought to present immediate safety or security concerns. Nonetheless, the continued presence of spent fuel at shutdown reactor sites is problematic and costly. Most obviously, it prevents these shutdown sites from being reclaimed for economically productive or otherwise desirable uses that would benefit the surrounding communities. Moreover, these communities were never asked about, and never contemplated or consented to, the conversion of these reactor sites into indefinite long-term storage facilities.

As a result, they generally also did not have an opportunity to negotiate for rights of participation or incentives and benefits of the sort that would likely be available to the host community of a dedicated storage facility. Finally, most of these shutdown reactor sites no longer have the capability to remove spent fuel from storage canisters for inspection if long-term degradation problems emerge that might affect the ability to transport the canisters. Consolidated storage sites can be developed to provide these capabilities.

Direct cost considerations alone provide a compelling reason to move stranded spent fuel as quickly as possible to a consolidated storage facility. This is because the cost attributable to storing spent fuel at plant sites increases dramatically once the reactor is shut down. Since the cost of loading fuel into dry storage casks has generally already been incurred at this point, continued storage involves little activity other than site security and monitoring. At an operating nuclear plant, security is already in place and only incremental effort is required to include the ISFSI within the plant's security umbrella. The same is true for the personnel needed to monitor the status of the fuel and perform any routine maintenance. When the rest of the site is shut down, however, these structures, systems, equipment and people are still needed to tend the spent fuel, and the cost is substantial. Recent studies find that the operation and maintenance costs for spent fuel storage at shutdown sites range from $4.5 million to $8 million per year, compared to an incremental $1 million per year or less when the reactor is still in operation. *Even assuming no further change in security requirements at shutdown sites, these cost estimates suggest that the savings achievable by consolidating stranded spent fuel at a centralized facility would be enough to pay for that facility.*[69] Consolidation would also allow any new safety or security measures that might be required in the future to be implemented more cost-effectively.

These cost advantages will only grow as increasing numbers of reactors reach the end of their operating lives, starting around 2030. Assuming a 60-year operating life, on average, for current plants, the number of shutdown sites could reach 30 by 2035 and 70 by 2050. Of course, subsequent life extensions beyond 60 years could push this curve farther into the future, but it is also possible that not all currently operating reactors will in fact have their lives extended to 60 years. In that case, the number of shutdown plants would increase more rapidly.

And while additional reactors have been proposed at some sites (which could delay full decommissioning of these sites for decades), it is impossible to know how many of these new units will eventually be built or how they would affect cumulative fuel storage costs.

Using the cost estimates cited previously, the added security and monitoring expenses associated with keeping stranded spent fuel at as many as 70 different shutdown reactor sites could be in the area of $350 to $550 million per year at today's costs.

In sum, equity and cost considerations together argue for moving as quickly as possible to transfer stranded spent fuel from shutdown reactor sites to consolidated storage. Given the significant direct benefits of transferring spent fuel from these sites, both for the surrounding communities and in terms of cost savings, the Commission recommends that *spent fuel currently being stored at shutdown reactor sites be "first in line" for transfer to a consolidated storage facility.*

5.2.2. Consolidated Storage Would Enable the Federal Government to Begin Meeting Waste Acceptance Obligations

Developing consolidated storage capacity would enable the U.S. government to begin fulfilling its legal obligations (described in section 3 of this report) with respect to the acceptance and removal of SNF from commercial reactor sites. In this way, it would also begin to address a large and growing source of financial and legal liability to the federal government and ultimately U.S. taxpayers.

The Commission has heard the suggestion that DOE should simply "take title" to the spent fuel that is currently being stored at reactor sites, even though DOE currently has no storage or disposal facilities to which it could transfer the spent fuel. The idea is that assuming management responsibility for this material would allow DOE to meet its contractual obligations under current law and end taxpayer exposure to further damage awards resulting from the federal government's failure to accept spent fuel on schedule.[71] Examining the case law, the Commission concludes that simply "taking title" would not change the ongoing taxpayer liability under DOE's contracts with utilities. Performance under existing contracts would require that DOE not only take title to the spent fuel, but also remove it from reactor sites. If DOE were to take title but leave the spent fuel where it is, DOE could not use the Nuclear Waste Fund to cover resulting maintenance and security costs.

A federal Court of Appeals has ruled that paying for at-reactor storage is not an allowed use of the Fund under the NWPA and under DOE's existing contracts with utilities. (For the same reason, the Fund cannot be used to cover damage payments to utilities for providing at-reactor storage.) Thus, the costs of at-reactor storage—regardless which entity has title to the spent fuel—must ultimately be paid by taxpayers, whether this occurs through damage payments to utilities from the federal Judgment Fund or through direct appropriations to DOE if DOE were to assume management responsibility for the spent fuel.

Moreover, taxpayers could remain liable for at-reactor storage costs even if the NWPA were amended to both allow DOE to take title to at-reactor spent fuel and allow the Nuclear Waste Fund to be used to cover at-reactor storage costs.

Because the explicit language of the Standard Contract limits the "costs" for which the fee can be collected to those costs set out in the NWPA, current contract holders may sue for damages on the theory that the Nuclear Waste Fund can be used only for purposes authorized under the NWPA as it existed when DOE and the utilities entered into these contracts.[72] The government could argue that the contract reference is to the Act as amended from time to time, and that paying for the cost of on-site storage effectively mitigates damages. How such litigation might be resolved, if it occurred, is difficult to predict—the outcome would depend on how legislation to authorize the "take title" approach was drafted and on the intricacies of federal contract law.[73]

It is reasonable to expect that taxpayers could ultimately remain liable for costs of at-reactor storage of spent fuel.[74]

Given the present situation, developing consolidated storage capacity could be the fastest and surest path for the federal government to begin performing under existing contracts and to ultimately achieve waste acceptance rates that can stop the further growth of taxpayer liability.

The existence of functional consolidated storage capacity would also change the federal government's ability to renegotiate contracts with utilities.

Table 1. Quantities of stranded spent fuel in storage at shutdown commercial U.S. reactor sites[70]

Plant	State	MTHM Stored at Site	MTHM in Pool Storage	MTHM in Dry Storage	Number of Casks	Estimated Casks	Total Casks (Actual Plus Estimated)	Average MTHM/ Cask
Big Rock Point	Michigan	58	0	58	7	—	7	8.3
Haddam Neck	Connecticut	412	0	412	40	—	40	10.3
Humboldt Bay	California	29	0	29	5	—	5	5.8
LaCrosse [a]	Wisconsin	38	38	0	0	5	5	7.6
Maine Yankee	Maine	542	0	542	60	—	60	9.0
Rancho Seco	California	228	0	228	21	—	21	10.9
Trojan	Oregon	359	0	359	34	—	34	10.6
Yankee Rowe	Massachusetts	127	0	127	15	—	15	8.5
Zion 1 & 2[b]	Illinois	1,019	1,019	0	—	61	61	16.7
Fort St. Vrain	Colorado	15	0	15	NA[*]	—	NA	—

[a] Testimony to Commission indicates target completion in 2012.

[b] Decommissioning contract entered with Energy Solutions. NAC MAGNASTOR canister will be used with capacity of 36 elements per cask. Target schedule for completion is 2013.

[*] Fort St. Vrain spent fuel is in vault storage.

Note: Some shutdown plant sites also have GTCC waste stored in dry casks.

Importantly, work toward a consolidated storage facility can begin immediately under the existing provisions of the Nuclear Waste Policy Act, which authorize the federal government to site and design a MRS facility and obtain construction authorization.[75]

Further legislative action would not be required prior to the designation of a MRS facility site (and potentially not until the construction phase), at which time Congress would need to amend the NWPA to allow construction to go forward independent of the status of a permanent repository.

Meanwhile, NRC regulations for independent spent fuel storage installations have already been developed and have been used to approve several types of storage technologies.[76]

(For example, it took DOE between two and five years to obtain NRC licenses for dry cask and dry vault spent fuel storage facilities at the INL.)

As with developing disposal capability, the critical challenge for consolidated storage will be finding a site or sites. Because the technical requirements for this type of facility would be less demanding than for a repository, finding a suitable location with an accepting host community may be less difficult, particularly if it is accompanied by attractive incentives.

The Commission has heard testimony indicating that potential host communities, states and tribes would be willing to participate in an open process that engages affected constituencies from the outset and gives them actual bargaining power. Nevertheless, the potential difficulty of siting consolidated storage and the need for a thoughtful approach to this task must not be underestimated.

Our specific proposals for a new approach to siting radioactive waste facilities in general are discussed in section 6.

5.2.3. Consolidated Storage Would Provide Flexibility to Respond to Lessons Learned from Fukushima and Other Events

A consolidated storage option (which would consist of dry storage, wet [pool] storage, or both) would provide flexibility to respond to changes in regulation or practice that might result from a fuller assessment of the events at Fukushima and of other unexpected and potentially disruptive events. While no determination has been made that current at-reactor storage arrangements in the United States are not adequately safe, access to consolidated storage would be very helpful if, for example, the decision were made to reduce inventories of spent fuel in reactor pools. In that case, having one or more consolidated storage facilities with the pool capacity to accept relatively young spent fuel would allow nuclear plant operators to focus on reducing the heat load in reactor pools by preferentially removing the hotter spent fuel, should that be determined to be the best approach.[77] After adequate additional cooling, the fuel could then be transferred to dry storage in a staged way.[78]

A consolidated storage facility with wet storage capacity would also provide the capability to remove even relatively hot, recently discharged fuel from reactor pools on relatively short notice and with minimum operational demands on reactor operators. The ability to move hot fuel off site could simplify the management of a post-accident situation at a reactor by, for example, removing an important potential source of risk and freeing up pool space for other purposes (e.g., storing radioactive debris).[79] As Fukushima has shown, completely unexpected problems can arise suddenly. At present, the United States lacks any capability to receive spent fuel in emergency situations, although DOE's standard contract with utilities would theoretically allow for the waste acceptance "queue" to be re-prioritized in such situations.

Finally, consolidated storage could enhance the safety and security of the overall waste management system simply because facilities for this purpose could be located where there is a much lower probability of extreme events (unlike reactors, for example, a storage-only facility need not be located near a large source of water), where the risks of broad-based population exposures in the event of a disaster are lower, and where local conditions are conducive to effectively monitoring and managing security risks.

5.2.4. Consolidated Storage Would Support the Repository Program

The Commission has concluded that siting and developing one or more consolidated storage facilities would improve prospects for a successful repository program.

First, the technical and institutional experience gained by siting, testing, licensing, and operating a consolidated storage facility, as well as planning for and executing a concurrent transport program, would benefit repository development and operation,[80] especially because all the activities involved (apart from those uniquely associated with underground disposal) would be the same.

In addition, consolidated storage would provide the flexibility needed to support an adaptive, staged approach to repository development. This kind of approach was recommended as early as 1990 by the National Academies' Board on Radioactive Waste Management and is discussed in more detail in section 6 of this report. The main point for purposes of this discussion is that a consolidated facility would allow federal acceptance of spent fuel to proceed at a predictable, adequate and steady rate—both before a disposal facility is available and when it is in operation.

Even after a disposal facility is open, consolidated storage would act as a buffer and provide valuable redundancy for the system as a whole. It would, for example, allow utilities to continue to ship spent fuel away from reactor sites as scheduled even if a repository had to slow or cease operation for a period of time for any reason. Alternatively, it could accommodate a surge of shipments from reactor sites if that were necessary, while allowing emplacement at a repository to proceed at a steady, pre-determined rate. To provide this flexibility, a consolidated storage facility would ideally be incrementally expandable (with the acceptance of the host community) in terms of its total storage capacity and fuel handling and management capabilities.

Consolidated storage also offers opportunities to simplify repository operations. For example, by accumulating a substantial inventory of spent fuel in one place, the storage facility could take over some of the thermal management activities that might be required for efficient repository operation (e.g., blending hot and cool fuel assemblies to create a uniform thermal load for waste packages). A consolidated storage facility could even offer the option of packaging the waste for disposal before it is shipped to the repository, further simplifying operations at the repository site.

5.2.5. Consolidated Storage Offers Technical Opportunities for the Waste Management System

A federal facility with spent fuel receipt, handling and storage capabilities can support other valuable activities that would benefit the waste management system. These include long-term monitoring and periodic inspection of dry storage systems and work on improved storage methods. Many current dry cask systems lack instrumentation to measure key parameters such as gas pressure, the release of volatile fission products, and moisture. Some of this work can be done in laboratories, but key aspects require the ability to handle and open loaded spent fuel storage containers and examine the fuel. A consolidated storage facility with laboratory and hot cell facilities and access to a substantial quantity and variety of spent fuel would provide an excellent platform for ongoing research and development to better understand how the storage systems currently in use at both commercial and DOE sites perform over time.[81]

5.2.6. Consolidated Storage Would Provide Options for Increased Flexibility and Efficiency in Storage and Future Waste Handling Functions

Finally, a consolidated storage facility could provide flexible, safe, and cost-effective waste handling services (i.e., repackaging or sorting of fuel for final disposal) and could facilitate the standardization of cask systems. This in turn could reduce the need for extensive handling at many reactor sites and make it possible to use more cost-effective storage systems at a central facility.[82] Such facilities could also offer enhanced remote handling capabilities, thereby reducing the potential for worker exposures.[83] This capability could be particularly important if changes in the condition of the spent fuel over time make it necessary to open storage containers and repackage the fuel before moving it elsewhere for disposition.[84] Dry storage facilities at shutdown reactors without pools do not have any of the fuel handling and recovery capabilities that would be provided in a consolidated facility—in effect, these facilities are simply well-guarded parking lots for storage casks. If fuel at these sites needed repackaging, a new fuel handling facility would have to be constructed at considerable time and expense.

Considering current uncertainties about long-term degradation phenomena in dry storage systems, it would be prudent to initiate a planned, deliberate, and reliable process for moving spent fuel from shutdown reactor sites to a central facility before any issues arise and where problems can be dealt with much more easily and cost effectively than at multiple shutdown sites. The importance of consolidating inventories of spent fuel before there might be a need to reopen dry storage containers increases as the period of storage being contemplated increases.[85]

5.3. Practical and Strategic Considerations and Next Steps for Proceeding with Consolidated Storage

For all of the reasons discussed in the foregoing section, the Commission recommends that the U.S. government proceed to develop consolidated storage capacity without further delay. The Commission has also heard and considered arguments *against* proceeding with consolidated storage. Of these, the most important objection and one that will need to be thoughtfully addressed is the concern that any consolidated storage facility could become a *de facto* disposal facility and—by reducing the pressure to find a long-term solution—thwart progress toward developing the deep geologic disposal capacity that will ultimately be needed. This is not a new concern; it is why the 1987 NWPA Amendments explicitly tied the construction of an MRS facility to progress on a first repository and set capacity limits for the MRS facility so that it could not accommodate all the spent fuel in need of disposal.

Circumstances today, however, are different. Trust and confidence in the federal government's basic commitment and competence to deliver on its waste management obligations have all but completely eroded since 1987. Restoring that trust and confidence must be the government's first priority and is essential for getting all aspects of the nation's nuclear waste program back on track. In this context, demonstrating that it is possible to muster the policy direction, technical expertise, and institutional competence needed to site and operate one or more consolidated storage facilities (while also vigorously pursuing final disposal capability) would by itself be enormously valuable. Near-term progress on a consolidated storage facility would not only address a major source of political, legal and financial liability that will otherwise complicate efforts to move beyond the current impasse in the repository program, it would also provide practical benefits in terms of gaining experience and providing the system-wide flexibility needed to support an adaptive, staged approach to repository development.

In sum, the Commission concludes that progress on consolidated storage will have a positive impact and indeed could play a crucial role in the success of a revitalized disposal program. Other concerns we have heard about consolidated storage—primarily related to the costs and potential worker exposures associated with handling and transporting spent fuel twice, once to move spent fuel from reactor sites to consolidated storage and then to move the fuel a second time to a permanent repository—are outweighed, in our view, by the increased flexibility, handling advantages, and potential cost savings that consolidated storage capability would provide.

That said, we do not underestimate the practical difficulties of siting a consolidated storage facility, particularly in a context of great uncertainty about the future of the repository program. If anything, the history of past efforts to develop a MRS facility show the same pattern of siting challenges as the repository program—and a similar record of failure in overcoming them. On the other hand, experience with the Office of the Nuclear Waste Negotiator in the early 1990s also gives some grounds for hope. As quoted in a recent Massachusetts Institute of Technology report on nuclear waste storage issues, the first Nuclear Waste Negotiator, David Leroy, concluded that "the volunteer siting process can work provided that the negotiator is given the resources and time to negotiate the terms of an interim storage facility and benefit package," although he also recognized that "the lack of a proposed repository makes the process more difficult."[86] Because siting and process issues are so important, and largely common to both disposal and consolidated storage facilities, they are addressed by a separate set of Commission recommendations and discussed at length in a separate section (section 6) of this report.

The salient point for purposes of this discussion is that the challenge of siting one or more consolidated storage facilities cannot be separated from the status of the disposal program. Many states and communities will be far less willing to be considered for a consolidated storage facility if they fear they will become the *de facto* hosts of a disposal site. This means that a program to establish consolidated storage will succeed only in the context of a parallel disposal program that is effective, focused, and making discernible progress in the eyes of key stakeholders and the public. A robust repository program, in other words, will be as important to the success of a consolidated storage program as the consolidated storage program will be to the success of a disposal program. Progress on both fronts is needed and must be sought without further delay.

It should be emphasized that the development of one or more storage facilities does not require, or even imply, an irreversible commitment to any particular long-term plan for moving fuel to these facilities or performing any specific set of activities at these sites. All of the capabilities that would ultimately be desirable do not have to be developed at once, particularly since it is not clear at this time exactly what features will be needed over the many decades such a facility or facilities would be in operation. A storage facility or system of facilities can be developed in a stepwise manner, as the need for expansion of capacity and capability becomes clearer. Furthermore, the initial cost to site, design, and license a storage facility is relatively low (less than $100 million),[87] so that the money put "at risk" in giving future decision makers the option to proceed with construction and operation of a storage facility is small compared to the potential benefits from having that option available. Siting, licensing, building and operating a storage facility with even limited initial capabilities would substantially resolve uncertainties about the costs and time required for these activities, including associated transportation needs, thereby providing a firmer basis for future decision-making with regard to potential expansion.

Principal Objections to Pursuing Consolidated Storage

The BRC's Transportation and Storage Subcommittee heard and considered a number of arguments *against* proceeding with consolidated storage. Most centered on three concerns: (1) that expending resources and effort on consolidated storage could slow or halt progress toward developing a permanent repository; (2) that consolidated storage would necessitate further handling of spent fuel and high-level waste, potentially increasing safety and security risks; and (3) that consolidated storage could increase overall waste management costs.

The first of these concerns, that any consolidated storage facility could become a *de facto* disposal facility and siphon resolve and resources away from repository development, is a longstanding one. It is why the NWPA explicitly prohibits the construction of an MRS facility before construction authorization has been issued for a first repository. Based on the record of progress to date, the Commission believes that the benefits of moving forward on both fronts—consolidated storage and geologic disposal— at the same time outweigh the potential downside risks. But clearly the challenge of establishing positive linkages such that progress on storage does not undermine, but rather supports progress on repository development remains an important one.

The linkages that exist under current law clearly have not worked as intended. Similarly, the concern about increased handling risks as a result of potentially moving SNF and HLW twice, first from decentralized storage to consolidated storage and then from consolidated storage to a geologic disposal facility, is a legitimate but in our view manageable one. Clearly, there are trade-offs: some risks might increase but benefits to the system as a whole (such as the ability to learn early lessons by moving spent fuel from shutdown reactors to consolidated storage, and the creation of increased capability to respond to emergency situations, for example) could more than offset these impacts so that overall risk for the waste management system as a whole could decline.

A 2007 study by the American Physical Society specifically looked at transport risks related to moving spent fuel twice and concluded these risks would be small.[88]

Ultimately, we believe the many safety and security benefits that would come with having one or more well-equipped, consolidated storage facilities outweigh objections centered on extra handling risks.

The Transportation and Storage Subcommittee looked in some detail at the third issue, concerning cost. As discussed elsewhere in this section, it found potentially substantial cost savings associated with removing SNF from shutdown reactor sites and with accelerating the federal government's ability to begin accepting waste in fulfillment of its existing contractual commitments (and thereby avoiding further damage payments to utilities).

The Subcommittee also looked at estimates of the cost of providing consolidated storage based on eight studies of this subject published since 1985. The conclusion was that it would be impossible to arrive at a single point estimate of centralized storage costs given the large uncertainties involved.

The more important conclusion, however, was that the extra cost to site, design and obtain a license for a consolidated storage facility was likely to be in the range of $50 to $100 million. While appreciable, these are small levels of commitment from the perspective of the overall spent fuel management program.

At the same time, a wide variety of circumstances can be anticipated in which centralized storage facilities could prove invaluable. In these circumstances, savings on the order of billions of dollars are possible.

With these findings in mind, the Commission concludes that it would be prudent to pursue the development of consolidated storage capability without further delay, recognizing that there will be an opportunity to make course corrections later as needed.

Finally, it is important to stress that other major Commission recommendations concerning the need for a new waste management organization, reliable access to the Nuclear Waste Fund, and a new approach to facility siting apply equally to a consolidated storage program and are just as important to its success.

These high-level, cross-cutting recommendations are covered in later sections of this report. Recognizing that it will take time and new authorizing legislation to implement the Commission's most important recommendations and recognizing that DOE remains responsible and ultimately liable for the government's existing waste acceptance obligations in the meantime,[89] it is important to reiterate an earlier point: that sufficient authority already exists under the NWPA to begin laying the groundwork for consolidated storage without further delay, assuming Congress makes appropriations available for this purpose.

Specific steps that DOE could take in the near term include performing the systems analyses and design studies needed to develop a conceptual design for a highly flexible, initial federal spent fuel storage facility, assembling information that would be helpful to the siting process for such a facility, and working with nuclear utilities, the nuclear industry, and other stakeholders to promote the standardization of dry cask storage systems with an eye to facilitating later transport and consolidation in centralized storage and/or disposal facilities.

5.4. The Case for a New Approach to Prioritizing the Transfer of Spent Fuel from U.S. Commercial Reactor Sites

Once one or more consolidated storage facilities are available, future decisions about how to prioritize or sequence the transfer of spent fuel from operating commercial reactor sites to these facilities should be driven first by safety and risk considerations, and then by issues related to cost and other impacts. The Commission recognizes that existing contracts have created a "queue" in terms of federal commitments to accept spent fuel from specific utilities. Unfortunately, the existing queue was not set up to maximize efficiencies or to minimize the impacts of fuel handling and transportation. Hence, we believe it would be appropriate for DOE to re-visit the current schedule as it is already authorized to do under certain circumstances (for example, the Standard Contract authorizes DOE to give priority to spent fuel from shutdown reactors). Other changes to the current queue may require the Department and utility contract holders to re-negotiate some existing commitments. The Commission believes a more flexible approach would benefit all parties involved.

Under DOE's Standard Contract with utilities, priority for the acceptance of spent fuel is allocated to utilities according to the "oldest fuel first" or "OFF" principle. This does not mean that utilities would necessarily choose to ship their oldest fuel first, since they would have a contractual right to decide each year (subject to DOE's approval) which fuel to ship from which reactor (with the overall amount being determined by the OFF allocation). The current approach, however, has a number of shortcomings, particularly from the standpoint of maximizing the value of at-reactor storage as one tool in an integrated management system.

First, the current approach may limit the ability to use at-reactor storage as part of an integrated thermal management strategy. The ability to select which spent fuel is delivered for disposal at a permanent repository each year may avoid the need for additional storage to hold fuel that is too hot for immediate emplacement. However, since utilities can choose which fuel to deliver, they may prefer to send the hottest eligible fuel in their pools, assuming that the plants are still operating when waste acceptance begins. This may require more complex thermal management activities at the consolidated storage or disposal facility.

Second, the current system can add complexity and reduce efficiency in planning for shipments of spent fuel to a consolidated facility. For example, an analysis performed for the BRC[90] showed that accepting fuel based on the OFF priority ranking could result in spent fuel being shipped from an average of about 60 nuclear power plant sites each year, compared to fewer than 20 if priority can be given to spent fuel from shutdown reactor sites.

While a robust transportation management system would be needed in either case, the planning challenge for transporting spent fuel from an average of about 60 sites annually would be considerably more complex than in a scenario where shipments are coming from one-third as many sites or even fewer.

Third, accepting spent fuel according to the OFF priority ranking instead of giving priority to shutdown reactor sites could greatly reduce the cost savings that could be achieved through consolidated storage if priority could be given to accepting spent fuel from shutdown reactor sites before accepting fuel from still-operating plants. Figure 16, which assumes that a disposal or consolidated storage facility begins operating in 2030, shows that the difference in cumulative operation and maintenance (O&M) costs between use of OFF and shutdown reactor priorities could amount to billions of dollars.[91]

The magnitude of the cost savings that could be achieved by giving priority to shutdown sites appears to be large enough (i.e., in the billions of dollars) to warrant DOE exercising its right under the Standard Contract to move this fuel first. Although this action would disrupt the queue specified in the Standard Contract, as utilities continue to merge and a growing number of reactors reach the end of their operating licenses, every utility (or nearly every utility) will have one or more shutdown plants. In that context, giving priority to moving fuel from decommissioned sites is likely to be seen by all parties involved as being in everyone's best interest.

In sum, the Commission takes the view that a new, independent waste management organization should be directed (as part of enabling legislation) to take the lead in working on a cooperative basis with nuclear plant operators to identify measures that could reduce the overall costs and impacts of an integrated spent fuel management system. As part of this effort, the new organization should seek to renegotiate contracts as necessary to implement cost-saving and risk-reducing measures, while also recognizing the contractual rights of current waste owners as originally established under existing statutes, and as subsequently interpreted by the courts.

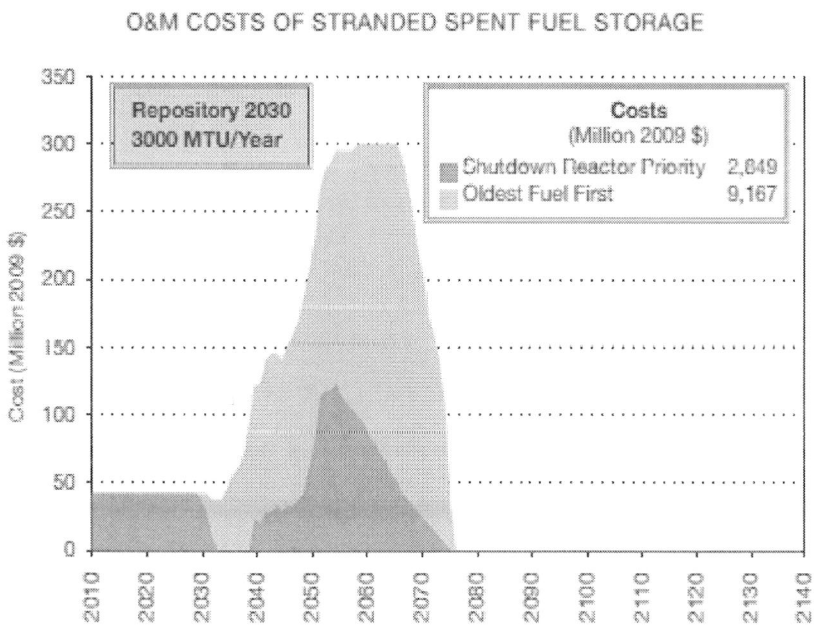

Figure 16. Operation and maintenance costs of stranded spent fuel storage.[92,93,94]

5.5. Safety and Security Considerations for Storage Systems

Safety and security are obviously paramount considerations in the storage and transport of SNF and HLW, under all circumstances and regardless of the type of site or facility involved. These are also issues that have drawn new attention in the wake of the disaster in Japan.

On March 23, 2011, NRC Chairman Gregory Jaczko directed the formation of a Near-Term Task Force to examine available information regarding the Fukushima disaster, and to determine whether changes should be made to ensure that the continued operation of existing reactors, and the licensing of new reactors, remain safe. On July 12, 2011, the Task Force released *Recommendations for Enhancing Reactor Safety in the 21st Century: the Near-Term Task Force Review of Insights from the Fukushima Dai-Ichi Accident.* The report found that it is unlikely a sequence of events similar to those experienced at Fukushima would occur in the United States, and concluded that "continued [reactor] operation and continued licensing activities do not pose an imminent risk to public health and safety." The Task Force went on to make 12 overarching recommendations to improve safety and to enhance the capability of reactor operators to react in the event of an emergency.

The Task Force report specifically looked at some issues related to spent fuel storage, an area of particular interest to the BRC. In its analysis of the Fukushima disaster, the report noted that when the station lost primary and backup power, operators were unable to monitor the water level and condition of the spent fuel pools at reactor units 1, 2, 3, and 4, and were also unable to run pumps that circulate water in the pools to keep the spent fuel cool. The Task Force observed that "operators were significantly challenged in understanding the condition of the spent fuel pools because of the lack of instrumentation or because of instrumentation that was not functioning properly." The loss of power and massive damage at the site also made it impossible for operators to add water to the pools as the water levels dropped, although they were later able to spray water into the pools using pumper trucks and high booms. The Task Force made specific recommendations for the NRC to consider that, in its view, would address "(1) the instrumentation to provide information about the condition of the pool and the spent fuel and (2) the plant's capability for cooling and water inventory management," in the event that a U.S. reactor suffered extended loss of AC power for whatever reason.

The report further observed that the four pools of concern "contained many fewer assemblies than typically stored in U.S. reactor unit spent fuel pools." In addition to the six pools adjacent to the reactors, the Fukushima facility also included a large storage pool away from the reactors themselves, which contained nearly 6,300 spent fuel assemblies. The report added that "U.S. reactor facilities do not typically have an additional spent fuel wet storage building like that at Fukushima Dai-ichi." Instead, many reactors have dry cask storage systems, which contain spent fuel that has been removed from the reactor for several years and can be cooled by air. The Fukushima plant had a small amount of fuel stored in nine dry casks in a separate building. It appears that the away-from-reactor storage pool and the loaded dry casks at Fukushima survived the disaster without suffering significant damage.

In a memorandum to NRC Commissioners dated October 3, 2011, NRC staff outlined recommendations concerning the prioritization of specific actions identified in the Task Force report.[95] The memorandum also noted that NRC staff had identified "a number of additional issues with a clear nexus to the Fukushima Daiichi event that may warrant regulatory action but which were not included with the [Near-Term Task Force] recommendations." Examining the transfer of spent fuel to dry cask storage was one of six actions that the NRC staff thought warranted "additional consideration for potential prioritization." NRC staff agreed at the time to provide NRC Commissioners with further recommendations concerning these additional actions within nine months, but also promised to promptly inform NRC Commissioners if the conclusion were reached that any of them warranted re-prioritization as a recommended near-term action.

Based on a review of the evidence to date, the Commission sees no unmanageable safety or security risks associated with current methods of storage (dry or wet) at existing U.S reactor sites. However, continued vigilance and careful attention to the lessons learned from Fukushima will be necessary to ensure that this remains the case, especially in light of the timeframes involved in establishing dedicated, away-from-reactor storage and disposal sites. Simply put, it will take years to more than a decade to open one or more consolidated storage facilities and even longer to open one or more disposal facilities. This means that storage of substantial quantities of spent fuel at operating reactor sites can be expected to continue for some time.

To provide effective oversight, regulatory authorities and nuclear plant operators, designers, and vendors must also be able to adapt quickly to new or unanticipated risks, such as emerged in the crisis at Fukushima. The post-accident response to that crisis is still ongoing and will continue for many months, which means that it will still be some time before a thorough investigation is complete and the resulting safety implications are fully understood.

Given the magnitude of the accident and its potential implications for future waste management policies, the Commission recommends that the NAS be asked to conduct an independent investigation of the events at Fukushima and their implications for safety and security requirements at SNF and HLW storage sites in the United States, once better information about the accident is available, including an analysis of the advantages and disadvantages of moving spent fuel from densely packed pools to on-site dry cask storage to facilitate low-density packing in the pools. This study would build upon the 2004 NAS study of storage issues (discussed in section 5.5.2 below) and would complement the other efforts to learn from Fukushima that have already been launched by the NRC and industry.

Besides a full investigation of events at Fukushima, the Commission has identified a number of priority areas for ongoing public and private efforts to ensure the continued safety and security of current storage arrangements. Specifically we urge continued work by the NRC, DOE, industry organizations such as EPRI, and others to explore fuel degradation mechanisms, identify unanticipated problems with extended fuel storage (i.e., unexpected corrosion rates), better understand the behavior of dry storage systems and their contents over time, investigate the feasibility of enhancing instrumentation in existing dry and wet storage systems, and promote the standardization of cask designs.

Similarly, we support ongoing efforts by the NRC to reexamine security requirements for storage sites and transportation and evaluate the need for enhanced security measures in the future.[96]

As part of this process the NRC should examine the advantages and disadvantages of options such as "hardened" on-site storage (HOSS) that have been proposed to improve security at existing sites (discussed further below). Obviously, any hardened system could be implemented more cost effectively at a consolidated storage facility than at existing sites due to economies of scale. Finally, continued vigilance and research is needed to stay abreast of evolving security risks and terrorism or sabotage threats, particularly as storage times increase and spent fuel becomes potentially more susceptible to theft or diversion.[97]

Specific issues and concerns with respect to the safety and security of storage technologies are summarized below.

5.5.1. Storage Security Considerations

Since the attacks of September 11, 2001, safety and security concerns specifically related to acts of terrorism or sabotage have received increased attention from agencies charged with regulating the storage and transport of nuclear materials.[98] Over the last decade, for instance, the NRC has issued more than 70 security and threat advisories to its licensees; in addition, starting in October 2001, the NRC initiated a series of classified studies that analyzed potential vulnerabilities and mitigation strategies at plants.

Under current NRC requirements, plant operators must demonstrate physical protection of pools through force-on-force testing involving simulated assaults in which the adversary is attempting to cause reactor or spent fuel damage.

Since late 2004, and as required by federal law since 2005, NRC-supervised testing is conducted at each operating power reactor once every three years (the operators conduct much more frequent tests on their own). This testing frequently includes simulated attacks on spent fuel pools. The NRC provides public and non-public reports to Congress every year on the results of these tests.[99]

The NRC is also primarily responsible for security requirements at ISFSIs. Like the security at an operating reactor, licensees must implement a "layered defensive strategy" that includes on-site protective forces with appropriate skills, weaponry, and other response equipment, and security systems. The strategy must include procedures to defend against physical attacks, insider threats, and cyber attacks. Security systems also provide for means to detect, assess, and communicate information about potential threats to local law enforcement authorities in the event of an attack.

Not surprisingly, security systems are tailored to the specific site, since relevant characteristics—such as the distance from storage facilities to the plant boundary—can vary from site to site. Licensees must conduct frequent performance drills and make internal assessments of force effectiveness; in addition, the NRC conducts its own periodic reviews of site-protective force training and force effectiveness.[100]

The NRC is currently undertaking a rulemaking to revise existing security requirements that apply to the storage of spent fuel at an ISFSI and to the storage of spent fuel and/or high level waste at a monitored retrievable storage installation (it will not address requirements that apply to storage in reactor pools). The rulemaking is intended to (a) examine the effectiveness of security orders imposed after the 9/11 terrorist attacks; (b) apply lessons learned from previous NRC inspections; and (c) ensure regulatory clarity and consistency between general and specific ISFSI licensees. The NRC issued a draft "regulatory basis" document in December 2009 and has received numerous comments on proposed technical approaches. Among other issues, the NRC is considering whether to require comprehensive "denial" capability on site—that is, sufficient security forces and weaponry for facility personnel to repel an attack on their own—or instead to require a detect/assess/communicate strategy that would rely on assistance from local, state and federal authorities.[101]

5.5.2. Storage Safety Considerations

The studies initiated by the NRC following 9/11 also addressed a number of issues directly related to the safety of pool storage, including the thermal response of fuel to fully drained and partially drained pool conditions, the structural response of spent fuel pools, options for mitigating spent-fuel heat-up or enhancing coolability, and confirmatory testing of analytical methods for calculating the thermal response of different types of fuel assemblies.

In February 2002, the NRC issued specific orders aimed at providing additional protection for spent fuel in pools based on the results of these initial studies.[102]

These orders are not available to the public, but they addressed strategies to restore or maintain core cooling and provide containment and spent fuel cooling under circumstances associated with the loss of large areas of the plant due to explosion or fire.

Additional guidance specifically related to pools was issued in 2004 and 2005 and subsequently incorporated into guidance for the 2009 Power Reactor Security Requirements final rule (74 FR 13926).

In 2003, an independent study of safety issues associated with the storage of spent fuel in reactor pools (by Robert Alvarez *et al.*)[103] raised concerns about the trend toward increased loading and higher-density spent fuel storage configurations in pools and the possibility that under certain conditions in which water is drained from a pool, the fuel could overheat and ignite the zirconium cladding, leading to large releases of radioactivity. (This possibility had already been identified by analyses performed for the NRC.)[104] The Alvarez report recommended that U.S. plant operators reduce their pool inventories and return to a more open storage configuration by transferring relatively older fuel to dry casks, which are passively cooled. The Alvarez study made other recommendations, such as installing emergency spray cooling systems and preparing to repair holes in spent fuel pool walls on an emergency basis, if called for.[105]

In a response to this study, the NRC argued that currently permitted, more densely arrayed pool storage could be carried out both safely and securely.[106] This position has continued to be questioned by advocates of lower-density pool storage, especially since the accident at the Fukushima Daiichi plant.

Prompted by conflicting public claims about the safety and security of commercial SNF storage at nuclear power plants,[107] the National Academies, at the request of Congress, completed an independent assessment of these issues in 2004 (an unclassified public report, titled *Safety and Security of Commercial Spent Nuclear Fuel Storage*, was published in 2006).[108]

The NAS study concluded that a successful attack on a fuel pool, though difficult, is possible and could result in a large release of radioactive material if it led to a propagating zirconium cladding fire. It also found, however, that additional analyses were needed to better understand these vulnerabilities and that a number of steps could be taken to reduce the likelihood of such a fire (including changing the configuration of hotter and cooler spent fuel assemblies in the pool and providing back-up cooling through spray systems). The NAS study further concluded that dry cask storage has inherent safety and security advantages over wet pool storage but is only suitable for older spent fuel (more than five years post-discharge). It did not, however, address the question of whether the transfer of spent fuel from pool to dry cask storage should be accelerated (the study simply notes that "further engineering analyses and cost-benefit studies would be needed before decisions on this and other mitigative measures are taken"[109]). The NRC has since taken actions to address the risks outlined in the NAS study. In February 2005, following completion of the classified version of the NAS study, NRC staff provided guidance for implementing the orders that had been issued in 2002, including best practices for mitigating losses of large areas of the plant and measures to mitigate fuel damage and minimize releases. The NRC subsequently conducted inspections at operating reactor sites to assure compliance with these orders.

In December 2006, the Nuclear Energy Institute (NEI) issued a document that provides guidance for implementing a set of strategies intended to maintain or restore core cooling, containment, and spent fuel pool cooling capabilities under the circumstances associated with the loss of a large area of the plant due to explosions or fire.[113] (The NRC endorsed this document as an acceptable means for developing and implementing the requirement for mitigation strategies.) The guidance related to pools includes adding make-up water to the pool and spraying water on the spent fuel. In addition to these measures, the industry has reportedly taken steps to implement the "checkerboarding" arrangement of hotter and cooler spent fuel assemblies in pools, as recommended in the NAS study.[114]

In 2002, a coalition of more than 150 national and local non-government organizations (NGOs) adopted a set of principles for at-reactor storage "based on the urgent need to protect the public from the threats posed by the current vulnerable storage of commercial irradiated fuel."[115] These principles include several points:

- Implement a low-density, open-frame layout for reactor fuel pools (which would involve accelerating the transfer of fuel older than five years to dry storage)
- Establish hardened on-site storage (HOSS – see text box) at reactors
- Provide for greater protection of fuel pools
- Require periodic review of HOSS facilities and fuel pools
- Provide dedicated funding to local and state governments to independently monitor and protect sites
- Do not reprocess spent fuel.

The recommendation to use HOSS, instead of conventional dry storage technology, at reactor sites is being considered as part of the NRC rulemaking that is currently underway to update nuclear plant security requirements. We believe the NRC rulemaking process is the appropriate venue for considering and assessing the technical merits of the HOSS concept at this time, since its principal objective is to increase resistance to terrorist attacks. Meanwhile, the question of whether steps should be taken to reduce the amount of spent fuel currently stored in reactor pools is distinct from the question of where and how the spent fuel should be stored if that were done. These issues are being reexamined by industry, the NRC, and others.

The Hardened On-Site Storage (HOSS) Concept

HOSS[110] dry cask or vault systems have been proposed to enhance the safety and security of spent fuel storage. As described by proponents, HOSS is the preferred end-point of a process that involves moving spent fuel from dense-packed cooling pools and into dry storage systems at reactor sites. The HOSS concept adds berms and reinforced concrete vaults and overstructures[111] to conventional dry storage systems with the intent of offering greater resistance to potential terrorist attacks using aircraft or conventional weapons.

Utilities and the nuclear power industry have generally not supported the HOSS approach to dry storage for a variety of reasons.

Industry representatives have suggested that the primary objectives of the HOSS approach are effectively already being met through a combination of robust cask systems (see section 5.1) and physical security measures. They believe that continued reliance on NRC requirements, which use a design-basis threat assessment methodology, will ensure facilities remain safe and secure by requiring tiered security forces, active and passive response systems, and conservative, robust technology designs.[112] The industry view is that the HOSS approach could increase risk rather than reduce it if the storage/vault system were to collapse under attack and then interfere with the cooling of the fuel.

6. A Consent-Based Approach to Siting and Developing Future Facilities for Nuclear Waste Management and Disposal

Having examined decades of experience in siting nuclear waste facilities in the United States and abroad, the Commission concludes that the United States needs to adopt a new approach to siting and developing nuclear waste management and disposal facilities in the future.

We believe siting processes for all such future facilities are most likely to succeed if they are:

1. Consent-based—in the sense that affected communities have an opportunity to decide whether to accept facility siting decisions and retain significant local control.
2. Transparent—in the sense that all stakeholders have an opportunity to understand key decisions and engage the process in a meaningful way.
3. Phased—in the sense that key decisions are revisited and modified as necessary along the way rather than being pre-determined.
4. Adaptive—in the sense that process itself is flexible and produces decisions that are responsive to new information and new technical, social, or political developments.
5. Standards- and science-based—in the sense that the public can have confidence that all facilities meet. rigorous, objective, and consistently-applied standards of safety and environmental protection.
6. (6) Governed by partnership arrangements or legally-enforceable agreements between the implementing organization and host states, tribes, and local communities.

The Commission recognizes that the NWPA and subsequent actions by Congress have established Yucca Mountain in Nevada as the site for a deep geologic nuclear waste repository, provided the repository license application submitted by DOE is found by the NRC to meet relevant requirements. The Commission takes no position on the Administration's request to withdraw the license application. We simply note that the U.S. inventory of SNF will soon exceed the amount that can be legally emplaced at Yucca Mountain until a second repository is in operation. So under current law, the United States will need to find a new repository site even if Yucca Mountain were to go forward. We believe the approach set forth here provides the best strategy for assuring continued progress, regardless of the fate of Yucca Mountain.

The remainder of this section discusses the basis of this Commission recommendation—including key lessons learned from past siting efforts—and elaborates on the details of the adaptive and staged approach we are recommending for siting new facilities.

6.1. Lessons Learned from U.S. Experience in Siting Nuclear Waste Facilities

The difficulty of siting any type of facility that handles, stores, or disposes of highly radioactive materials has been at the heart of the federal government's failure to deliver on its waste management obligations to date. Three examples from the U.S. experience are particularly instructive for future siting efforts: the currently suspended program to develop a geologic repository at Yucca Mountain in Nevada, the successfully completed and currently operating WIPP disposal facility for transuranic defense waste in New Mexico, and a series of thus far unsuccessful public and private efforts to establish an MRS facility for commercial SNF. Each of these experiences is summarized as part of the historical overview provided in section 3 of this report. In this section, we highlight lessons learned from these past siting efforts that helped inform the Commission's recommendations.

In the history of the U.S. nuclear waste management program, the contrasting experiences with Yucca Mountain and WIPP offer important insights. Yucca Mountain was singled out as the sole site to be considered for a first national geologic repository in the 1987 Amendments to the NWPA and the record since has been one of frequent regulatory and legal deadlock, extreme political controversy and strong state opposition, steadily escalating project costs, and delays measured in decades.

The problems that plagued Yucca Mountain from the outset and that have led to the current impasse are not hard to identify:

- Short-circuiting of the initial site selection process that had the effect of tainting all subsequent state-federal interactions over the project
- Lack of appropriated funds to complete project milestones on time
- Overly prescriptive requirements and rigid deadlines that made it difficult to respond to stakeholder concerns
- Inconsistent program leadership and execution. All of these flaws only served to exacerbate what was arguably the most important and most enduring problem of all—the fact that the project was strongly opposed, from the time Yucca Mountain was named in 1987 as the only site to be studied, by the majority of Nevada residents and by the state's political leaders. That the project suffered from protracted delays and has now been suspended—after an investment of more than 20 years and billions of dollars in resources—speaks volumes about the difficulty of siting a facility over the objections of the host community, state, or tribe and about the broader shortcomings of the U.S. program.

In stark contrast to Yucca Mountain, the WIPP facility in New Mexico has been operating successfully for more than a decade with broad local and state support, although that project too was often controversial, suffered numerous setbacks in the siting and licensing process, and took years longer to complete than originally planned. The crucial difference in the WIPP case was the presence—also from the outset—of a supportive host community and of a state government that was willing to remain engaged. Starting in the early 1970s and continuing to the present, elected officials and other local leaders in and around the WIPP site, particularly in the Carlsbad business community, made it very clear that they approved of the development and use of the facility to dispose of defense TRU wastes. This unwavering local support helped to sustain the project during periods when federal and state agencies had to work through disagreements over issues such as the nature of the wastes to be disposed, the role of different entities in providing oversight, and the standards that the facility would be required to meet.

Even so, the path to successfully licensing and opening WIPP was neither straightforward nor quick (see text box in section 3.4.2). On the contrary, it involved years of legal, regulatory, and political activity and complex negotiations between the State of New Mexico and the federal government. Ultimately, local support combined with other confidence-building measures proved sufficient to allay state concerns and allow the project to go forward. But no one could have designed the process that was ultimately followed ahead of time nor could that process ever be replicated.

Attempts to site an MRS facility in the 1980s and 1990s, by contrast, have had more in common with the Yucca Mountain experience in the sense that none of them— despite the availability of unspecified inducements under the 1987 NWPA amendments—succeeded in overcoming opposition at the state level. Outreach by the short-lived Office of the Nuclear Waste Negotiator in the early 1990s prompted a number of communities and tribes to express interest in being considered for a facility, but the program was closed down before any of those possibilities could be fully explored. A subsequent private initiative by several utilities to work directly with the Goshute Indian tribe to open a consolidated spent fuel storage facility on the tribe's Skull Valley Reservation in Utah resulted in the NRC issuing a license but likewise encountered strong state-level opposition and is still being litigated.

In sum, U.S. experience to date clearly underscores the inherent complexity and difficulty of siting nuclear waste facilities, particularly in the face of state-level opposition. At the same time, the record, along with input received from a number of parties during the BRC's deliberations, provides grounds for optimism that it can be done. The WIPP example, in particular, represents an affirmative demonstration that with adequate patience, flexibility, and political and public support, success is possible.

6.2. Experience with Nuclear Waste Facility Siting in Other Countries

In designing a new approach to siting, the United States can also look to a substantial body of experience in other countries. All of the countries the Commission studied (see appendix C) provided useful insights for the U.S. program going forward. Sweden and Finland are furthest along in selecting and developing a repository site, while other countries—like France and Canada—have also made substantial progress (of these countries, Canada provides perhaps the closest analogue to the United States in terms of political structure). In addition, Spain recently selected a site for a consolidated storage facility. Overall, the experience of these countries provides strong support for the Commission's conclusion that a transparent, consent-based approach built on a solid understanding of societal values has the best odds of achieving success in siting, constructing, and operating key waste management facilities.[116,117]

In Finland, plans to develop a geologic disposal facility for SNF at the island of Olkiluoto have the support of the host community, Eurajoki (which initially vetoed its selection as a repository site).[118] Finland's efforts to site a deep geologic repository and undertake associated environmental impact assessments began in 1983, when the government issued a major policy decision on the management of SNF and on the schedule and process to be used for selecting a final repository site.[119] The siting process entailed three steps. First, a country-wide screening study was undertaken between 1983 and 1985. This was followed, from 1986 to 1992, by preliminary site investigations. In the third phase, from 1993 through 2000, detailed site investigations and environmental impact assessments were conducted for four sites.

All four sites were found to be technically suitable for the final disposal of SNF, but local support for a repository was strongest in the communities of Eurajoki and Loviisa where nuclear infrastructure already existed. Of these two sites, a larger area for surface support facilities was available at Olkiluoto. In addition, because of two existing reactors at Olkiluoto, a large portion of the country's spent fuel inventory was already on the island.

In 1999, Posiva Oy (the company responsible for managing spent fuel in Finland) applied to the Finnish government for a decision-in-principle to go forward with a repository at Olkiluoto. At that point, the government requested statements on Posiva Oy's application from the municipality of Eurajoki and from the relevant regulatory authority. Eurajoki's municipal council voted in favor (by 20 votes to 7) and the Finnish government followed with a positive decision-in-principle in December 2000. After further discussion, Finland's Parliament overwhelmingly ratified the government's decision (by a vote of 159 to 3) in May 2001. Detailed site characterization studies at Olkiluoto began in 2004 with the construction of an underground research tunnel. A license application for the facility is now planned for 2012 with an anticipated start date for repository operations in 2020.

The Swedish waste management company, SKB, is likewise moving forward with the development of a geologic repository for spent nuclear fuel with the consent of the host municipal government. Between 1977 and 1985, SKB identified a number of "investigation areas" in different parts of the country. Such areas were selected for further studies on the basis of existing geological data as well as an assessment of the ease of getting permission by the land-owner to carry out such investigations (including borehole drillings). This approach gradually met more and more opposition. In 1985, SKB decided to stop these investigations, partly as the result of a governmental request. At that time, geological information had been collected from about 15 locations. An overall conclusion was that it is possible to find sites that meet the stipulated geological requirements for a deep geological repository in most parts of Sweden.

In early 1992 SKB initiated a new siting process. The process started with a letter from SKB to all Swedish municipalities (about 290) explaining SKB′s task to find a site for a repository for spent fuel and inviting interested municipalities to voluntarily apply. SKB's invitation resulted in two municipalities agreeing to a feasibility study. These feasibility studies were followed by referendums in both municipalities to ascertain public opinion regarding further participation in the siting process. In both cases, the referendums resulted in a rejection of further participation.

At that point, SKB conducted further feasibility studies and identified five potentially promising sites. Of these, SKB approached the three geologically appropriate communities that already housed nuclear facilities. In 2001, the government approved SKB's proposal to undertake a detailed investigation of these three sites: (1) the existing Forsmark nuclear site near the municipality of Östhammer, (2) Oskarshamn, which was the site of an underground nuclear research laboratory constructed in the early 1990s and (3) an area in the northern part of Tierp. A few months later, the municipal councils in Östhammer and Oskarshamn consented to further investigations, while Tierp opted out (importantly, either Östhammer or Oskarshamn could have vetoed its selection as a disposal site for HLW).[120] Ultimately, this process worked. Of the two remaining options, Forsmark—which already hosts a large nuclear power plant and an operating repository for short-lived low- and intermediate-level radioactive waste—was ultimately selected in 2009 because it offered better geology. In March 2011, SKB applied to the Swedish government for permits to construct a repository in Forsmark.

A unique feature in the Swedish approach is that, before the final site decision was made, there was an agreement that the community not selected would receive a larger amount of compensation than the community that was selected. The rationale was that the community selected to host the repository would realize additional economic benefits, in the form of construction activity, infrastructure investments, permanent jobs to operate the repository, and ancillary development (e.g., research and fabrication facilities, etc.). The value of these benefits to the local economy was estimated at about $300 million.[121] Ultimately, the community near Forsmark will receive approximately 25 percent of this estimated value for hosting the repository, while the community at Oskarshamn, which was not selected, will receive the remainder—approximately 75 percent of the estimated benefits—for participating in the siting process. At this point, the anticipated start date for repository operations is 2025.

France established its Agency for Radioactive Waste Management (ANDRA) in 1991 to develop a strategy and perform research on managing the high-level and intermediate-level long-lived radioactive waste generated by that country's nuclear reactor fleet (prior to this time the process was largely controlled by the industry and the national government). Lines of authority and decision-making responsibility were further clarified with the passage of the 2006 Planning Act, which established the decision in principle to develop a geological repository, to be located at a site and in a geological formation that had already been studied through an underground laboratory. To date, there has been community support for the siting process: local governments in the Meuse/Haute-Marne region volunteered to host an underground site-characterization program and can expect to benefit from a series of measures designed to support local development, including a dedicated tax on basic nuclear installations, along with additional projects. More recently, ANDRA signed a contract with a joint venture of two engineering companies to conduct industrial design work for a deep geological repository for France's high- and intermediate-level radioactive waste. The first conceptual study phase is to be conducted in 2012 and will lead to a public consultation that will take place in 2013.[122]

Canada's Nuclear Waste Management Organization (NWMO) was formed in 2002 after the failure of a decades-long, technically-oriented effort to establish a repository. A commission chartered in the 1990s to review the Canadian program concluded that while the program had conducted the scientific and technical aspects of the program well, it did not enjoy public confidence and had not provided for "social safety." This review led to legislation that established the NWMO.

The NWMO has adapted many lessons from the Finnish and Swedish experience to its approach to nuclear waste management in Canada and pioneered a number of novel steps in its approach as well. The very first step taken by the NWMO was to ask how its attempt to develop a repository would be any different from those of the past. The conclusion was reached that the NWMO should first seek to understand the deeply-held values of citizens, and only then review its options in light of that citizen input.[123] After several extensive iterations with Canadian citizens and stakeholder organizations, the NWMO has explicitly adopted a phased, adaptive approach they call Adaptive Phased Management.

This deliberate, transparent, and highly engaged process has led nine communities to volunteer to engage the NWMO in the earliest stage of discussion and information gathering, prior to considering whether to have surveys conducted. Canada went through an evaluation of its program by an external commission more than a decade ago and fundamentally restructured its approach as a result.

Canada's provincial-level government in some ways mirrors the intermediate level of government comparable to U.S. state government, which does not exist in Finland, France, or Sweden. Canada's progress to date thus provides additional insights and enhances confidence in the siting process we are recommending.

Spain provides the most recent example of a successful consent-based siting process for a nuclear waste facility—in this case, a consolidated storage facility for spent fuel from that country's eight operating and two shutdown reactors.[124] In December 2004 a resolution supported by all parties in the Spanish Parliament called on the government to put an end to dispersed spent fuel storage at multiple reactor sites by developing a central storage facility for the spent fuel, as well as for a small quantity of solidified high-level waste due to be returned from France.

In 2006, an inter-ministerial commission of the national government was established to define siting criteria for the facility and to develop and supervise a transparent, democratic, and participatory siting process. In the same year, the commission initiated an information campaign aimed at municipalities in the country (the siting process had no formal role for the autonomous communities, i.e. the large regional governments such as Valencia and Andalucia that are analogous to states). ENRESA, the national waste management organization that would be responsible for designing, constructing, and operating the facility, supported the commission's siting process by performing technical studies and providing information to stakeholders.

The proposed facility–described as a technology park– includes not only the storage facility itself (which will also accept intermediate-level radioactive waste from nuclear power plant decommissioning), but also other facilities intended to support local and regional development, including several laboratories.

In December 2009, the government issued a call for proposals from communities interested in hosting the facility, and by the end of February 2010 eight communities with potentially qualified sites were accepted as candidates. Following an evaluation of the proposals, the commission proposed a preferred candidate site (Zarra, in Valencia) in September 2010, but the government did not formally endorse the recommendation pending efforts to gain consensus at the autonomous community level. After delays due in part to an acceleration of the schedule for national elections, in December 2011 the new government announced selection of a site in the autonomous community of Castilla la Mancha that had been one of the four top-ranked candidates and enjoyed broader support beyond the host community than had been the case with Zarra. The entire siting process, from establishment of the interministerial commission in 2006 through site selection at the end of 2011, took less than six years.

In addition to hearing from leaders of the Canadian, Finnish, French, and Swedish programs and visiting facilities in Finland, France, and Sweden, several Commission members had an opportunity to travel to Japan, Russia, and the United Kingdom to hear firsthand from leaders of those countries' nuclear waste management programs. As an element of these fact-finding trips, members heard from local government officials and from a variety of non-governmental organizations and other stakeholder groups. In contrast to the U.S. situation, these officials and others expressed a high degree of confidence in the site identification and selection processes used to locate a repository and in the institutions responsible for implementing and overseeing those processes. Although the countries we visited were in various stages of the siting and licensing process, they stressed that several elements were critical in establishing a foundation for trust:

- A clear and understandable legal framework
- An opt-out option for the local affected community, up to a certain point in the process
- The availability of financing for local governments and citizen organizations for conducting their own analyses of the site and siting issues
- Compensation for allowing the investigation and characterization of the proposed site
- A concerted effort to promote knowledge and awareness of the nuclear waste issue and plans for addressing it through mechanisms such as:
    - Seminars, study visits, and reviews conducted by the local government
    - Information to and consultation with local inhabitants
    - Socioeconomic studies and evaluations of impacts on local businesses

- Openness and transparency among and within the implementing organization, the national government, local governments, and the public.

How these elements might be included in a new approach to siting future facilities for nuclear waste and spent fuel management and disposal in the United States is the subject of the next section.

6.3. Key Elements of a Phased, Adaptive Approach to Siting and Developing Facilities

Based on the history of waste management efforts at home and abroad, the Commission concludes that the United States must commit to a new, more flexible and more adaptive approach to siting and developing facilities in the future. "Learning by doing" has produced substantial improvements in the reliability, safety, and performance of commercial nuclear reactors in the United States. It has also contributed to an impressive track record of safe transport and handling with respect to the transfer of defense TRU wastes to the WIPP facility in New Mexico. Compared to the prescriptive approach used in attempting to develop a repository for spent fuel and HLW at Yucca Mountain, other nations—notably Sweden and Finland—appear to be proceeding with less controversy using an adaptive, staged management approach (recognizing that some other nations using an adaptive approach have not yet succeeded in identifying repository sites).

The notion that such an approach could produce better outcomes for this nation's nuclear waste management program also is not a new one. In a comprehensive 2001 report on the status of efforts to provide for the disposition of HLW and spent fuel,[125] the NAS concluded that "geological disposal remains the only long-term solution available" and recommended that national waste management programs "should proceed in a phased or stepwise manner."[126]

The Commission concurs strongly with the NAS recommendation. In our view, moreover, the events of the last decade only bolster the case for a phased, adaptive approach because they demonstrate that without political buy-in and trust, progress in the long and demanding process of finding a resolution to our nation's waste management challenges will be extremely difficult to sustain.

Of course, the first requirement in siting any facility centers on the ability to demonstrate adequate protection of public health and safety and the environment.

As part of a phased, adaptive approach,[127] the Commission recommends that starting early in the process of exploring a potential site the waste management organization begin to develop a "safety case" that collects in one document the wide range of relevant technical and other information (including information on legal, financial, and managerial aspects of the waste management system) that together provides a basis for confidence in the safety of the proposed facility at the proposed site.[128]

The articulation of a safety case provides a way to communicate important information to decision- makers, stakeholders, and the public; it also helps to promote a broader and more accurate understanding of the scientific, technical, and other bases for decisions about the facility, including ultimately the licensing decision. The purpose of the safety case would not be to expand on requirements already included in the existing licensing process,[129] but rather to make the rationale for decisions about the facility accessible and understandable to the public and to a wide range of decision makers beyond the audience of regulatory experts who are already familiar with the full range of arguments being considered as part of the process.[130]

To support the consent-based siting process we have recommended, the safety case should (1) be easily accessible to all concerned stakeholders and to local, tribal, and state government representatives and (2) should strive to make clear and explicit all the assumptions and evidence that have been considered as part of building the case for confidence in the long-term performance of the proposed facility at the proposed site. In addition, the safety case should be updated as needed to provide an input to decisions throughout the facility development process. It should also be updated periodically after the facility begins operation if agreements with local communities, tribes, or states require a periodic revalidation of the facility's ability to meet safety requirements.

Siting New Nuclear Waste Management Facilities – Getting Started

First, the Environmental Protection Agency and the Nuclear Regulatory Commission should develop a generic disposal standard and supporting regulatory requirements early in the siting process. Generally-applicable regulations are more likely to earn public confidence than site-specific standards. In addition, having a generic standard will support the efficient consideration and examination of multiple sites.

Once the new waste management organization is established it should:

- *Develop a set of basic initial siting criteria* – These criteria will ensure that time is not wasted investigating sites that are clearly unsuitable or inappropriate.
- *Encourage expressions of interest from a large variety of communities that have potentially suitable sites* – As these communities become engaged in the process, the implementing organization must be flexible enough not to force the issue of consent while also being fully prepared to take advantage of promising opportunities when they arise.
- *Establish initial program milestones* – Milestones should be laid out in a mission plan to allow for review by Congress, the Administration, and stakeholders, and to provide verifiable indicators for oversight of the organization's performance.

One important implication of pursuing an adaptive staged approach is that the focus is on initial operation of a repository rather than on rapidly disposing of a large inventory of waste.[131] This follows from the NAS description of the characteristics of a successful geologic repository program[132] as one in which:

- A geologic site and engineered system, judged to be technically suitable using the particular country's accepted regulatory, public, and political processes, have been identified

- Operational and long-term safety aspects are made consistent with the current scientific understanding of repository systems, safety features are reviewed; and the necessary licenses are granted
- An ongoing long-term monitoring and observation program designed to substantiate the current scientific understanding of the safety aspects of the repository system is in progress
- Sufficient societal consensus is achieved to allow operations to begin and continue
- Initial waste emplacement has taken place with plans for reversibility
- All necessary safety and security measures are set up to place additional waste, if decided
- Procedures and funding arrangements are agreed to for either:
    - Backfilling (if used), closing, and sealing the repository (if technical and societal confidence in the long-term isolation properties continues), or
    - Maintaining capability for long-term control and monitoring, and capability for treating wastes, if waste retrieval is necessary for technical or societal reasons.

It is very important to recognize that these requirements in turn imply a need for substantial buffer storage capacity in the waste management system so as to decouple the program's ability to accept waste from the emplacement of that waste in a repository for disposal. This in turn would provide the flexibility needed to develop repository capacity in a more gradual and stepwise manner. The need for buffer capacity is addressed by the Commission's recommendation concerning the expeditious development of one or more consolidated storage facilities for SNF, as discussed in section 5 of this report.

6.4. Specific Steps in an Adaptive, Staged Facility Siting and Development Process

Experience in other countries and from the WIPP facility in the United States suggests that an adaptive, phased, and ultimately consent-based process should start by encouraging expressions of interest from a large variety of communities that can offer a potentially suitable environment for the type of facility under consideration. The waste management organization should also be able to approach communities that it believes can meet the siting requirements. As communities engage the process, the implementing organization must be flexible enough not to force the issue of consent while also being fully prepared to take advantage of promising opportunities when they arise.

Throughout, meaningful consultation with stakeholders to inform them of the status of the siting process and make needed adjustments (much as was done by the NWMO in Canada) will be critical to building credibility and confidence in the implementing organization.

Prior to launching a consent-based siting process, the implementing organization should develop a set of basic, initial siting criteria designed to ensure that time and resources are not wasted to investigate sites that are clearly unsafe, unsuitable or inappropriate for waste facility development. At the same time, it will be important to communicate with local communities and stakeholders about the nature of the risks involved in hosting a facility and about options for addressing and managing those risks. As the siting process continues and as various candidate sites pass initial screening criteria, additional sets of criteria should be applied to eliminate all but the most suitable sites for further characterization. These additional criteria might include geologic features, anticipated socioeconomic effects, transportation access and impacts, costs, and other important considerations. Obviously, as a candidate site is characterized in greater and greater detail it will be necessary to demonstrate not only that the preliminary criteria are satisfied, but that all applicable environmental, health and safety, and other requirements set forth by the responsible regulatory authorities can be met.

The Commission takes the view that any future site, provided it has met all regulatory requirements and has been selected with local- and state-level consent should require no additional approval, including congressional approval.[133] Likewise, after a disposal facility enters operation, any modification or expansion of the facility's mission should be consent-based. This approach is consistent with an overall framework that gives the new implementing organization authority—subject to congressional oversight—to make binding agreements with regard to developing key parts of the nuclear waste management system. As with other details of establishing a new management approach and a new implementing organization, the specific requirements for moving forward with a particular site would have to be set forth in new legislation.

The Commission also recommends that pilot, test, and demonstration facilities (including an *in situ* research and demonstration laboratory) be co-located with new waste management facilities, as appropriate, wherever feasible. This will make it possible to conduct tests aimed at improving operational efficiency and safety and signal a continuing commitment to R&D to reduce residual uncertainties.[134] These facilities have also been used as excellent public communication tools in Sweden and France, for example, to explain to the interested public exactly how a repository operates.

The National Academics' 2003 *One Step at a Time* report identified seven key attributes of adaptive staging:

1. Commitment to systematic learning. Project managers intentionally seek, are open to, and learn from new knowledge and stakeholder input. Stages are designed specifically to increase available scientific, technical, societal, institutional, and operational knowledge.
2. Flexibility. Project managers are able and willing to reevaluate earlier decisions and redesign or change course when new information warrants.
3. Reversibility. Project managers are able to abandon an earlier path and reverse the course of action to a previous stage if new information warrants.
4. Transparency. The decision-making process and the basis for decisions are documented and accessible in real-time and plain language to all stakeholders.
5. Auditability. Documentation for the basis of decisions is complete and made available to all interested parties for review purposes.
6. Integrity. Technical results are accurately and objectively reported and all uncertainties, assumptions, and indeterminacies are identified and labeled.
7. *Responsiveness*. Project managers seek and act on new information in a timely fashion.

Finally, the Commission recognizes that reasonable milestones for major phases of program development and implementation are important to keep the program focused and ensure that it is moving forward. The Finnish waste management program demonstrates the usefulness of milestones as a mechanism to help sustain steady and meaningful progress. Since an adaptive phased approach requires both clear programmatic planning and flexibility, we recommend that the implementing organization establish reasonable time horizons for the major stages of the program. As one example, the implementing organization might contemplate a range of, say, 15 to 20 years to accomplish site identification and characterization and to conduct the licensing process.

A notional timeframe for siting and developing a consolidated storage facility would presumably be shorter, perhaps on the order of 5 to 10 years. The implementing organization will be responsible for setting overall and intermediate milestones for each stage of the process. Of course, there will be unforeseen developments that could cause siting to take a longer or shorter period of time.

This is why the program requires flexibility. Program milestones should be laid out in a regularly updated mission plan (as discussed in chapter 7) to allow for review by Congress, the Administration, and stakeholders, and to provide verifiable indicators for external oversight of the organization's performance.

Any needed changes would be presented in mission plan revisions for review as appropriate.[135]

6.5. Support for Participation

A noteworthy feature of the Swedish repository program is that funds from the nuclear waste management organization are set aside to be awarded to NGOs involved in the siting and repository development process. These funds are used by the NGOs to investigate technical and other aspects of the nuclear waste management program.

In the course of the Commission's deliberations, many participants emphasized the importance of citizen participation. For example, a letter from the South Carolina Governor's Nuclear Advisory Council and others pointed out that "citizen participation results in better and quicker decisions that are accepted by the larger public."[136]

For a complicated and technically-involved issue like the development of a nuclear waste repository, the inability of citizens and citizen groups to access the necessary technical expertise can be a major barrier to participation (see further discussion of this issue in section 6.6). In a large country like the United States, sheer distance can also be an issue; important meetings, conferences, and other events are regularly held in far-flung locations, and travel and lodging expenses can be beyond the means of individuals and groups who would otherwise wish to participate.[137]

Perhaps even more important, states, tribes, and affected communities—in order to gain trust and confidence in the decisions taken by the waste management organization—must be empowered to meaningfully participate in the decision-making process. This means being in a position to evaluate options and provide substantive input on technical and operational matters of direct relevance to their concerns and interests.[138]

Features of Adaptive Staging

Every first-of-a-kind, long-term, and complex project develops in stages. With time, stages and schedules are inevitably revised in light of experience and knowledge gathered along the way. However, many national repository programs, including the U.S. program, have run afoul of rigid milestones for commencing full-scale waste emplacement.

By contrast, adaptive staging entails a flexible approach where the overall direction to be taken and its end points are outlined at the beginning and all parties, including stakeholders, acknowledge that the program can be revised as it progresses. Adaptive staging is less "error-prone" than a rigid approach and it allows the current generation to manage waste using the best available knowledge without foreclosing options if future generations decide to take a different approach.

It is important to emphasize that these elements should not be implemented in a way that causes continual delay. Certainly, an adaptive staged approach may result in higher initial costs and a slower pace of waste emplacement in the beginning. But the point is to implement a process that is ultimately more efficient—both in terms of cost and time—because it corrects potential problems before they become expensive and time consuming. Finally, an adaptive staged approach implies continued investment in new learning, including support for science and technology development that can improve the performance of the whole waste management system.

In sum, the Commission believes that a new U.S. waste management organization should adopt the Swedish practice and set aside funding for participation by citizens, citizen groups, and other NGOs. The availability of funding should be widely announced and reasonable criteria should be established against which to evaluate applications for financial support.

6.6. The Role of States, Tribes, and Communities in an Adaptive, Consent-Based Siting Process

It has long been accepted that host states, tribes, and local governments should play an important role in siting nuclear waste management and disposal facilities.[139] As one early study put it: "If the federal government is to make progress toward a permanent solution of the radioactive waste problem, it cannot go it alone—citizens will insist on assurances (other than federal assurances) that proposed actions will not involve undue risks to the host states."[140]

In the debates leading up to the original NWPA of 1982, Congress considered a wide range of options for formalizing the host states' role in repository siting—from merely providing for consultation to giving host states a complete veto over proposed projects within their borders. Ultimately, the formula adopted in the NWPA included provisions for "consultation and cooperation," combined with some state oversight rights and the ability to veto a proposed site. The state veto, however, was subject to congressional override— an option that was exercised when Congress overrode Nevada's veto of the Yucca Mountain site in 2002.[141]

As we noted in our brief review of lessons learned from the U.S. experience so far, states have generally resisted—in some cases very strongly—efforts to site HLW and spent fuel disposal and away-from-reactor storage sites within their borders.[142] By contrast, some local governments and tribes have viewed these facilities more positively—and in some cases have supported them strongly—primarily on the basis of anticipated job creation and economic development benefits. Indeed, some of the most supportive communities have been those with a long history of hosting nuclear facilities. Tribal and local support, however, has not usually been sufficient to overcome state-level opposition. This suggests that to be successful, a new waste management organization must find ways to address state concerns while at the same time capitalizing on local support for proposed facilities.

What those concerns might be and how the tensions inherent in the federal–state and federal–tribe relationship might be successfully navigated in different siting contexts is impossible to anticipate in advance. Clearly, locating and constructing facilities for the management and disposal of SNF and HLW will require complex and possibly lengthy negotiations between the federal government and other relevant units of government. In these negotiations, it will be important to define the roles, responsibilities, and authorities of host state, tribal, and local governments both throughout the siting and licensing process and once a facility is operational.[143] In addition, host jurisdictions should have the option to enter into partnership arrangements or other legally-binding, court-enforceable agreements with the implementing organization to ensure that all commitments concerning the development and subsequent operation of waste management facilities are upheld. A similarly consent-based approach should be used in the future in deciding whether modifications to the scope or mission of an existing facility are appropriate and acceptable.

Beyond engaging in substantive negotiations and binding agreements with other units of government as part of the facility siting and development process, the Commission believes that states and tribes should retain—or where appropriate, be delegated—direct authority over aspects of regulation, permitting, and operations where oversight below the federal level can be exercised effectively and in a way that is helpful in protecting the interests and gaining the confidence of affected communities and citizens. Such authorities could be included in legally-enforceable agreements or partnerships if such arrangements are negotiated between the implementing organization and states, tribes, and/or local communities that agree to host a waste management facility. We recognize that defining a meaningful and appropriate role for states, tribes, and local governments under current law is far from straightforward, given that the Atomic Energy Act of 1954 provides for exclusive federal jurisdiction over many radioactive waste management issues.[144] Nevertheless, we believe it will be essential to affirm a role for states, tribes, and local governments that is at once positive, proactive, and substantively meaningful and thereby reduces rather than increases the potential for conflict, confusion, and delay. At the same time, host state, local, and tribal governments have responsibilities to work productively with the federal government to help advance the national interest.

Several commenters have expressed a desire to see the Commission explicitly define the point at which potential host state, tribal and local governments could no longer unconditionally (that is, without cause) "opt out" of a facility siting process. These commenters correctly note that the level of state, tribal and community acceptance of a proposed waste management facility can and likely will fluctuate over time. The Commission believes that defining the point at which the right to unconditionally opt out expires must be part of the negotiation between affected units of government and the waste management organization. In our view, however, the right to opt out without cause should expire no later than the time when a license application for a proposed facility is submitted.

We believe this approach makes sense given that, under the process we have recommended, the potential host community, tribe, and state would have had to consent to be considered for a waste site, with full knowledge of the relevant safety standards and siting criteria. Further, the host state and affected tribal and local governments would have had to agree to the terms of site study and what was to be built prior to the submission of a license application.

When studies were complete, a license application would be prepared, and the Commission believes the host state and affected tribal and local governments should be given the opportunity to sign off on it before submittal. After that time, the state and other units of government would only be allowed to opt out "for cause"—such as bad faith on the part of the facility operator.

Formal agreements, of the type we have recommended elsewhere, would be in place to cover this situation.

It is worth noting that in the context of the fundamentally consent-based facility siting and development process we are recommending, negotiations with host states, tribes, and local governments would obviate the need for a state-level veto, just as the veto/override provisions of the NWPA would not have applied to a repository or MRS facility sited through the voluntary Nuclear Waste Negotiator process established in the 1987 NWPA amendments. Meanwhile, legislation to establish a new waste management organization and associated funding reforms (discussed in detail in the next two chapters) must make it clear that the organization has the responsibility, authority, and resources to negotiate and comply with enforceable commitments.

Consent

Another question highlighted in numerous comments to the BRC is the question of how to define "consent." Some stakeholders, for example, have suggested that consent within a state could be measured by a state-wide referendum or ballot question. On the other hand, the WIPP facility was sited, opened, and has been operated without the state's elected leaders employing such consent-measuring mechanisms. The Commission takes the view that the question of how to determine consent ultimately has to be answered by a potential host jurisdiction, using whatever means and timing it sees fit. We believe that a good gauge of consent would be the willingness of the host state (and other affected units of government, as appropriate) to enter into legally binding agreements with the facility operator, where these agreements enable states, tribes, or communities to have confidence that they can protect the interests of their citizens.[145]

Here, as in other aspects of facility siting, it is instructive to look again to the WIPP experience, since that project was controversial at the state level for many years despite strong local support from the Carlsbad business community. After years of delay and state–federal disagreements, an important development came when Congress required EPA (not DOE) to certify that the facility met applicable standards for waste disposal, including requirements under the Resource Conservation and Recovery Act (RCRA) for the disposal of mixed hazardous and radioactive waste.[146] This meant that the State of New Mexico retained authority to regulate mixed waste at WIPP and that the New Mexico Environment Department had to issue a Hazardous Waste Facility Permit for the repository. Even though the state did not have direct regulatory authority over the radioactive components of the waste being brought to the facility,[147] this development was very important in terms of giving state officials and residents beyond the local community confidence that the facility was safe. Similarly, DOE's decision to work cooperatively with Carlsbad and the Western Governors' Association to develop a safe transportation program for WIPP was extremely helpful in addressing transportation-related concerns. The resulting Western Governors' Association WIPP Transportation Safety Program Implementation Guide includes many procedures that would otherwise be considered "extra-regulatory" and could not be mandated by the states without federal consent. And finally, the establishment of the federally-funded, university-housed Environmental Evaluation Group was important for gaining the trust of state officials and the local community because it provided an independent and credible source for technical information and review of the WIPP project.[148]

Trust, in fact, is often the core issue whenever different parties are involved in a complex adjudicatory process— and it can be especially difficult to sustain when much of the power or control is viewed as being concentrated on one side. In a recent news article, former Governor Michael Sullivan of Wyoming pointed to a lack of trust as one of the central issues that led him to veto a proposed monitored retrievable storage facility in Wyoming in 1992.149 The WIPP example suggests that having some degree of direct state- or local-level control (in the WIPP case, this was possible through RCRA) can be helpful in instances where faith in federal agencies is lacking. In some cases, states have pursued formal agreements with the federal government that can be enforced in the courts, if necessary. In 1995, for example, the State of Idaho entered into an agreement with DOE and the U.S. Navy that allows DOE to ship a limited quantity of spent fuel (primarily from research reactors and from the Navy's nuclear-powered fleet) to INL for storage over a 40-year period. The agreement also obligates DOE to move all spent fuel into dry storage by 2023 and to remove all naval spent fuel from Idaho by no later than 2035. If DOE fails to meet any of the agreement milestones at any point, DOE is subject to fines of $60,000 per day and the State of Idaho may ask the U.S. District Court to halt any further spent fuel shipments to INL. The State of Washington recently entered into a similar agreement with DOE concerning the storage of wastes at Hanford.

The same issues of trust, consultation, and control arise in the context of the federal government's interactions with Indian tribes, another important stakeholder group in the context of nuclear waste management decisions.[150] In fact, because many existing and proposed nuclear sites are either on or near tribal lands, tribal governments have been involved in nuclear technology and nuclear waste issues for decades. The 1982 NWPA requires consultation with states and affected Indian tribes and specifically addresses the participation of tribes in repository siting decisions. In the wake of the 1987 NWPA amendments, several tribes expressed interest in exploring the possibility of hosting nuclear waste facilities on at least an interim basis. As was the case with local communities, however, these expressions of interest generally met with opposition at the state level.

Intergovernmental relationships will require careful attention as the U.S. nuclear waste management program is revived. Experience shows that an unwilling state government can successfully stand in the way of tribal efforts to site nuclear waste management facilities and the Commission believes it would be unrealistic to attempt to locate a facility on tribal land in the face of determined state-level opposition. Yet unlike local communities or state governments, tribes have a unique "government-to-government" relationship with the United States.[151] Their right to make their own laws and be governed by them is limited only by their status as dependent domestic nations and by federal law. Therefore, the federal organization tasked with managing the waste problem will be required to work with federally-recognized tribes on a government-to-government basis.

Legally, states have a very limited role in Indian affairs. They do not have the power to regulate Indian tribes or tribal lands unless such powers are delegated to them by the federal government. Since 1975, moreover, federal policy has supported tribal self-determination. This means that meaningful consultation with tribal governments is required in the development of federal policies and practices that may impact tribal lands, people, or resources.[152] The existing State and Tribal Government Working Group (STGWG) provides an example of one mechanism for facilitating regular consultation between states and tribes and the federal government. Established in 1989 at the request of 10 state governors, the group grew to include 15 states and 10 tribes who would meet with DOE to discuss the federal government's cleanup activities at facilities that have been or are still part of the nation's nuclear weapons complex. STGWG now meets twice annually. As with states, some precedent also exists for giving tribes a degree of regulatory control over specific facilities or operations in the nuclear waste management system. In 1991, the Shoshone-Bannock Tribe attempted to stop the shipment of commercial spent fuel across its reservation in Idaho. A lawsuit resulted and while the courts concluded that federal law (in this case, the Hazardous Materials Transportation Act) did not allow the tribes to ban spent fuel shipments from crossing their land, it did allow them to develop regulations for those shipments.

In sum, whatever the specific authorities and resources of a given community, state, or Indian tribe, experience shows that determined opposition at any level of government can at a minimum significantly complicate and delay, and in many cases defeat, best efforts to site a facility. In this context, it is difficult to overstate the importance of support for a facility or site at the state, tribe, and local level (obviously, public acceptance is not the only criterion; to be considered, any site must also meet safety and technical criteria and other requirements). Support from Congress—for the new waste management organization and its activities as well as for participating states, tribes, and communities—is obviously also important. In the case of WIPP, Congress engaged with the siting process over a period of many years and at several critical junctures congressional intervention, far from undermining the process, helped build trust, resolve issues and ultimately achieve success.

6.7. Benefits to Host States, Tribes, and Communities

In addition to conducting a process that is consent-based, transparent, and responsive to tribal, state, and local governments' need for meaningful input and control, it will be important to demonstrate that the decision to host a facility can deliver real benefits (economic and otherwise) to the tribe, state, and local community.[153] Affected states, tribes, and communities will reasonably expect incentives for helping to address the important national issue of nuclear waste management. To be most effective, such incentives must be provided in ways that are generous, creative, and attentive to their symbolic content.

Besides financial incentives, benefits could include local preferences in hiring and in the purchase of goods and services by the waste management facility, infrastructure investments (such as new roads or rail lines[154]), as well as the opportunity to host co-located research and demonstration facilities or other activities that would generate new employment opportunities and make a positive contribution to the local and regional economy.[155] For example, Spain's effort to find a volunteer host for a storage facility for spent fuel and a small amount of HLW included a technological research laboratory to deal with waste processing, waste forms, disposal of HLW as well as spent fuel, etc. as an integral part of the facility. Eight volunteer communities for the integrated storage/research facility were identified and a final site was selected in December 2011 (see section 6.2).

As noted in section 7.4.1, we recommend that the responsibilities of the new waste management organization include promoting the social and economic well-being of communities affected by waste management facilities. The Commission also recommends that the benefits provided by the current NWPA[156] be modified and expanded to give the waste management organization greater flexibility to promote economic development. Specifically, the Commission believes that the level of benefits currently specified in Section 171 of the NWPA is inadequate. Accordingly we recommend that the NWPA be amended to authorize a new federal corporation (described in the next chapter) to negotiate substantial benefits—potentially well above the amounts currently contemplated in Section 171—to state and local governments, communities, tribes, or other organizations as appropriate. The specific uses of these funds and the metrics that would determine their amounts should be an element of negotiation between the federal government and local communities and governments interested in hosting facilities, but we envision that benefit payments could be used for a wide range of uses, including for economic development purposes. All such payments should be subjected to external, independent auditing.

In addition to locating waste management-related activities in the affected state and community, these states and communities could also be given preference in the siting of other federal projects (provided they are otherwise suitable to host those projects). Section 174 of the NWPA, titled "Other Benefits—Considerations in Siting Facilities," already specifies that the Secretary of Energy "in siting Federal research projects, shall give special consideration to proposals from states where a repository is located." This approach can provide additional benefits to host communities and states without requiring new appropriations or increasing the cost of already planned programs or projects. The Commission recommends that this provision be expanded to include states that host any waste management facilities sited by the new waste management organization and to clarify that the special consideration applies to research, development, and demonstration facilities (not research contracts) that receive federal funding, including any federal matching funds.

In addition to incentives and benefits, neighbors and others impacted by nuclear waste management facilities need assurance of reasonable compensation for real costs. The Commission believes that the framework for evaluating and providing compensation for direct impacts in the current NWPA is workable and should be left alone.[157]

Experiences in Sweden, Finland, and elsewhere have shown that it may not be possible or even advisable to specify incentives and compensation up front; rather, in keeping with an adaptive approach, these determinations are best left to the discretion of the implementing organization and potential host governments—including communities surrounding the host community. These stakeholders will be in the best position to determine what incentives are both appropriate and in their best interests.

Finally, it is important to recognize that Congress may ultimately have a role in providing or approving benefits and compensation for hosting nationally-needed nuclear waste facilities, particularly since some benefits—such as transfers of federal land to host states, tribes, or communities to compensate for land withdrawn for waste facilities—may be beyond the waste management organization's authority and could require legislation.

7. A New Organization to Lead the Nation's Waste Management Program

> Having examined the history of U.S. nuclear waste policy over the last 60 years, the Commission concludes that a new, single-purpose organization is needed to develop and implement a focused, integrated program for the transportation, storage, and disposal158 of nuclear waste in the United States.

This is one of the Commission's central recommendations, and it cross-cuts all aspects of the new strategy we propose for managing the back end of the nuclear fuel cycle in the United States. We believe that new institutional leadership for the nation's nuclear waste program is clearly needed and that a new organization offers the best opportunity to establish—from the outset—the track record of consultation, transparency, accountability, and scientific and technical credibility needed to re-establish trust with the public and key stakeholders.

We urge that legislation to establish this new institution be enacted soon, because the sooner a new institution can take over the waste management program, the sooner it can begin restoring trust and building the relationships and the credibility necessary for success over the many decades the nuclear waste program will operate.

The remainder of this chapter elaborates on the rationale for establishing a new waste management organization and discusses related issues and design decisions, including the form and structure of a new organization, key attributes, scope of responsibility, governance and oversight issues, stakeholder participation, and the transfer of contracts and liabilities. The critical issue of funding is covered in the next chapter of this report.

7.1. The Rationale for a New Waste Management Organization

For the last 60 years, the DOE and its predecessor agencies have had primary responsibility, subject to annual appropriations and policy direction by Congress, for implementing U.S. nuclear waste policy. DOE is a large cabinet-level agency with multiple competing missions, a budget that is dependent on annual congressional appropriations, and top management that changes with every change of administration, and sometimes more frequently than that.

Clearly, multiple factors have worked against the timely implementation of the NWPA and responsibility for the difficulties of the past does not belong to DOE alone. Nevertheless, the record of the last several decades indicates that the current approach is not well suited to conducting a steady and focused long-term effort, and to building and sustaining the degree of trust and stability necessary to establish one or more disposal facilities and implement other essential elements of an integrated waste management strategy. These considerations lead the Commission to agree with a conclusion that has also been reached by many stakeholders and long-time participants in the nation's nuclear waste management program: that moving responsibility to a single purpose organization—outside DOE—at this point offers the best chance for future success.

For example, a new organization dedicated to the safe, secure management and ultimate disposal of high-level nuclear waste can concentrate on this objective in a way that is difficult for a larger agency that must balance multiple agendas or policy priorities. A new organization will be in a better position to develop a strong culture of safety, transparency, consultation, and collaboration.[159] And by signaling a clear break with the often troubled history of the U.S. waste management program it can begin repairing the legacy of distrust left by decades of missed deadlines and failed commitments.

Finally, while the Commission recognizes that it will never be possible or even desirable to fully separate future waste management decisions from politics, we believe a new organization with greater control over its finances could operate with less influence from short-term political pressures. We do not propose that a new organization be less accountable for its actions—on the contrary, effective oversight by Congress and by a strong, independent regulator remains essential. But with greater control over year-to-year budgets and operations, we believe a new organization could more easily maintain the program-level continuity and mission consistency that has often been lacking at DOE.

From an implementation standpoint, this is clearly among the most difficult recommendations advanced by the Commission. Nevertheless, it is also one of the most important, since even the wisest policies are likely to fail without an institutional structure that is capable of implementing them.

7.2. Options for Structuring a New Waste Management Organization

Proposals to establish a new waste management organization are not new. In 1982, the original NWPA directed DOE to study alternative approaches for constructing and operating civilian radioactive waste management facilities, including, specifically, the feasibility of establishing a private corporation for these purposes. More recently, legislation introduced in the 110th and 111th sessions of Congress[160] would have amended the Atomic Energy Act of 1954 to create a new federal corporation (called the "United States Nuclear Fuel Management Corporation") that would "assume responsibility for the activities, obligations, and use of resources of the federal government with respect to SNF management." Over the nearly three-decade period between the original NWPA legislation and this recent proposal, alternative means for financing and managing the nation's HLW program have been extensively studied but never implemented.

Though it is clear to the Commission from its study of this history that a new waste management organization could take a number of forms, we conclude that a federal corporation chartered by Congress offers the most promising model. This is also the organizational form proposed in recent legislation and recommended by an independent advisory committee (the Alternative Means of Financing and Managing or "AMFM" Panel) in 1984.[161] We believe that an independent federal corporation with a well-defined mission, access to adequate resources, ability to make binding contractual commitments, and subject to rigorous external oversight is more apt to achieve the combination of attributes discussed in this section.[162] The Tennessee Valley Authority (TVA), which was established in 1933 to promote resource development in the Tennessee Valley region, may provide a useful existing example of such a federally-chartered, mission-oriented corporation.

Key Attributes f a New Waste Management Organization

How a new waste management organization behaves and delivers on commitments is more important than what specific organizational form it takes. In presentations, public comments, and written submissions to the Commission, stakeholders and experts repeatedly stressed that actions and behavior, more than policies or promises, would be key to restoring trust in the nation's waste management program and in the institutions responsible for operating that program.

Policy makers should therefore consider what design features—including what organizational structure and operational ground rules—would foster the behaviors and attributes most critical to the new organization's success:

- Mission orientation—A well-defined, stable mission, and the organizational capability to focus resources, personnel, and attention on that mission, without being diverted by other priorities.
- Performance—Ability to achieve and sustain high standards of technical, managerial, and craft performance through a skilled workforce supported by a high-reliability, safety-oriented culture.
- Integrity—The intent to be truthful, honest, accurate and open in conducting the program and to place ethical considerations and public well-being at the center of decision making.
- Empowerment—Sufficient authority and independence from political micromanagement to be able to implement the mission.
- Continuity—Stability in terms of organizational structure, culture, and leadership, particularly at the senior levels.
- Flexibility—The ability to anticipate and adapt to new challenges, including sufficient organizational independence to do so.
- Transparency—A clear, open, and transparent decision-making process.
- Participation—Straightforward paths for involvement by all interested parties, with adequate staff and funding dedicated to outreach.
- Responsiveness—The willingness and ability to respond effectively to the concerns and expectations of diverse stakeholders and constituencies.
- Funding—Assured financing to accomplish the mission.
- Accountability—Mechanisms to assure responsible action and to ensure effective oversight by Congress, independent regulators, financial and technical reviewers, and the public.
- Constancy—Commitment to behavior that builds trust and confidence, most importantly by delivering on promises, contracts, obligations, and deadlines.

Two of these attributes—flexibility and responsiveness—are particularly important for program success. Not coincidentally, they are also supported by most of the other attributes listed. Flexibility is needed because the program must operate over very long timeframes in which major changes in technology, institutions, and societal values are inevitable but frequently unpredictable.

The capacity to adapt will be essential. Responsiveness means the ability of the new organization to continually understand and reflect the values of stakeholders and the broader public. Finally, accountability to Congress, to other oversight bodies, to key stakeholders, and to the public is also critical to gaining and sustaining trust, as is a consistent commitment to transparency and communication about how decisions are being made and how competing values and interests are being balanced.

Compared to simply creating a new single-purpose federal agency (even one housed entirely outside DOE), we believe a corporate organization will also (a) be less susceptible to political micromanagement, (b) have more flexibility to respond to changes in external conditions, and (c) have a greater ability to manage costs and schedules.

More important than the specific form of the new organization, however, is that it possess the attributes, independence, and resources to effectively carry out its mission. While a corporate structure appears to the Commission to offer particular advantages, previous studies have concluded that a number of different organizational forms could also get the job done.

Striking the right balance of independence and accountability is the key challenge, whether a new waste management organization is structured as a federal corporation or takes some other form. In any case, Congress must provide clear policy direction, exercise ongoing oversight, and establish the necessary funding mechanisms but should leave control of operational decisions and resource commitments for implementing the policy direction to the new organization.

Those decisions and commitments, and indeed the performance of the organization as a whole, would, of course, be subject to policy, safety, security, technical, and financial review by appropriate government agencies and Congress. We recommend that a board of directors be appointed by the President and confirmed by the Senate (for staggered seven-year terms). The fundamental role of the board would be to provide management and fiduciary oversight and operational direction.

Members of the board should be selected to provide a range of perspectives and expertise and to ensure that key interests are represented.[163]

In addition to an engaged and highly competent board of directors, a new waste management corporation will need the leadership of a strong chief executive.

It will therefore be critically important to define the position and powers of the CEO in terms that will attract candidates with exceptional management, political, and technical skills and experience. Under both the original AMFM Panel proposal and recent legislative proposals, the CEO would be appointed by the corporation's board of directors. The Commission supports this approach.

Other important questions concerning the scope of responsibilities for the new organization, oversight, and stakeholder participation are taken up in the next sections, while the critical issue of funding is discussed in the next section.

Example of How The Nuclear Waste Management Corporation's Board of Directors Might Be Structured

Size: Eleven members, including the CEO.

Appointment: appointed by the President by and with the advice and consent of the Senate (with the exception of the Board-appointed CEO who serves *ex officio*).

General qualifications: To be eligible to be appointed as a member of the Board, an individual shall — (1) be a citizen of the United States; (2) with the exception of the CEO, not be an employee of the Corporation; (3) make full disclosure to Congress of any investment or other financial interest that the individual holds in the energy industry; and (4) affirm support for the purposes of the Corporation.

Composition: The composition of the board is intended to ensure the ability to exercise sound, professional managerial and financial oversight of the functions of the waste management corporation.

The President and the Senate, in making these appointments, will consider individuals who bring that expertise, ability, and experience. Nominees should be selected from organizations contributing to the Nuclear Waste Fund, state public utility commissions, the environmental non-governmental organization community, representatives of workers involved in the construction or operation of radioactive waste management facilities, and others with demonstrated knowledge and experience relevant to the conduct of the activities of the organization, including technical, scientific, social science, community and stakeholder relations, siting, and public health and safety functions. Not more than six of the members shall be from any one political party.

Chairman: Selected by the Board.

CEO: Selected by the Board in its sole discretion, and serves as an *ex officio* member of the Board. Compensation for members who are not federal employees would be provided through an annual stipend (e.g. TVA) or through compensation at a specified daily rate (e.g. proposed United States Nuclear Fuel Management Corporation Establishment Act) Term of office: Seven years, renewable once. Staggered terms, member would continue to serve until his/her successor is appointed.

Subcommittees and consultants: The Board is authorized to form subcommittees and hire consultants and advisors as needed.

7.3. Scope of Responsibilities for a New Waste Management Organization

The Commission's strong view is that to be successful, a new waste management organization must be clearly focused on issues of direct relevance to its primary mission, which is the safe management and disposal of SNF and high-level radioactive wastes.

Specifically, we recommend that the scope of the organization be limited to those functions already assigned to the government in the NWPA, as amended, including:

Responsibility for siting, obtaining licenses for, constructing, operating, and ultimately closing facilities for the disposal of civilian and defense HLW and spent fuel.

- Responsibility for providing consolidated storage capacity for commercial spent fuel, either by directly siting, obtaining licenses for, constructing, and operating storage facilities or by contracting with private entities that have established and licensed such capability with the agreement of the host state, tribe, and other affected units of government.
- Responsibility for the transportation of commercial spent fuel once it has been accepted from utilities for disposition.
- Responsibility for conducting non-generic RD&D activities related to storage, transportation, and geologic disposal, as well as non-generic R&D on the social dimensions of nuclear waste management.[164]
  (Responsibility for generic research in alternative disposal methods and advanced fuel cycle and waste form options should remain with DOE and private industry. In other words, it should continue to be funded by general appropriations and by industry.)

We include a lead role in defining and funding related social science research among the new organization's responsibilities because we believe public acceptance and policy preferences will continue to have an important, if not decisive, influence on nuclear materials management policies in the future.

Current understanding of these attitudes and preferences is inadequate, and in any event social attitudes and preferences will undoubtedly change with time as views on safety, energy security, environmental protection, and other issues also change. Targeted social science research can help improve understanding of the public's concerns and provide the foundation for an informed consideration of social issues in the research agenda and in waste management decisions.

By contrast, research and other activities related to the development of new reactor and fuel cycle technologies should, in our view, be left to DOE and industry. That said, authorizing legislation should give the new organization clear direction concerning the need to work with industry and DOE to ensure that waste management considerations are integral to future reactor and fuel cycle developments and that the waste management system will have the flexibility to support such developments.[165,166]

The Commission notes that none of the past studies of organizational options for waste management have recommended broadening the scope beyond storage, transportation, and disposal; in addition, most countries that have confronted this question have opted to separate institutional responsibility for waste disposal and advanced fuel cycle facilities.

For example, France, which is one of the principal nations actively engaged in nuclear fuel reprocessing and recycling, has separated responsibility for waste management from other fuel cycle functions and has given that responsibility to an independent organization (ANDRA), distinct from the government agency (CEA) that is responsible for reactor and fuel cycle RD&D.

Late in our review we heard comments from several states that host DOE defense waste in support of leaving responsibility for defense waste disposal with DOE. These states generally agreed with the proposal in the Commission's draft report to establish a new organization to manage civilian wastes, but believe the government can more effectively meet its national security obligations and cleanup commitments if responsibility for defense waste disposal remains with DOE.

The Commission also heard from interested parties, such as NEI, who provided credible arguments for why the original commingling decision should be sustained. Whatever one's view of the pros and cons of the current policy, a decision to move responsibility for defense wastes to a new organization (versus leaving that responsibility with DOE) would have major implications for the scope of responsibility for the new organization, as well as for key questions of funding, governance, and Congressional oversight.

Since a 1985 decision by President Reagan that a separate permanent repository for defense high-level waste was not required,[167] DOE has planned to dispose of all high-level waste and spent fuel from national defense activities and from DOE's own research activities in a repository for commercial waste developed under the NWPA.

Any investigation of whether the United States should consider reversing the 1985 decision to commingle defense and civilian waste for disposal will require both a re-examination of the factors that were required by section 8 of the NWPA to be evaluated as part of the presidential decision, and an assessment of facts and factors that have changed since the presidential decision. These facts and factors include:

- The sharp shift in focus at DOE from the production of materials for nuclear weapons to the cleanup and disposal of legacy wastes from the Cold War.
- The establishment of legally-binding site clean-up commitments that require DOE to remove defense wastes from some sites where they are currently stored by 2035.
- The current lack of statutory authority to develop a repository at a site other than Yucca Mountain under the Nuclear Waste Policy Act.
- Successful development and operation of a geologic repository (WIPP), with a mission explicitly limited to the disposal of transuranic waste only from defense nuclear activities.
- Our recommendation to establish a new organization outside of DOE to develop and operate repositories under an amended NWPA.

We are not in a position to comprehensively assess the implications of any actions that might affect DOE's compliance with its cleanup agreements, and we did not have the time or the resources necessary to thoroughly evaluate the many factors that must be considered by the Administration and Congress in making such a determination.

The Commission therefore urges the Administration to launch an immediate review of the implications of leaving responsibility for disposal of defense waste and other DOE-owned waste with DOE versus moving it to a new waste management organization.

This review should include an assessment of issues associated with the disposition of DOE-owned wastes from non-defense sources (e.g. a portion of the high-level waste now stored at West Valley, New York, and a variety of wastes now in storage at INL such as damaged fuel from the Three Mile Island Unit 2 reactor).

The implementation of other BRC recommendations, however, should not wait for the commingling issue to be resolved. Congressional and Administration efforts to implement our recommendations can and should proceed as expeditiously as possible.

7.4. Governance/Oversight Recommendations for a New Organization

This section turns to the issue of accountability in a new organization. As we have already noted, considerations of independence and accountability are fundamentally intertwined and must be carefully balanced. Put another way, a new waste management organization will only be entrusted with substantial operational and financial autonomy if Congress and the American public are confident that safeguards are in place to ensure that the organization behaves responsibly and uses public resources wisely to achieve national policy objectives. For this reason, all analyses and proposals involving new institutional leadership for the nation's waste management program, starting with the AMFM Panel report in the 1980s, have paid considerable attention to issues of governance, oversight, and accountability.

7.4.1. Congressional Oversight

Congress would play a central role in ensuring the accountability of a new waste management organization in several ways. First, Congress would define—through enabling legislation—the mission, structure, responsibilities, and powers of the new organization.[168] Specifically, Congress must define:

- The national nuclear waste policy framework within which the organization must operate
- The institutional form of the new organization
- Financial resources and funding mechanisms for the new organization
- The roles of state, local, and tribal governments in siting waste management and disposal facilities, including the nature of public funding for state, local, tribal and other stakeholder participation
- The organization's responsibility to promote the social and economic well-being of communities affected by waste management facilities,169 as well as the general nature of incentives to be provided and the manner in which states, tribes, and localities are to be funded during the siting process.

In addition, the organization should be required to prepare regular reports to Congress on its activities, expenditures, and progress. Review of these reports, along with periodic oversight hearings and Senate confirmation of the new organization's board of directors would be the chief mechanisms through which Congress would exercise oversight.[170]

In defining these responsibilities, we encourage Congress to adopt an approach that provides the new management entity with as much latitude as possible in accomplishing its mission.

While Congress would provide a policy framework at the outset, some mechanism for facilitating later adjustments or course corrections (after the initial policy direction is specified in law) may be desirable.[171] One option would be to use the mission plan already required in the NWPA as a vehicle for ongoing Congressional oversight. The new waste management organization could submit a mission plan describing its planned activities, schedules and milestones, and supporting budget to DOE and Congress on a regular basis (e.g., every three to five years). The mission plan could also lay out a timeline or milestones specifying when and for what purposes the organization will require access to the corpus of the Nuclear Waste Fund held by the Treasury (the Commission's recommendations related to the nuclear waste fee and Fund are discussed in chapter 8). If desired, legislation establishing the new organization could include an expedited process similar to that provided by the Congressional Review Act (CRA) through which Congress could veto a proposed mission plan revision by passing a joint resolution, subject to presidential veto.[172] This approach would allow substantial congressional control over changes of direction without requiring that legislation be passed to approve such changes whenever they are needed.

7.4.2. Management Oversight

In many of the proposals for a new organization advanced to date (including by the original AMFM Panel, the proposed legislation cited above, and this Commission), a first layer of accountability below Congress is provided by a board of directors. This would provide a degree of ongoing management oversight and control that is not normally present with a typical federal agency program; it would also be particularly appropriate for an organization that is engaged in a business-like, fee-for-service activity such as managing high-level nuclear waste. The board of directors would have the usual powers granted such bodies: it would set broad policies and objectives (within the statutory framework set by Congress); select top managers, establish the management structure, and define personnel policies; approve annual budgets; and report to external stakeholders on the performance of the organization. This approach appears to be the norm in other nations' waste management programs. Having looked at organizational arrangements for radioactive waste management in 12 other countries, the Commission found that in all but two cases the implementing organization is overseen by a board of directors or supervisors.[173]

7.4.3. Independent Regulation

A new waste management organization would be subject to the same federal and applicable state health, safety, and environmental regulations as a private corporation. Currently, regulatory responsibility for various aspects of nuclear materials management is divided among several federal agencies: EPA and the NRC are responsible for radiological health and safety; EPA is responsible for other environmental impacts; the DOT is responsible for transportation safety (other than certifying transportation cask designs); the Occupational Safety and Health Administration (OSHA) and the Mine Safety and Health Administration are responsible for worker safety; and the NRC, DOT, and others (through the implementation of Department of Homeland Security [DHS] standards) are responsible for security. In addition and as discussed in chapter 6, the waste management organization should be authorized to accept additional state and/or tribal oversight and regulatory requirements as a part of any legally binding agreement or partnership arrangement negotiated with states, tribes, or local communities that consent to host a waste management facility.

7.4.4. Scientific and Technical Oversight

Many proposals for an independent waste management organization provide for broad independent technical oversight in addition to, and separate from, any specific health and safety or environmental standards that might apply to the waste management facilities built and operated by the organization. Independent scientific and technical oversight of the nuclear waste management program is essential and should continue to be provided for out of nuclear waste fee payments. The existing NWTRB currently provides this type of wide-ranging technical oversight.

Independent reviews of key aspects of the program on an *ad hoc* basis by independent organizations such as the NAS can also be useful in providing guidance and enhancing public confidence in the technical competence of the organization's work. The waste management organization should therefore be given the authority and responsibility to implement programs and procedures aimed at facilitating such independent reviews, including authority to fund such activities, where appropriate.

Assuring the relevance, quality, and comprehensiveness of the organization's scientific and technical work is important to program excellence. It is also necessary to earn the confidence of the scientific community and larger public. A rigorous, open, and documented peer review process appropriate to the different types of work products developed by the new organization (e.g., peer review mechanisms for research would differ from those for engineering design) can play a major role in providing this assurance, in conjunction with a rigorous quality assurance program. Peer review provides one mechanism by which outside experts can critique analyses, studies, or proposals put forward by the waste management organization. Such evaluations can be used as management tools for verifying or validating the assumptions, results, and conclusions of the organization's work.

In sum, besides ensuring that interested parties and stakeholders have timely access to data and analyses, the new waste management organization should encourage and support the peer-reviewed publication of all work that is important to its activities, including site characterization work as well as analyses aimed at demonstrating the safety and suitability of plans for repository design and operations.

The organization should also encourage and support its staff and the external research teams it funds—not only in publishing results in recognized professional journals, but also in delivering presentations and papers at scientific and technical conferences, as well as participating in national and international meetings. This will allow the organization's work to benefit from full exposure to the broader scientific community and other interested stakeholders. We envision that a robust peer review effort will not substitute for, but rather will augment, the oversight provided by relevant regulatory authorities, the NWTRB, and other outside sources of expertise, while also improving decision-making by bringing other relevant work to the attention of the organization.

7.4.5. Financial Oversight

With greater budget control and assured access to the NWF, the new organization must also be subject to independent financial oversight to ensure that public resources are being used appropriately in support of waste program objectives.

Beyond a board of directors, most proposals provide for additional oversight in the form of independent audits of the new organization's finances along with reviews by the Government Accountability Office (GAO).

The NWPA already requires an annual GAO audit of the activities of DOE's OCRWM, as well as a comprehensive annual report by OCRWM on its activities and expenditures and an annual report to Congress from the Secretary of the Treasury (after consultation with the Secretary of Energy) on the financial condition and operations of the NWF. These requirements could simply be extended to the new organization (except that the organization would not report to Treasury through DOE). A mechanism for Congress to review regular updates of the organization's mission plan and budget would provide an additional vehicle for overseeing the organization's use of funds.

Particular attention must be paid to which entity has authority over the level of the nuclear waste fee. Under current law, the Secretary of Energy is required to make adjustments to the fee, as necessary, to ensure recovery of the full costs of managing and disposing of commercial SNF. The AMFM Panel recommended that a "Waste Fund Oversight Commission" be established for the specific purpose of ensuring that NWF fees are being used cost-effectively and to approve or disapprove proposed changes to the level of the fee. In its 2001 update of the AMFM study, DOE instead recommended that the Federal Energy Regulatory Commission (FERC) serve this purpose. Giving authority to review and approve fee increases to an independent organization with suitable expertise and staff would enhance confidence that the increases are just and reasonable and are not simply the result of ineffective use of the program's resources. Since the FERC already exists and deals with rate issues, the Commission recommends that it be used for this function.[174] As it determines how to carry out this new responsibility, we encourage FERC to consider the development of a "joint board" with state commissioners as provided for under Section 209 of the Federal Power Act.

7.5. Stakeholder Participation

The NWPA states that "state and public participation in the planning and development of repositories is essential in order to promote public confidence in the safety of disposal of such waste and spent fuel." The Commission agrees and recommends that legislation to establish a new waste management organization include appropriate mechanisms to facilitate and support constructive stakeholder participation. Such mechanisms should address two distinct areas of stakeholder participation: interaction with national stakeholder groups and interests and interaction with states, communities, and tribes that would be directly impacted by particular facilities or operations. Because providing for extensive stakeholder participation will require a significant commitment of staff and resources, enabling legislation should specify that this is an authorized use of the NWF and related activities should be covered in annual reports and long-term plans.

The list of stakeholders with an interest in the overall direction and conduct of the national waste management program is a long one. It includes, among others, utility companies; public utility commissions; taxpayers; states, tribes, and local communities that might be affected by waste facilities or activities; public interest groups; the nuclear industry; DOE; the U.S. Navy; the academic community; and the non-proliferation and nuclear security policy community.

Not all of these stakeholders could be represented on the board of directors of a new waste management corporation, nor would this necessarily be appropriate given that the primary role of the board of directors is not to represent all stakeholder views, but rather to carry out fiduciary responsibilities for management oversight. To provide an ongoing conduit for input from the full range of interests noted above, a larger and more widely representative stakeholder advisory committee should be established. This committee would report to the waste management organization's CEO and/or board of directors (similar to DOE's Environmental Management Advisory Board).[175] This would not supplant direct interactions between the waste management organization and various stakeholders or interest groups, but it would ensure that the organization regularly hears the full range of perspectives represented by these different groups. Ongoing dialogue with a stakeholder advisory committee can help the organization develop broadly acceptable policies and plans and identify areas of disagreement that remain to be resolved.

Of the activities the waste management organization will be involved in, siting will likely draw the most intense stakeholder attention and concern. The Commission therefore considered an option in which a different entity or authority—one not charged with developing and operating waste management facilities—would undertake siting as a separate function. Ultimately, the Commission concluded that siting should remain under the auspices of the same waste management organization, for several reasons. First, siting decisions will have a major impact on storage and disposal operations, and siting decisions and criteria must meet operational and design standards. Most crucially, the same waste management organization must be accountable on an ongoing basis for living up to all commitments made during the site selection, characterization, and approval process.

Nevertheless, it will be important to recognize siting as a unique function of the organization for which active engagement with a broad range of stakeholders and other experts will be particularly critical. Throughout the siting process the waste management organization will need to operate with a high degree of independence and objectivity to maintain credibility with the wide range of stakeholders that will be involved. The Commission therefore recommends that a special subcommittee of the stakeholder advisory committee be established to provide specific guidance on the siting process. The special subcommittee would provide a conduit and focal point to ensure that stakeholder input on these issues is given serious consideration and acted on as appropriate. Members of this subcommittee could include representatives from the full stakeholder committee supplemented by other individuals with relevant expertise. Whether separate siting subcommittees should be established for consolidated storage facilities as distinct from disposal facilities is a question that should be decided by the new waste management organization.

Finally, it will be important for members of the general public to have an opportunity to review and comment on the ongoing activities of the waste management program. Requiring that the organization develop and regularly update its mission plan (as discussed above) and make that plan available for public comment would provide a mechanism for soliciting broad-based input.

7.6. Interactions with Affected States, Tribes, and Local Governments

States, tribes, and local communities that are potential or actual hosts of waste management facilities[176] have a special interest in being involved in the process of evaluating potential sites and developing and operating the facilities proposed for these sites. As the siting process narrows to consider specific locations, interactions with potential community, state, and tribal hosts will take on increasing importance. The NWPA makes extensive provisions for coordinated planning and consultation with affected states and Indian tribes. For example, section 116 of the NWPA requires OCRWM, after it has approved a site for characterization or upon request, to seek to enter into and negotiate consultation and cooperation (C&C) agreements with eligible states and affected tribes. The purpose of this type of agreement is to specify the procedures that will be followed in areas of mutual concern, such as public health and safety, environmental and socio-economic impacts, access to technical data and expertise, joint surveillance and monitoring of project activities, public education, resolution of conflicts and off-site concerns, financial assistance, and notification of waste shipments.

These provisions in the NWPA were modeled on the 1981 C&C agreement between DOE and the State of New Mexico for the WIPP facility; they apply to all types of waste management facilities, although section 116 (the section containing these provisions) is focused on repositories. Although C&C agreements would be negotiated with state or tribes, it will also be important for the waste management organization to engage directly with local communities early and often throughout the process. Ultimately the range of issues that could come up in negotiations with potential host communities, states, and tribes is very wide; a few examples from past siting processes include environmental monitoring and testing; authority to issue needed water, waste discharge and construction permits; emergency response agreements; research and education agreements; and economic impact assistance payments. Clearly all levels of government must be involved from the first phases, but how the siting process unfolds and in what order different agreements are struck between different parties is not something that can or should be dictated in advance. This is also why the attributes described previously, including flexibility, responsiveness, and transparency, will be so important to the success, not only of a given siting process but of the waste management program more broadly.

In this context, it is notable that the NWPA's current consultation and cooperation provisions apply only to relations between the federal government and state or tribal governments, and do not extend to local governments.[177] In visits to Sweden and Finland,

Commission members saw first-hand how close involvement with the local host community was critical in winning acceptance for waste management facilities. Local involvement is likely to be critical in the U.S. context also. When a community task force in Oak Ridge, Tennessee evaluated a DOE proposal to site a MRS Facility in the area, the task force made its support for the facility conditional on the adoption of specific measures to enhance local authority. These included provisions for C&C agreements directly between DOE and units of local government (as well as between DOE and the state) and granting preferred status to local governments in interactions between the state, DOE, and NRC regarding the MRS facility.[178] We therefore recommend that the waste management organization's authority and responsibility to negotiate binding agreements with host states and tribes be extended to also include local host governments.

7.7. Transfer of Contracts and Liability to a New Organization

Transferring responsibility for nuclear waste management to a new organization raises the difficult question of how to handle existing liabilities under DOE's current contracts with utilities. (Earlier chapters have discussed the litigation currently underway with respect to the breach of these contracts.) Congress will need to give careful consideration to the treatment of existing contractual liabilities in legislation to establish a new waste management organization.[179] A core question will be how to pay for damages accrued until federal facilities are available. A federal court has ruled that the NWF cannot be used to pay damages because at-reactor storage is not an allowed use of the Fund under the NWPA and DOE contracts with utilities, even if the federal government were to take title to the spent fuel at reactor sites. As a result, damages are now being paid out of the Judgment Fund, which receives a permanent indefinite appropriation from the Treasury. It will therefore be important to clarify responsibility for contracts and associated liabilities going forward.

7.8. Near-Term Steps

The Commission strongly believes that new institutional leadership is critical to getting the nation's nuclear waste management program on track. But we recognize that it could take several years for a new organization to be authorized, funded, staffed and fully launched. In the meantime, it will be important for DOE to keep the program moving forward through non-site specific activities, including R&D on geological media and work to design improved engineered barriers.

For instance, DOE's Office of Used Nuclear Fuel Disposition Research & Development is implementing the Used Fuel Disposition Campaign. The objectives of the Campaign are to identify alternatives and conduct R&D on transportation, storage, and disposal options for SNF from existing and potential future nuclear fuel cycles. This program and other non-site-specific generic activities, such as support for and coordination with states and regional state government groups on transportation planning, should be continued.[180]

To ensure continued progress, we urge the Secretary of Energy to task a senior official with sufficient authority to coordinate all of the DOE elements involved in the implementation of the Commission's recommendations.

8. FUNDING THE WASTE MANAGEMENT PROGRAM

To succeed, a new waste management organization must have the resources needed to implement an effective program.

Since 1983, nuclear utilities and their ratepayers have been paying a nuclear waste fee into a Nuclear Waste Fund (NWF) in the Treasury. The Fund is dedicated to covering the cost of disposing of commercial radioactive waste, but for reasons discussed below the money in the Fund is effectively unavailable for its intended purpose. The Commission believes that *the success of a revitalized nuclear waste management program will depend on making the revenues generated by the nuclear waste fee and the balance in the NWF available when needed and in the amounts needed to implement the program.*

The Commission spent considerable time on this issue. The remainder of this chapter details our specific recommendations for implementing the funding reforms that are required to support a revitalized U.S. waste program.

8.1. Background

The 1982 Nuclear Waste Policy Act created a "polluter pays" funding mechanism[181] to ensure that the full costs of disposing of commercial HLW would be paid by utilities (and their ratepayers), with no impact on taxpayers or the federal budget. Nuclear utilities are assessed a full-cost-recovery user fee on every kilowatt-hour of nuclear-generated electricity as a *quid pro quo* payment in exchange for the government's contractual commitment to begin accepting commercial spent fuel for disposal beginning by January 31, 1998. The fee is collected from utilities that own or operate nuclear power plants; generally it is passed on to utility ratepayers.

The fee was initially set at 1 mill (0.1 cents) per kilowatt-hour (where it still is); however, the Act requires the Secretary of Energy to review the adequacy of the fee annually and adjust it as needed to ensure that going forward the government can recover the full costs of waste management and disposal. In recent years, the fee has generated approximately $750 million in annual revenues; the total amount collected through 2010 amounted to just over $16 billion.

Fee revenues go to the government's Nuclear Waste Fund, which was established for the sole purpose of covering the cost of disposing of civilian nuclear waste. (Costs for disposing of defense nuclear wastes are paid by taxpayers through direct appropriations from the Treasury that do not pass through the Nuclear Waste Fund.) The unspent balance in the Fund is allowed to accumulate and accrue interest with the idea that it will be available as needed to fund program expenditures in future years.

The current unspent balance in the Fund (known as the "corpus") totals nearly $27 billion, including interest. Federal appropriators are supposed to be able to access the Fund when and in the amounts needed to implement the waste program without facing competition from other funding priorities.

The clear intent of Congress in establishing a self-financing mechanism based on contractually-obligated user fees was to "provide an assured source of funds to carry out the programs and...eliminate...annual budgetary perturbations in an ever more constrained

Federal budget," while at the same time ensuring that "the Federal budget will not be burdened by repository program expenditures" (see text box).

Views on the Intent of the Nuclear Waste Fund

Senator James McClure, chairman of the Senate Committee on Energy and Natural Resources and floor manager of the Senate nuclear waste policy legislation:

> "By establishing a 1 mill-per-kilowatt-hour user fee on nuclear generated electricity, this bill for the first time would provide a direct financial linkage between the beneficiaries of nuclear power and the cost for interim management and ultimate disposal for nuclear wastes...This funding mechanism would provide an assured source of funds to carry out the programs and would eliminate not only annual budgetary perturbations in an ever more constrained Federal budget, but the too often repeated shifts of policy direction under succeeding administrations. The nuclear waste policy, programs and required financing would be statutorily fixed and quite predictable under this approach." Congressional Record-Senate, December 20, 1982, pp. S15655 - S15656

Congressman Morris Udall, Chairman of the House Committee on Interior and Insular Affairs and key sponsor and manager of nuclear waste legislation in the House:

> "The cost of the waste disposal program will be borne by the generators of the waste. The program will be financed up-front by nuclear utilities, so that the Federal budget will not be burdened by repository program expenditures. Utility payments will be made into a Nuclear Waste Trust Fund set aside exclusively for repository development purposes... The Nuclear Waste Trust Fund will be isolated from other Federal programs, and will not be used to finance any activities other than repository development." Congressional Record-House, September 30, 1982, p. H8163

American Nuclear Energy Council, Edison Electric Institute, and Utility Nuclear Waste Management Group:

> "The central concept of the financing plan which we support is premised on complete cost recovery of all reasonable facility costs. ...While the electric utilities do not endorse the precedent of collecting a tax, we recognize that nuclear waste management is a unique Federal responsibility resulting from joint effort of the government and industry to utilize nuclear energy for the public benefit.
>
> Such a financing arrangement is not viewed as a precedent, but rather an innovative mechanism for ensuring the financial viability of a successful long-term Federal waste management program.
>
> ...Again, we must emphasize that the full payment for reasonable costs of storage and disposal of commercial spent fuel and radioactive wastes will be paid by the utilities and will be included as part of the cost of the nuclear fuel." Joint statement submitted to the House Committee on Science and Technology on October 5, 1981.

Congressional oversight through the annual appropriations process would ensure that expenditures from the Fund would be made prudently and for their intended purposes. But the Fund was clearly designed to ensure that the waste program's needs and schedules determined its funding, rather than allowing federal budget constraints to limit the program's progress.

Indeed, the Nuclear Waste Policy Act's provisions for an expanded and accelerated repository program and its direction to DOE to assume contractual obligations for accepting waste on a defined schedule demanded an assured funding source to support the activities needed to meet these legal obligations.[182]

8.2. Current Treatment of the Nuclear Waste Fund in the Federal Budget

8.2.1. A Case of Unintended Consequences and Constraints

The Fund has not worked as intended to insulate the nation's civilian nuclear waste management program from the vagaries of the federal budget process while at the same time insulating the federal budget from the costs of the waste program.

A series of actions by successive administrations and Congresses (see text box below) has made the approximately $750 million in annual fee revenues and the unspent $27 billion balance in the Fund effectively inaccessible to federal budgeters and appropriators, forcing them to take money away from other federal priorities to fund activities needed to meet contractual waste management obligations.

As a result, waste management needs have had to compete with other priorities in DOE's annual budget request and in the Congressional appropriations process (figure 17), subjecting the program to exactly the sort of "budgetary perturbations" that the funding mechanism was intended to avoid.

Senator Bennett Johnston, then Chairman of the Senate Committee on Energy and Natural Resources, pointed out the problem in 1994:[183]

> "We thought we had provided a guaranteed funding source for the waste program when we created the Nuclear Waste Fund in 1982.
>
> The Waste Fund consists of money paid by electric ratepayers for the sole purpose of funding this program
>
> ...Unfortunately, the Waste Fund has become entangled in budget rules adopted in recent years to combat the deficit. The unintended consequence of these rules had been to put most of the Nuclear Waste Fund out of reach of the very program for which the money is being collected."

In other words, a program that was intended to be fully self-financing now has to compete for limited discretionary funding in the annual appropriations process, while the contractual user fees intended to prevent this from happening are treated just like tax revenues and used to reduce the apparent deficit on the mandatory side of the federal budget (which deals with expenditures and receipts that are not subject to annual appropriations).

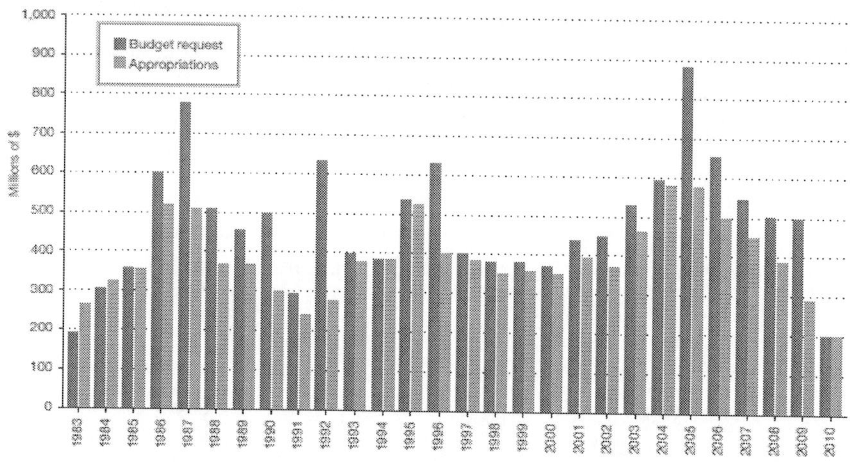

Figure 17. Nuclear waste program: budget requests versus appropriations. [184]

The Layering of Budget Constraints on the Nuclear Waste Fund

Since the establishment of the NWF in 1982, Congress enacted several budget control acts that dramatically reduced the funding flexibility originally envisioned in the NWPA:

- The Balanced Budget and Emergency Deficit Control Act of 1985, also known as Gramm-Rudman-Hollings (GRH), made the NWF subject to a government-wide deficit-reduction process. In implementing GRH, the Office of Management and Budget (OMB) "split" the NWF; fee receipts were placed on the "mandatory" side of the budget (dealing with activities controlled by permanent laws rather than annual appropriations), where they are treated like tax revenues and used to offset mandatory spending; while expenditures were placed on the "discretionary" side (dealing with activities controlled by annual appropriation acts), where they are subject to the deficit reduction process.
- The 1987 amendments to GRH placed appropriations from the NWF under the spending cap applicable to all domestic discretionary programs, even though the NWF was self-financed. This had the effect of forcing spending for the NWF to compete with other spending programs, which did not have dedicated funding sources. As a result, OMB also dropped its historical practice of setting separate budget planning targets for the NWF, forcing it to compete against other DOE programs within a single DOE budget target for domestic discretionary spending.
- The Budget Enforcement Act of 1990 (BEA) set new caps on discretionary spending and established new pay-as-you-go (PAYGO) requirements to ensure that the net effects of legislative changes affecting mandatory spending were budget neutral.
- In the Conference Report accompanying the Omnibus Budget Reconciliation Act of 1990, spending from the NWF was included in domestic discretionary appropriation accounts for fiscal year (FY) 1991, and was therefore subject to the spending cap set in the BEA.

- The 1997 Amendments to the Balanced Budget Act extended caps on discretionary spending and PAYGO requirements for mandatory spending accounts through FY 2002.

This layering of budget requirements seriously eroded the NWF's funding capability in two ways:

- It imposed annual spending and revenue controls on a fund that was designed to finance a 125-year program on a life-cycle cost basis; and
- It made the NWF dysfunctional by creating separate and unrelated rules applicable to the revenue and spending components of the Fund.

The overall effect, in short, has been to prevent the NWF from being used for its intended purpose. Under PAYGO requirements, increased funding for the waste management program must be offset by cuts in other programs within the annual discretionary appropriations caps. The original NWPA requirement for annual appropriations from the NWF was intended to ensure that Congress retained control over the program; its purpose was never to limit the funding needed to implement the program.

Source: Alternative Means of Financing and Managing the Civilian Radioactive Waste Management Program, U.S. Department of Energy, August 2001, DOE/RW-0546, pp. 12-13.

These problems have also materially contributed to the failure of the federal government to meet its contractual obligations and to the government's large and growing exposure to financial liabilities for resulting damages—damages that will have to be paid by taxpayers.

We discuss this issue in detail in section 8.5, but for now it is worth pointing out that the damage payments being awarded to compensate utilities for the costs of continued at-reactor storage of spent fuel that was supposed to have been accepted by the federal government do nothing to advance the objective of providing for the permanent disposition of the fuel. Meanwhile, the unspent balance of fee revenues and interest accumulating in the Waste Fund represents a large and growing liability for taxpayers that must be paid at some point in the future. Because DOE's contracts with utilities create a legal obligation, those funds can and must eventually be used only for the purpose for which they have been collected; the Treasury bills in the Fund that were issued as IOUs for the fee receipts and interest must at some point be redeemed either by future tax revenues or by borrowing from other sources that in turn must be repaid.

8.2.2. Disadvantages of the Appropriations Process

Even if competition with other programs for limited discretionary funding were not an issue, the current statutory requirement that makes use of the NWF subject to appropriations has led to unforeseen difficulties caused by the appropriations process itself. Although the current system assures Congress explicit and extensive year-to-year oversight and control as intended by the NWPA, it has clearly proven to be a poor mechanism for financing a very long-term and complex effort. First, the annual appropriations process creates substantial funding uncertainty, which can make it difficult for the implementing agency to make and honor longer-term commitments, retain staff expertise, and exercise independent judgment about programmatic priorities and resource allocation. Second, Congress has increasingly failed to pass appropriations bills in a timely manner in recent years, forcing federal agencies to operate on continuing resolutions for extended periods of time while coping with the delayed availability of requested funds.

A 2005 report on the management and funding of nuclear waste management programs in the 11 member nations of the International Association for Environmentally Safe Disposal of Radioactive Materials (EDRAM)[185] noted that all these nations have applied the principle that waste producers should pay for the management of their wastes. Where EDRAM members differ is in how they estimate, collect, and manage waste management fees. The United States stands out as the only nation where the national legislature directly controls, on an annual basis, the expenditure of funds collected for nuclear waste management purposes.[186]

8.3. Fixing the Funding Problem

The federal government's failure to deliver on its statutory waste management obligations to date and the fact that the Waste Fund and fee are not working as intended have prompted the National Association of Regulatory Utility Commissioners, along with some nuclear utilities and the NEI, to pursue legal action against DOE aimed at suspending the collection of nuclear waste fees until such time as a new waste management plan for the country is in place. The outcome of this and other pending legal actions remains uncertain at present, but they underscore the growing frustration among state regulators, nuclear utilities, and consumer advocates about the continued lack of progress toward a durable waste management solution. In fact, there is a growing sense of outrage that the only aspect of the waste management program that has been implemented in full and on schedule is the part that involves collecting fees for a contractually required service that the federal government has never managed to deliver.

The Commission concludes that for the waste management program to succeed, the nuclear waste funding mechanism must be allowed to work as intended so that the ability to implement the waste program is not subject to unrelated federal budget constraints. If that is not done, key recommendations of the Commission will be undermined— e.g., efforts to develop both storage and disposal facilities will be in conflict rather than mutually supportive and commitments to provide benefits to host communities over the life of the program will lack credibility. Fixing this problem requires extricating the nuclear waste fee and NWF from the web of budget rules that have made these user-provided resources effectively unavailable to federal budgeters and appropriators, forcing them to take limited discretionary funds away from other federal programs in order to pay for the activities needed to meet federal waste management contractual obligations and thereby put an end to growing taxpayer financial liability for failure to meet those obligations.

The Commission also concludes that a new waste management organization bound by a well-defined mission should be entrusted—subject to an appropriate level of oversight by Congress and relevant regulatory authorities—with greater autonomy and control of its budget over multiple year periods than is possible under the annual appropriations process, just as the TVA has control of the use of its receipts from electricity sales (subject to congressional oversight). This kind of authority is crucial, among other reasons, to allow the new organization to negotiate meaningful, enforceable, and ultimately *credible* commitments with other parties—including with the communities, states, and tribes that will be most directly affected by its activities. Fixing the current funding problem requires removing waste program funding decisions, to the extent they concern activities related to the civilian wastes for which the nuclear waste fee is being paid, from dependence on the annual federal budgeting and appropriations process, while ensuring appropriate oversight by Congress and other third-party agencies.

The Commission recommends that this transition be accomplished in two stages:

1. Near-term non-legislative actions that would allow full access to future waste fee revenues subject to appropriations control but independent of competition with other funding needs.
2. Legislative action as part of the establishment of an independent waste management organization that would allow it to function as an autonomous self-financed entity like TVA or the Bonneville Power Administration, with full control of the use of its revenues subject to congressional and other independent oversight and with access to future fee receipts and, eventually, the current corpus of the Nuclear Waste Fund.

8.3.1. Near-Term Non-Legislative Action to Increase Access to Fee Revenues

The Commission recognizes that legislative action to create a new waste management organization with full access to the nearly $27 billion balance in the NWF will be difficult in the current political and budgetary climate, despite the fundamental equity arguments for this action. Therefore, we urge the Administration to take prompt action aimed at enabling appropriators to use the annual nuclear waste fee revenues for their intended purpose, free from competition with other spending priorities, while stopping further additions of surplus revenues to the NWF until such access has been guaranteed. We believe this can be accomplished by adopting a combination of measures that are already allowed under existing legislation.[187]

Specifically, the Administration should (1) change the way in which the nuclear waste fee is collected so that only an amount equal to actual appropriations from the NWF is collected each year, with the remainder retained by utilities in approved trust funds to be available when needed for future use, and (2) work with the congressional budget committees and the Congressional Budget Office to reclassify the fee receipts from mandatory to discretionary so that they can directly offset appropriations for the waste program.[188] Taken together, these steps would make the nuclear waste program funding mechanism work essentially as Congress intended in the NWPA, at least for future fee revenues. Each is discussed further below.

Change the Timing of Nuclear Waste Fee Collections

Under the current approach, the entire 1 mill/kwh fee is collected from contract holders each year (the total collected amounts to approximately $750 million per year) and deposited in the Treasury, independent of the sum actually appropriated from the Fund for use by the waste management program. This annual revenue stream is counted in the federal budget baseline as an offset to mandatory spending, which raises the criticism that the fee is simply being used to reduce the budget deficit instead of for its intended purposes. This criticism becomes more acute as the gap between annual fee payments and appropriations from the Fund widens.

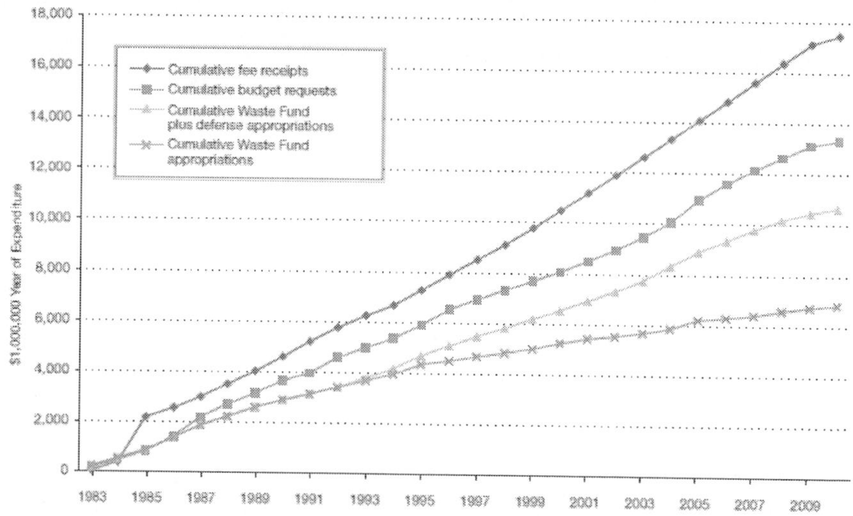

Figure 18. Cumulative nuclear waste fees, budget requests, and appropriations.[189]

Figure 18 shows the large and growing gap between cumulative nuclear waste fee receipts (not including interest on the NWF balance) and appropriations from the NWF. The longer annual fee payments continue to accumulate in the Fund, the greater the budgetary and political difficulty of restoring the Fund to its intended purpose will be.

To stop the flow of waste fees to an inaccessible account in the Treasury, to put an end to the perception that the fee is simply being used to reduce the federal budget deficit, and to take the first crucial step towards making future fee revenues accessible to appropriators, the Administration should adopt a modified version of an approach proposed by the Secretary of Energy in 1998 as part of a litigation settlement concept.[190]

The key element of that proposal was to change the timing of fee payments into the NWF through administrative action so as to match the annual flow of cash into the Fund with actual spending from the Fund in support of nuclear waste management activities. Specifically, DOE proposed to offer to amend its contracts with utilities to allow utilities to retain the portion of the 1 mill/kwh fee that exceeded the annual appropriations level. As soon as the federal government began to accept waste, utilities would pay the deferred fees plus interest at the Treasury rate.[191] The modified approach proposed here would require each utility to place the unused fee receipts in an irrevocable trust account at an approved, third-party financial institution, allowing the money to be withdrawn only for the purpose for which the trust account was created, at the time and in the amounts needed to fund the federal waste management program. This would make the "waste disposal trust accounts" similar to the decommissioning "sinking funds" most utilities use to meet NRC requirements that they provide assured funding for reactor decommissioning. Funds in those accounts can only be used for decommissioning. By analogy, if a similar irrevocable trust account were created for NWPA purposes, the licensee could only pay out the money to the waste management organization as required to meet program needs. This approach would make the utility waste trust accounts collectively serve the function that the Nuclear Waste Fund was supposed to, providing a source of funds in reserve that can be used in years in which the waste program's funding needs exceed the total annual fee receipts.

A key feature of this proposal is that it would be accomplished using the Secretary of Energy's existing authority under the NWPA to establish procedures for the collection and payment of the fees.[192] Under current budget rules, any legislative action that has the effect of reducing NWF receipts to the U.S. Treasury will be subject to "pay as you go/cut-as-you-go" or "PAYGO/CUTGO" requirements.[193] This means that new revenues or budget cuts will be needed to cover the change in funds flowing to the Treasury resulting from new legislation. However, any changes to fee revenues resulting from non-legislative action under existing law would have no PAYGO/CUTGO impact.[194] At the same time, by ending the practice of counting revenues from the entire 1 mill/kwh fee in the federal government's budget baseline, this step would substantially ease the PAYGO/CUTGO burden associated with subsequent legislative action to transfer fee receipts to an independent organization.[195] Furthermore, tying annual fee collections to actual appropriations for the waste program would strengthen the rationale for reclassifying fee receipts as a discretionary offsetting collection, which is the second step required to implement our recommendations for interim funding.[196]

Reclassify Waste Fee Revenues from Mandatory to Discretionary

The above-described step of splitting fee collections does not, by itself, address the problem that appropriations from the Fund are subject to caps on discretionary spending, because the fee receipts have been placed on the other side of the mandatory/discretionary spending firewall where they are not directly available to appropriators. A second step is needed to move the receipts to the discretionary side so they can be used by appropriators to fund the waste program without reducing funds available for other discretionary programs. *To implement this approach, the Administration should work with the appropriate congressional authorities to re-classify waste fee receipts from mandatory to discretionary offsetting collections so that they can directly offset appropriations for the waste program.* Combined with the previous step that would tie annual fee receipts to actual appropriations levels, this would enable a funding process similar to that used to fund the NRC (i.e., where funding is provided primarily by user fees that are set at the level of annual budgetary authority established in appropriations bills).

DOE's 2001 analysis of alternative means of financing and managing the waste program, which was prepared at the request of Congress, specifically considered this option and concluded it would be feasible. Current practice would require OMB to seek the concurrence of the Congressional Budget Office and the congressional budget committees for this reclassification. In addition, appropriations language would be required to credit the fee to waste management appropriations; indeed, we urge the Administration to include such language in its FY 2013 budget proposal.[197]

Implications of the Proposed Approach

The two-step approach we propose would accomplish several things:

- It would reduce PAYGO/CUTGO challenges for future legislative action to give a new organization access to the nuclear waste fee and Fund by lowering the baseline projection of fee receipts for federal budget purposes and by slowing the continued build-up of the corpus of the Fund.

- By eliminating surplus collections, it would address the concern of utilities and public utility commissions about the misuse of the fee and Fund to reduce the annual deficit instead of for the purposes of the NWPA. Instead, the surplus fee revenue would go into approved third-party trust accounts that would be available when needed to meet the operational costs of disposal, when program expenditures can be expected to exceed fee receipts.
- It would facilitate adequate appropriations for the program in the near term by giving appropriations from the Fund (up to the amount of revenue generated annually by the 1 mill/kwh fee plus any additional amount obtained from balances in the utility trust accounts) a net budgetary impact of zero, since the appropriation would be directly offset by the collection of an equal amount in fee revenues. As noted above, a similar approach is already being used to fund the NRC.
- Finally, it would demonstrate the federal government's determination to make the funding mechanism established in the NWPA work as originally intended.

There are also several things this two-step action would *not* do:

- It would not reduce Congress's oversight role in the budget process for the waste program. Under current practice, OMB would seek the concurrence of the Congressional Budget Office and congressional budget committees for reclassifying fee receipts, appropriations language would be needed to credit fee receipts against appropriations, and congressional appropriations committees would continue to control the annual level of program funding through the appropriations process. Legislation will be required to remove this funding from the annual budget process while retaining an appropriate degree of external oversight of program spending, as recommended earlier.
- It would not increase access to the corpus of the NWF. This must be accomplished in subsequent legislation since DOE's existing contracts with utilities create a legal obligation for the federal government to ultimately expend these funds for the waste management purposes for which they were collected.
- It would not adversely impact the discretionary funding of any single program or agency since the changes would occur on the mandatory side of the budget, although it would—by removing projected fee revenues from the budget baseline—lead to a very small percentage increase in the federal government's nominal annual budget deficit.

We understand that nearly 30 years of interpretation and application of general budget concepts and practices have led to the current treatment of the waste fee receipts and program expenditures. But the application of general concepts and practices to unique situations can sometimes have unintended and perverse results—as they have in this case. We cannot believe that anyone intended the current situation: the government is in default on a contractual obligation to dispose of spent fuel from nuclear utilities; the user fees being paid to the government to finance the activities needed to meet that obligation are used to offset the deficit, while expenditures for those activities are constrained under limits on discretionary appropriations; and all the while, taxpayer liabilities resulting from failure to meet the government's contractual obligations continue to grow. The Financial Report of the

United States Government for FY 2011 reports that these liabilities totaled $49.1 billion—including both the unpaid damages for non-performance and unspent Nuclear Waste Fund fees and interest.[198]

We believe that this situation must be changed, in order to put an end to the continuing damage to taxpayers, nuclear utilities and their ratepayers, and the credibility of the federal government's commitment to meet its statutory and contractual obligations. To do so, we believe that budget policy leaders in the Administration and Congress can and should act in the same bipartisan spirit of cooperation that characterized passage of the NWPA to make whatever reinterpretations of, or even exceptions to, the decades of budget interpretations and practices that will be needed to make the waste management funding mechanism work as originally intended.

We recognize that there may be concerns that the actions to give full access to the nuclear waste fee and Fund the Commission recommends might set precedents that would have broader implications for other federal programs. However, we believe that the current circumstances— involving a highly unusual contractual arrangement mandated by the NWPA and the existence of growing taxpayer liabilities for failure to comply with the terms of that arrangement—are so narrowly drawn that any precedents that are set would have at most very limited implications elsewhere.[199]

8.3.2. Legislative Action to Provide Budgetary Autonomy (Subject to Oversight)

The above-described steps would enable appropriators to fund a restart of the waste program from future fee receipts without taking funds from other programs. However, growing delays and uncertainties in the overall federal appropriations process will continue to make long-term planning and commitments difficult; and eventually access to the unspent balance in the Nuclear Waste Fund will be needed. Legislation to establish a new waste management organization should give the organization the same authority to use its revenues to carry out its civilian nuclear waste obligations independent of annual appropriations (but with congressional oversight) as is now given to the Tennessee Valley Authority and Bonneville Power Administration.

As noted earlier, legislation that has the effect of reducing nuclear waste fee receipts to the U.S. Treasury or increasing projected spending from the NWF will be subject to PAYGO/ CUTGO requirements, depending on when the changes will occur. The Commission recognizes that there have been numerous unsuccessful legislative proposals to increase access to the fee revenues and the NWF while addressing such requirements.[200] Nonetheless, access to the corpus of the NWF will ultimately be needed to meet the fluctuating revenue demands of the waste management program going forward. This will include covering years when costs peak— for example during the construction of waste management facilities. That the balance in the NWF (including accrued interest) would be fully accessible when and as needed was a fundamental premise underlying the commitments made in the NWPA—that premise must be restored. Anticipating that the near-term non-legislative actions proposed above may be able to provide adequate funding for a restarted waste program for the next decade or perhaps longer, the Commission recommends that legislation establishing a new waste management organization include a defined schedule of payments to transfer the balance of the Fund to the organization over a reasonable future time period, starting 10 years after the organization is established.[201]

As we have already noted, our recommendations for separating the NWF from the congressional budget process are in no way intended to imply a diminished need for rigorous program oversight. On the contrary, we believe these budget and funding reforms—to be acceptable to Congress and the public—must be coupled with strong provisions to ensure that the waste program is being implemented effectively and is making appropriate use of the NWF and fees with which it has been entrusted.

Finally, as discussed above, the Commission is aware that efforts to fix the use of the NWF could be caught up in broader questions such as the treatment of trust funds in the federal budget more generally. However, DOE has testified to Congress that proposals to correct the treatment of the waste fee and Fund are unlikely to create wider precedents beyond similar contractual fee-for-service situations (if any exist).[202]

8.4. Paying for the Defense Waste Share

The preceding discussion has addressed only the portion of waste program costs that are attributable to the management of commercial waste and that are paid for through the nuclear waste fee and NWF. Since current policy presumes that national defense wastes will be disposed of in a repository developed pursuant to the NWPA, a portion of the costs of the program are paid directly by appropriations from the national defense side of the federal budget.[203]

Using a methodology for allocating costs between government-managed nuclear materials and commercial wastes that was first published in 1987,[204] DOE's 2007 Fee Adequacy Assessment estimated the defense share of total program costs at 19.6 percent for 2007.[205] (The defense share adjusts each year as assumptions change.)

Steady progress on implementing a disposal solution will require that appropriations for the defense share are made as needed to pay the full cost of defense waste disposal (note that, in the absence of a disposal facility, the GAO has estimated that continued storage of defense wastes at DOE sites will cost well over a billion dollars through 2040[206]). Historically, appropriations from the defense side of the waste management budget have not been nearly as constrained as those from the civilian side. Since the inception of the program through the end of FY 2010, defense appropriations (in nominal dollars) amounted to $3,756 million compared to $6,837 million from the NWF, or just over 35 percent of total waste program appropriations. By comparison, the defense share of total program cost over the life of the repository was estimated at 19.6 percent in 2007. In the last ten fiscal years, defense appropriations have represented over 61 percent of total appropriations for the waste program.[207]

Given this history, it would not appear that measures are needed to ensure adequate appropriations for the defense share of repository costs in the future.[208] However, once it becomes necessary to fund the construction of a repository (whether that repository is for commingled civilian and defense wastes or for defense wastes only), consideration might be given to mechanisms like multiyear appropriations that are sometimes used with large defense procurements (e.g., for the construction of an aircraft carrier) to ensure that expensive and complex projects can be completed in a timely and cost-effective manner.

8.5. Dealing with Ongoing Litigation

For reasons discussed in other chapters, DOE was unable to begin accepting commercial spent fuel by January 1998, as required under the Standard Contract. DOE and utilities have been engaged in protracted litigation since then over the Department's failure to perform its obligations,[209] as shown in table 2. Some 78 lawsuits have been filed, dozens of lawsuits have yet to be tried, some utilities have reached settlements with the government, and courts have reached judgments in other cases that find DOE in "partial breach" of its contracts. This means taxpayers must pay damages incurred by utilities as a result of DOE's failure to accept fuel,[210] even as DOE remains obligated to do so in the future.

DOE currently estimates that total damage awards to utilities could amount to $20.8 billion if the federal government begins accepting spent fuel in 2020. DOE has previously estimated that liabilities could increase by hundreds of millions of dollars annually if the schedule for starting acceptance slips beyond 2020. DOE and the Department of Justice (DOJ) note a significant development in 2008 that substantially affects future damage estimates. The Court of Appeals for the Federal Circuit ruled in one case that DOE was obligated to accept spent fuel at higher rates than were used in the settlements on which previous damage estimates were based, and directed the trial court to apply these higher rates in determining damages.[211] The new rates have been assumed for future settlements and decisions in the most recent liabilities estimate, and are largely responsible for the increase over the 2010 estimate of $16.4 billion.

To date, damages and judgments in the amount of $2 billion have been paid from the taxpayer-funded Judgment Fund, which is overseen by the DOJ. The Judgment Fund is being used because a federal court ruled in *Alabama Power Co. v. United States Department of Energy*, 307 F.3d 1300 (11th Cir. 2002), that the government could not use the Nuclear Waste Fund to pay for damages incurred as a result of DOE's delay. In addition, DOJ has incurred $188 million in costs for the cases it has litigated through 2011 and more cases are expected in the future. Because DOE is only in "partial breach" of the contracts, utilities can only file for actual damages incurred as of the date of filing. As a result, utilities must re-file periodically—at least every six years because of the statute of limitations—to recover additional damages after the previous claim was filed. For this reason, a steady stream of lawsuits can be anticipated until either (a) DOE has accepted enough waste to "catch up" with the amount it should have accepted on the schedule determined by the courts or (b) DOE has negotiated settlements with all contract holders that would allow damages to be paid without further litigation.

The litigation that has already occurred over the federal government's failure to meet its existing waste acceptance obligations has been expensive, time-consuming, not conducive to resolving the current impasse in the nation's nuclear waste management program, and detrimental to the full and open communication among parties needed for integrated planning concerning spent fuel management.

Because most of the major recurring issues have been resolved in litigation and the outcomes are now more predictable, moving toward a simplified claims process for the purpose of settling existing lawsuits has been suggested,[213] and since February 2011, the Department of Justice has executed 13 additional settlements resolving claims covering 25 reactors and has authorization to enter another settlement covering four reactors.[214] Settling current and pending lawsuits as quickly as possible would reduce unnecessary litigation costs, make it possible to assess the cost impacts of changing current spent-fuel acceptance priorities more reliably, and facilitate more open communication and coordination between the waste management organization and contract holders.

The Commission therefore urges all parties to continue to work to conclude these proceedings in a fair manner, either through settlement agreements or through another process, such as mediation or arbitration, consistent with the precedents set by past court decisions.

Table 2. Status of Doe-Utility Standard Contract Litigation (as of December 2011)[212]

Standard contracts	76
Reactors covered by contracts	118
Cases filed through 2010	78
second-round	(12)
Claims	$6.4 billion
Voluntarily withdrawn	7
settled	23
separate settlement agreements	21
Reactors covered by settlements	65
Final judgments	24
not appealable	(13)
on appeal	(11)
Pending before the trial court	24
DoJ trials through 2010	30
litigation costs through 2010 (experts and support; no DoJ or Doe staff)	$188 million
DoJ trials expected 2011 through 2012	up to 6
Amount of judgments on appeal	$509 million
Payments for final judgments and settlements to date	$2 billion
estimated total damages (if acceptance starts in 2020)	$20.8 billion
estimated increase for each year slippage	up to $500 million

9. TRANSPORTATION ISSUES

The ability to safely and securely transport nuclear materials is critical to implementing a comprehensive and workable waste management system.

Overall, the set of standards and regulations that currently exists to govern the transport of spent fuel, high-level waste, and other nuclear materials has functioned well and the safety record for past shipments of these types of materials has been excellent.

Indeed, states have been working cooperatively with DOE for many years to plan for shipments, often through agreements with regional groups of states and in ways that involve radiological health, law enforcement, and emergency response personnel.

However, past performance and mere compliance with current regulations does not guarantee that future transport operations will match the record to date or that these operations will be conducted in a way that inspires public confidence, particularly as the logistics involved expand to accommodate a much larger number of shipments. This chapter discusses the Commission's specific findings and recommendations with respect to the transport of spent fuel and nuclear waste in the context of the broader waste management strategy we are proposing.

9.1. Background and Context

In 2006, the National Academies issued a report titled *Going the Distance: the Safe Transport of Spent Nuclear Fuel and High-Level Radioactive Waste in the United States.*[215] The report concludes that there are "no fundamental technical barriers" to the safe transport of such materials, but it made a number of recommendations to improve safety, communicate risk, and conduct planning and other activities in preparation for a large-scale transport campaign for spent fuel.

Many of these recommendations have since been adopted, at least in part, by federal agencies such as the NRC, DOE, and the U.S. Department of Transportation (DOT).[216] Table 3 below summarizes the key NAS study recommendations and their current status.

The Commission believes that other NAS recommendations that have not yet been implemented, for whatever reason, should be revisited and addressed as appropriate.[217]

Table 3. Recommendations of the NAS Going the Distance Report and Their Current Status

Recommendation	Current Status (as of January 2012)
undertake full examination of spent fuel transport security by independent, cleared technical experts.	members of the BRC and staff with appropriate clearances have been briefed by nRC and Doe on transportation and storage security analyses undertaken since 2006. However, the BRC does not believe this constitutes the "full examination" recommended by the nAs study and we are not aware of any other efforts that would satisfy the nAs recommendation.
Be proactive in formally assessing and managing "social risks." Expand transportation external Coordination (teC) working Group to include this issue, establish external risk advisory group, potentially under nwtRB auspices.	Research on social risks and risk perception has been ongoing, but the Commission is unaware of any specific changes that have been implemented by Doe or other agencies as a result. The teC working Group (now defunct) did not expand its scope to address this issue. the nwtRB has periodically examined public perceptions of transportation risks but has not established an external risk advisory group.
the nRC should analyze very long duration fires, and implement regulatory controls to reduce the chances of a spent fuel shipment being involved in such a scenario.	the nRC has made a practice of studying real-world fires and analyzing how casks would perform under such conditions. The nRC has also worked with the Association of American Railroads to establish a "no pass" rule for tunnels that would be used to transport spent fuel, effectively precluding the possibility that other trains with flammable materials would be in a tunnel at the same time. this would prevent a long-duration fire of any significant size.

Table 3. (Continued)

Recommendation	Current Status (as of January 2012)
Full-scale package testing should continue to be used as part of package performance evaluation. testing to destruction should not be required.	the nRC had planned to implement a Package Performance study that would involve a full-scale cask and would not "test to failure." the project was never begun due to lack of funding and the eventual cancellation of the yucca mountain project. the study would have involved use of a transportation, Aging and Disposal (tAD) canister in an overpack; initial design work for this canister has been completed but none have been fabricated.
Doe should continue to ensure systematic involvement of states and tribal governments in decisions about routing and scheduling for current spent fuel shipments.	Doe has continued to involve states and tribes in transportation planning, and has established a national transportation stakeholders' Forum for that purpose. After the yucca mountain project was cancelled, Doe's em program and nRC's spent Fuel storage and transportation office provided funding to state regional groups (albeit at reduced levels).
Dot should ensure states rigorously comply with requirements for sound risk assessments in designating routes.	Dot has developed regulations for determining "preferred routes" for highway shipments, and the nRC reviews routes for security. Doe follows the same requirements for its shipments. The Commission is unaware of any recent or proposed campaigns where a state has attempted to re-route spent fuel shipments using impermissible assessments or practices.
Mostly rail has clear advantages; doe should complete the Nevada rail line and examine how to reduce the need for cross-country truck shipments by expanding intermodal service.	the yucca Mountain project had formally established a "mostly rail" policy before the program was cancelled. construction of the Nevada rail line never commenced. Federal Railroad administration (FRa), doe and some state agencies began a pilot project to examine near-reactor infrastructure and related intermodal issues, but that work too has been halted.
doe should ship oldest fuel first to a repository or storage facility. conduct a "pilot" campaign by shipping fuel from shutdown reactors first	this recommendation was never implemented, but is consistent with the commission's recommendation to ship fuel from shutdown reactors first.
doe should identify and make public its suite of preferred routes as soon as practicable to support state, tribal and local planning and preparedness, following the research reactor fuel program's process of involvement.	the yucca Mountain project began development of a formal route assessment process based on doe's established practices, but this effort too was halted. doe programs continue to consult with states and tribes on routing issues.
immediately implement section 180(c) of the nWPa to provide funding and technical assistance to corridor states and tribes.	this recommendation was never implemented, but it is consistent with the BRc's recommendation that funding should be provided to implement section 180(c) early in the planning process
Federal agencies should develop clear and consistent guidance on what and how information about transportation should be protected, and commit to open access to information that does not need such protection.	doe and other agencies did develop a joint transportation classification guide for yucca Mountain shipments, which have not commenced. doe programs continue to follow their own security requirements.

Recommendation	Current Status (as of January 2012)
doe should fully implement its dedicated train decision before large-quantity shipments begin.	the yucca Mountain project did issue a formal decision to use dedicated train service for its shipments. the BRc is unaware of any recent or proposed spent fuel rail shipments that would not involve use of dedicated trains.
doe and congress should transfer responsibility for spent fuel transportation to an outside entity.	this recommendation was never implemented, but is consistent with the BRc's recommendation that a new entity is needed to manage transportation (and everything else) related to snF and hlW.

9.2. Regulatory and Technical Issues

Several transportation-related regulatory and technical issues received particular attention in the course of the Commission's deliberations. For example, the BRC's Transportation and Storage Subcommittee heard testimony that DOE's plans to use its own self-regulating authorities under the Atomic Energy Act sharply undercut credibility in the proposed transportation program. The existing regulatory framework for commercial transportation—which features extensive oversight and involvement by the NRC, mode-specific administrations of the DOT,[218] and state and tribal officials—is proven. Consistent with the recommendations articulated in chapter 7 of this report, the Commission believes that a new waste management organization should be subject to independent regulation of its transport operations in the same way that any private enterprise performing similar functions would be—in other words, the new organization should not receive any special regulatory treatment. This will help assure regulatory clarity and transparency.

Another transport-related issue that will need to be addressed concerns the fact that the NRC has not yet granted a license for the transport of higher burnup fuels, which are now commonly being discharged from reactors.[219] It will be necessary to reexamine current regulations and to develop a technical basis for assigning burnup credit to ensure that these higher-burnup fuels can be transported when needed. In addition, spent fuel that has been stored for extended periods may be degraded and may require additional handling and preparation before it can be transported.

Finally, numerous parties have suggested that expanded full-scale testing of transportation casks (in addition to computer modeling) could be useful in enhancing public confidence in transport safety. Full-scale testing is part of the testing methodology used by the NRC in its integrated evaluation program. The NAS *Going the Distance* study endorsed the NRC's approach and recommended that full-scale cask testing, as well as other accepted methodologies, should continue to be used for technical reasons. In 2005, the NRC approved a staff proposal for the full-scale testing of a rail cask (like one shown in figure 19)—of the kind expected to be used in transporting spent fuel to a HLW repository—in a scenario involving a collision with a locomotive traveling at high speed followed by a hydrocarbon fire. DOE supported the proposed Package Performance Study and suggested combining it with an emergency response exercise to maximize the benefits of the study. Plans to provide NRC with needed funding in 2009 did not materialize because of budget constraints (the estimated cost of the study was approximately $15 million) and uncertainties about the Yucca Mountain project. The Commission's view is that funding for the proposed test, if it has independent value, should be provided from the Nuclear Waste Fund so that the NRC can update these plans and proceed with those tests the NRC determines to be most useful.

With regard to transportation security, the NRC has existing security regulations and orders in place and is currently undertaking a separate rulemaking to codify further transportation security requirements.[220]

The proposed protective strategy for transportation includes several elements:

- Advance planning and coordination with states
- Increased notifications and communications before and during shipment
- Continuous and active shipment monitoring
- Use of armed escorts over the entire shipment duration (previously, armed guards had been required only in highly populated areas)
- Background investigations of personnel with access to Safeguards Information.

In its *Going the Distance* report, the NAS noted that "[m]alevolent acts against spent fuel and HLW are a major technical and societal concern." However, the report authors were unable to perform an in-depth analysis of transportation security due to informational constraints (primarily lack of access to classified materials).[221] Accordingly, the committee recommended that experts with full access to all relevant information conduct an independent assessment of security risks before any large-scale campaign to ship materials to a disposal or consolidated storage facility is launched.

In subsequent discussions with the NRC's Office of Nuclear Security and Incident Response, BRC Commissioners and staff reviewed the additional analyses NRC has conducted following the release of the NAS report and others developed since that time.[222] We found that the NRC has taken reasonable actions to respond to the vulnerabilities that have been identified to date and we expect the current NRC rulemaking process to be sufficient to ensure that any needed future changes will be made appropriately.

Figure 19. Casks being transported by rail.

9.3. The Role of State, Tribal and Local Governments and the Importance of Early Planning for Future Transport Needs

Extensive planning and preparation for transport arrangements will be required even if only the 2,800 metric tons of spent fuel currently being stored at shutdown reactors are slated for initial transfer to consolidated storage.[223] The Commission has heard testimony indicating that advance planning timeframes on the order of a decade could be required to plan and coordinate a transport strategy and to establish the institutional and physical infrastructure to conduct a large-scale shipping operation.[224] Historically, some programs have treated transportation planning as an afterthought. No successful programs have done so.

This lead time is important from a purely logistical standpoint because some critical elements of infrastructure and equipment do not currently exist and will need to be designed, fabricated, tested and licensed before significant amounts of waste can be moved. For example, the Association of American Railroads requires the use of cask cars, buffer cars and escort cars with special safety features for future rail shipments of SNF. No such cars currently exist, and developing and placing this type of equipment into service will require between five and seven years.[225] The current commercial fleet of licensed casks is quite small and is primarily limited to legal-weight truck (LWT) casks. While a significant transportation campaign could begin using trucks and current LWT casks, a sizeable fleet of rail rolling stock will be needed to move larger quantities of SNF (such as those currently loaded in dual-purpose containers in dry storage at reactor sites). Moving significant quantities of spent fuel from reactor sites will require substantial new, specially-designed and dedicated equipment; it will also require infrastructure modifications and improvements at existing reactor sites, which vary widely in terms of their condition and accessibility. These logistical issues, which must be addressed sooner or later in any event, do not present large technical challenges, but they are nonetheless complex and will take time to resolve.

Substantial lead time is also needed to ensure that planning and institutional arrangements are in place and tested by the time major shipments commence. This means early efforts are needed to undertake planning activities that involve state, tribal and local officials. To that end, DOE should complete the development of procedures and regulations for providing technical assistance and funds (pursuant to section 180(c) of the NWPA) for training local and tribal officials in areas traversed by spent fuel shipments. Although the final destination of the material to be shipped (whether for storage, recycling or disposal) is not known, every origin site is known.

DOE has a well-established practice of working with state and regional groups and other organizations to coordinate and provide technical assistance for transportation. Future programs should build upon these proven approaches.

In particular, DOE has for many years supported cooperative agreements with state regional groups, or SRGs, to partner with local authorities through whose jurisdictions radioactive materials will be transported. Collaboration through the SRGs has proved important, not only because states have primary responsibility for protecting the health and safety of their citizens, but because they share (and sometimes disagree about) common concerns.

Bringing corridor jurisdictions together under the auspices of these groups allows issues to be identified and resolved by all parties. It also means the shipper and carrier do not have to negotiate individually with jurisdictions that may have inconsistent or even conflicting priorities. States have extensive experience with transportation issues and important roles to fulfill with respect to issues such as routing, inspections, training, emergency preparedness, communications, public information, and security for radioactive materials and other hazardous shipments.

The WIPP facility in New Mexico, for example, provides a longstanding and highly successful model for partnering with states to achieve shared success in addressing issues related to the transport of nuclear materials. Beginning with the Western Governors' Association (WGA) and later expanding to include other SRGs, states worked with one another and DOE over a period of years to develop inspection protocols, training programs, information products and other areas of transportation planning. The goal of such efforts, according to one policy statement issued by the WGA, was to achieve "the safe and uneventful transport of radioactive, radioactive materials, and spent nuclear fuel" by (among other things) conducting "early coordination and effective communications with state, tribal, and local governments."[226]

On occasion, DOE agreed to go beyond regulatory requirements when doing so was reasonable and prudent, and would enhance overall safety. For example, the Commercial Vehicle Safety Alliance developed an enhanced inspection standard to ensure that trucks carrying waste to WIPP, and spent fuel to other sites, were "defect free" before departure, and inspected again upon arrival.

Over time, inspectors in "downstream" corridor states became confident that WIPP trucks were among the safest on the highway and would waive en-route inspections, allowing them to spend more time examining other trucks whose equipment was deficient and posed significant safety risks.

This program, which was originally a voluntary "extra-regulatory" measure, was later incorporated into regulations. Overall it lowered the time-in-transit and related costs for WIPP shipments.

In a paper examining the WIPP transportation record and safety program after ten years of operation, several participants on the WGA working group noted that:

> Working Group members recognized that there were three key elements to achieving the Governors' objectives of safe and uneventful transportation and public acceptance of the program: accident prevention; effective emergency response if there were an accident; and a successful public information program. They also recognized that a cooperative effort by federal, tribal, state and local governments was necessary to achieve the objectives.227

This basic approach to collaborative planning has been applied to other campaigns with the Southern States Energy Board and the Northeastern and Midwestern offices of the Council of State Governments. Spent fuel transportation programs (including those supporting research reactors and the Navy) continue to follow the same basic approach, with modifications as appropriate (for example, Navy SNF shipments are classified). The Commission believes that while DOE has frequently been criticized for its management of the civilian nuclear waste program more generally, the Department's record of planning in cooperation with concerned stakeholders to address transportation issues has for the most part been quite successful. Any new entity charged with managing spent fuel and waste in the future should emulate and build on this success.

9.4. Next Steps

Under Section 180(c) of the NWPA, DOE is required to provide funding and technical assistance for the training of public safety officials to states and tribes whose jurisdictions would be traversed by shipments of spent fuel to storage facilities or to a repository. Over the course of more than a decade and several stops and starts, DOE's Office of Civilian Radioactive Waste Management (OCRWM) developed a policy for implementing a 180(c) program. The BRC commissioned a paper on the background and history of the 180(c) program in an effort to develop recommendations for the future implementation of this or a similar initiative.[228] The results of this paper, in combination with the recommendations put forward by the NAS in 2006 and comments received by the BRC in the course of our deliberations, suggest that at least two actions are needed in the near term.

First, early implementation of the 180(c) program as currently defined in the NWPA should be initiated by DOE and should be supported by the Nuclear Waste Fund, *even before* any potential storage or disposal site is identified. DOE should (1) finalize procedures and regulations for providing technical assistance and funds for training to local governments and tribes pursuant to Section 180(c) of the NWPA and (2) begin to provide such funding, independent from progress on facility siting. While it would be premature to fully fund a technical assistance program before knowing with some certainty where the destination sites for spent fuel are going to be, substantial benefits can be gained from a modest early investment in planning for the early transport of spent fuel from shutdown reactor sites. Consistent with the Commission's recommendation that spent fuel from shutdown reactor sites be "first in line" for acceptance at a consolidated storage facility, initial routes from those sites can be easily identified, and pilot training programs for emergency responders along those routes should begin without further delay. This would be consistent with the recommendation of the *Going the Distance* study that DOE initiate transport "through a pilot program involving relatively short, logistically simple movements of older fuel from closed reactors to demonstrate its ability to carry out its responsibilities in a safe and operationally effective manner."

Second, legislation to implement the broader waste management strategy being recommended by the BRC should include amendments to Section 180(c) that would expand the authority and responsibility of the new waste management organization to include authorities equivalent to those given to DOE in the WIPP Land Withdrawal Act with respect to transportation to WIPP, including:

- A program to provide information to the public about the transportation of spent fuel or high level waste to or from a repository or storage facility. [WIPP LWA Sec. 14(c)(1)(D)(iv)]
- Authority and direction to assist states, tribes, and local governments, through monetary grants or contributions in-kind (subject to appropriation) in acquiring equipment for responding to an incident involving shipments covered by the law. [WIPP LWA Sec. 14(c)(2)]
- Broad authority and direction to provide in-kind, financial, technical, and other appropriate assistance (subject to appropriations) to states and tribes whose jurisdictions would be traversed by shipments of spent fuel to interim storage or to a

repository, for the purpose of transportation safety programs related to such shipments that are not otherwise addressed in the law. [WIPP LWA Sec. 14(d)]

The Wipp Transportation System a Decade of Safe, Secure Shipments of Radioactive Waste

In March 1999, the Waste Isolation Pilot Plant (WIPP) in New Mexico received its first shipment of transuranic (TRU) radioactive waste. WIPP has received over 10,200 shipments as of January 2012. The experience of the WIPP transportation system provides grounds for confidence that nuclear waste can be transported across the nation safely and securely. The system was designed by DOE and includes multiple coordinated elements aimed at assuring safe and secure transport.

The Transport Container – All waste is transported in packages approved for use by the NRC. Several different types of shipping containers have been developed to enable shipment of different types of waste. All packages must meet NRC and U.S. Department of Transportation (DOT) radiation limits.

The Drivers and Carriers – DOT sets standards for drivers of trucks that carry hazardous cargo. DOE agreed to go beyond these requirements for its WIPP drivers and carriers. WIPP drivers must meet or exceed experience, licensing and training qualifications, and maintain good driving records. Once hired, drivers are also instructed in defensive driving, adverse weather, road hazards, and mountain driving, in addition to extensive WIPP relevant training, and are subject to stringent penalties if they deviate from specific procedures.

Drivers work in pairs to ensure that the truck and payload are attended at all times and that drivers are rested while driving. WIPP drivers must stop and check their trucks and payload every 150 miles or three hours en route.

The Shipping Network and Emergency Preparedness and Response Systems – DOT regulations require radioactive materials to be shipped on the interstate highway system unless states designate other routes. WIPP shipment protocols and routes were developed through cooperative efforts between states, tribal governments and DOE.

Prior to departing a TRU waste site, state police inspect WIPP trucks to Commercial Vehicle Safety Alliance Level VI standards, the most rigorous in the commercial trucking industry. WIPP drivers notify state officials two hours before entering each state and WIPP trucks are subject to inspections at each state port of entry.

The states and DOE have agreed on procedures to monitor weather and road conditions so that shipments can avoid hazards. Shipments will not depart DOE facilities if they are likely to encounter severe weather along the route. If unexpected bad weather or road conditions are encountered, procedures for the selection and use of safe parking areas have been developed. Designated federal, state and tribal officials can also monitor the shipments.

While designed to prevent accidents from occurring, the WIPP transportation system also has extensive measures in place to address emergency response in the event a shipment is involved in a serious accident. Specific plans and procedures for dealing with an accident are in place throughout all routes used in the transportation system; these plans cover notification, incident command, and response procedures.

In addition, more than 26,000 trained emergency response professionals are in place along the routes. In coordination with DOE, the states have developed a WIPP-specific training regimen for emergency first responders; this regimen has been incorporated directly into hazardous materials training programs for fire fighters, police and emergency medical staff along the routes. In 1994, the National Academy of Sciences projected that WIPP's planned shipping program would be "safer than that employed for any other hazardous material in the U.S.." Experience to date bears out this assessment.

10. REGULATORY ISSUES

Regulation is an essential element of a safe, secure, environmentally responsible and ultimately effective nuclear waste management strategy.

The federal government has sole authority to regulate SNF and high-level radioactive waste. Under current law, the two agencies with primary responsibility for regulating facilities or activities related to radioactive waste management are the NRC and EPA.[229] Storage facilities for spent fuel and HLW are regulated and licensed by the NRC, while disposal repositories are subject to both EPA and NRC regulation. Specifically, EPA is responsible for issuing "generally applicable standards for protection of the general environment from offsite releases from radioactive material in repositories." These standards apply to the management and storage of waste during the operational period, as well as to the performance of a disposal facility during the post-closure period (i.e., after waste is no longer being actively emplaced).

The NRC, meanwhile, is charged with issuing "requirements and criteria" to be used in approving construction, operation, and closure of repositories. These criteria, which may not be inconsistent with the standards issued by EPA, must require a repository to use a system of multiple barriers and must include any restrictions on the retrievability of the emplaced waste that the NRC deems appropriate. In addition, the NRC is responsible for regulations dealing with nuclear materials safeguards and security and with protection of facility workers from radiological exposures. Other categories of worker protections are the responsibility of OSHA.

Finally, the DOT has direct regulatory responsibility for important aspects of the systems and practices used to transport radioactive wastes, while the Department of Homeland Security and other agencies play a role in addressing security and counter-terrorism-related issues involving nuclear facilities and materials.

This chapter discusses a number of regulatory issues that will have important implications for the future storage and disposal of SNF and HLW. We focus particularly on regulations for disposal facilities, as this is the area that presents the most challenging regulatory issues.

The discussion in this chapter reflects current arrangements under which authority for establishing regulations and evaluating compliance is held by the federal regulatory agencies; as discussed in section 6.6, we recommend that state and tribal governments have the opportunity to negotiate important roles in aspects of regulation, permitting and operational oversight.

10.1. Issues and Challenges in Regulating Storage and Transportation

As noted in chapter 3 of this report, the NRC recently extended its "Waste Confidence Decision" to up to 60 years after the termination of an operating reactor's license (with extensions up to 60 years).[230] Several states have since filed suit against the NRC over this finding.[231] In the meantime, the NRC has also begun researching the potential environmental impacts of storage over even longer timeframes—more than one hundred or even several hundred years. It is important to emphasize, however, that even if the NRC finds that storage can be safely implemented over these very extended timeframes, this would not mean that deferring disposal for additional decades to (in the worst case) centuries would be justified or would make sense—in either cost or risk management terms.

In June 2010, the NRC launched a comprehensive review of regulations related to extended storage and transport including, specifically, the adequacy of existing mechanisms for ensuring safe and secure storage and transportation for extended periods beyond 120 years.[232]

This review is expected to be complete in 2017. A newer and unanticipated challenge involves developing an appropriate regulatory response to the events that occurred at Japan's Fukushima Daiichi nuclear power station following the March 2011 earthquake and tsunami. The NRC and other agencies, such as the IAEA, are in the early stages of conducting in-depth investigations of the crisis; in addition, the Commission is recommending a separate NAS study of Fukushima. As a result of these efforts, new storage-related regulatory requirements may be deemed necessary and appropriate and if so, should be implemented as expeditiously as possible.

More generally, the primary regulatory challenge for storage facilities (given a realistic appraisal of the time likely to be necessary to open and load one or more geologic disposal facilities) remains ensuring their performance over extended periods of time (120 years or more).

This will require a better understanding of degradation mechanisms that could, over multiple decades, potentially affect the integrity of spent fuel or its cladding. It also requires better information about environmental conditions and the state of spent fuel inside existing dry storage systems.

As noted in chapter 5, because these systems generally lack instrumentation, knowledge of key parameters such as (but not limited to) gas pressure, the release of volatile fission products, and moisture is limited to nonexistent for most dry cask installations. Some of these issues will be addressed as part of the Extended Storage Collaboration Program that EPRI has launched—in conjunction with the NRC, DOE, the Nuclear Energy Institute, individual utilities and dry storage system vendors—to research the technical basis for long-term dry storage of SNF.

The current regulatory system for assuring the safety and security of nuclear waste shipments, meanwhile, has functioned well to date. As discussed in chapter 5 of this report, however, the challenge will be to ensure that the current system can keep up in terms of managing health and safety risks and providing adequate physical security if the quantity and volume of waste shipments—including shipments of higher-burnup fuels—increase substantially in the future.

A separate NRC rulemaking is currently underway to codify further transportation security requirements for future nuclear waste shipments.

10.2. Issues and Challenges in Setting Regulatory Standards for Disposal Facilities

Regulating facilities for the disposal of HLW and spent fuel presents unique challenges because of the extraordinarily long time periods over which these materials present health, safety, security, and environmental concerns.

In its 2006 *Safety Requirements* report, the IAEA elaborated on the basic aims of geological disposal:[233]

- To contain the waste until most of the radioactivity, and especially that associated with shorter lived radionuclides, has decayed
- To isolate the waste from the biosphere and to substantially reduce the likelihood of inadvertent human intrusion into the waste
- To delay any significant migration of radionuclides to the biosphere until a time in the far future when much of the radioactivity will have decayed
- To ensure that any levels of radionuclides eventually reaching the biosphere are such that possible radiological impacts in the future are acceptably low

The task for regulators is to translate these general aims into specific standards or technical performance requirements that must be met before a facility can be licensed. Different countries have taken different approaches to this task with the result that regulatory requirements for disposal facilities vary around the world. Increasingly, however, there is some convergence in these requirements, particularly as international organizations such as the Nuclear Energy Agency (NEA)[234] and the IAEA have published recommendations or guidance in this area.

In the United States, there are currently two sets of federal regulatory standards for high-level radioactive waste disposal facilities one set that was developed specifically for Yucca Mountain and another, earlier set that would, under current law, apply to all other sites (this earlier, generic set of standards was essentially complete by the time Congress directed the development of Yucca Mountain-specific standards in 1992; see further discussion in the text box).[235]

Because the thinking about repository regulations evolved considerably during the development of the Yucca Mountain requirements, the Commission concludes that the generic regulations that would currently apply to all other sites will need to be revisited and revised in any case. In addition, the Commission has heard a range of views both about broader reforms to the current U.S. regulatory framework for geologic disposal facilities and about specific changes to existing repository requirements. We have addressed some of the broader reform questions, but have not attempted to develop specific recommendations concerning the appropriate form and stringency of regulatory standards for disposal facilities. Resolving these issues will involve societal value judgments that should be mediated through the normal regulatory development process. In that process, EPA, NRC, and other agencies can and should draw from an extensive literature and considerable regulatory experience to make appropriate determinations for assuring safe and secure nuclear waste disposal in this country.

The remainder of this section briefly reviews some of the most important and controversial technical and policy issues to be resolved in setting performance standards for disposal facilities, before offering some general principles to guide the development of future regulations in the United States.

10.2.1. Timeframe

Since long-term protection of human health is one of the core functions of deep geologic disposal, quantitative limits on the public's future exposure to radioactivity are typically included in disposal facility standards. These limits may take the form of a dose-based or risk-based standard (the two are essentially equivalent in practice) that limits the exposure to individuals resulting from radiation releases from the repository. Alternatively they can take the form of a release-based standard that limits the amount of radioactive material that is allowed to escape the repository (see text box describing U.S. disposal facility regulations).

A critical regulatory question then centers on the timeframe over which compliance with these numeric limits must be demonstrated. This has been a controversial question in the past, because the long-lived nature of the radiological hazard posed by SNF and HLW creates an inherent tension between the objective of protecting future generations on the one hand, and the practical difficulties of making very long-term projections about human and natural systems on the other hand. In the United States, the EPA initially proposed a compliance timeframe of 10,000 years for the proposed Yucca Mountain repository; however, this limit was later increased to 1 million years.[236] Many individuals have told the Commission that it is unrealistic to have a very long (e.g., million-year) requirement for demonstrating compliance in a traditional regulation; the Commission agrees. Other countries have taken different approaches to this issue: some have opted for shorter timeframes (a few thousand to 100,000 years), some have developed different kinds of criteria for different timeframes, and some have avoided the use of a hard "cut-off" altogether and have instead opted to require a demonstration that the proposed facility is at very low risk for catastrophic disruptions that could lead to large-scale releases of radioactivity. Some countries, such as Finland and Sweden, have more stringent regulations for the first few thousand years after repository closure, compared with the period from 1,000 years to 100,000 or 1,000,000 years. In doing so, they acknowledge the fact that uncertainties in predicting geologic processes, and therefore the behavior of the waste in the repository, increase with time.

10.2.2. Compliance Methodology

As critical as the form and stringency of the standards to be applied to a disposal facility is the decision about what approach or methodology will be used to determine whether they have been met. An integral part of EPA's standards for geologic disposal facilities is a requirement (embedded in the standards themselves) that compliance be demonstrated using a quantitative performance assessment to project repository performance over very long time periods: 10,000 years in the case of WIPP and up to 1,000,000 years for Yucca Mountain.[237] Over the last decade or more, however, there has been increasing attention worldwide to compliance methodologies that integrate quantitative and qualitative lines of argument to show that a repository will remain safe after our ability to monitor a repository is lost.[238,239] This shift has been motivated in part by increasing recognition of the inherent limitations of quantitative projections over geologic time periods.[240]

Instead of focusing on comprehensive calculations of projected dose levels to populations hundreds of thousands of years or more in the future, for example, the safety analysis used to support regulatory demonstrations of compliance might use such calculations for an initial period for which they would be most defensible, and then follow the evolution of troublesome radionuclides in the given geologic environment over the long term using other existing and compelling scientific knowledge.[241] For example, Finnish regulators require quantitative assessment where possible, but also call for use of complementary considerations when quantitative analyses are not feasible or are too uncertain.[242]

Performance assessment is valuable as a systematic method for organizing the understanding of a geologic repository.[243, 244] It is also valuable for focusing the information used to support a compliance demonstration.[245] Furthermore, NRC regulations stipulate that additional factors (such as the demonstration of the effectiveness of multiple barriers, empirical observations, and a performance confirmation program) are to be considered in a licensing decision.[246]

Nonetheless, the heavy emphasis on quantitative performance assessment in U.S. disposal facility safety standards may lead to a focus on showing numerical compliance that could obscure understanding of the actual operation of the disposal facility system247 and divert attention from the overall strategy being employed to safely dispose of nuclear waste.248 While EPA recognized this concern when it extended the compliance period to 1 million years for its Yucca Mountain standard in compliance with a judicial decision,249 the standard itself makes no distinction between the roles of the performance assessment in demonstrating compliance for the first 10,000 years and for the very long term—use of performance assessment is an integral part of the standard for both time periods.

The need for flexibility in demonstrating compliance with any very long-term quantitative performance standard might be better met if performance assessment were included as one of a set of methods to be used in formulating the overall safety case for compliance, rather than being incorporated in the statement of the standard itself.[250]

10.2.3. Standard of Proof for Compliance Demonstrations

The "standard of proof" required for compliance demonstrations should be viewed as integral to a long-term performance standard. While EPA disposal facility regulations (both generic and Yucca Mountain-specific) require the use of quantitative performance assessments to show compliance with quantitative standards, they also state that unequivocal proof of compliance is neither expected nor required because of the inherent limitations of such assessments.[251]

Thus, licensees must demonstrate a "reasonable expectation" of compliance with standards for the post-closure period. The NRC adopted EPA's approach of applying a "reasonable expectation" standard to the post-closure period, while using a "reasonable assurance" standard for the operation of facilities during the pre-closure period (consistent with the NRC's practice for other licensed operating facilities subject to active licensee oversight and control).

10.2.4. Other Protection Requirements

Protection of the natural environment (along with, but distinct from, human health per se) is widely accepted as an important objective of geologic disposal; however, there has been less convergence internationally around how to assess this objective and develop appropriate criteria.

A recent (2010) NEA review of regulatory developments pertaining to geologic disposal describes a number of national and international efforts—some ongoing—to develop ways of accounting for the long-term protection of flora and fauna.

Current U.S. Disposal Facility Regulations

"GENERIC" EPA AND NRC REGULATIONS

EPA standards for all sites other than Yucca Mountain are defined under 40 CFR Part 191, "Environmental Radiation Standards for Management and Disposal of Spent Nuclear Fuel, High-Level and Transuranic Radioactive Wastes" (with additional implementation and compliance criteria for WIPP found in Part 194). This regulation was first issued in 1985, remanded by a federal court for reconsideration of certain provisions, and reissued in 1993 to apply only to geologic disposal facilities other than Yucca Mountain (see below).

The core of Part 191's disposal standard is a "containment" requirement designed to protect human populations by limiting the cumulative releases of key radioactive isotopes over the 10,000-year period following closure of a disposal facility. Compliance is to be demonstrated by use of quantitative performance assessments that take into account "all significant processes and events" to show that there is a "reasonable expectation" (not absolute proof) that cumulative releases for a number of specific isotopes will have a low likelihood (less than one chance in 10 for low releases and less than one chance in 1,000 for higher releases). The EPA regulation also includes an individual protection requirement, which stipulates that for 10,000 years there should be a reasonable expectation that no member of the public will receive an annual dose greater than 15 millirems (150 microsieverts), considering only the undisturbed performance of the repository (rather than all significant processes and events, as required for the containment standard). NRC regulations for all sites other than Yucca Mountain are defined in 10 CFR Part 60, "Disposal of High-Level Radioactive Wastes in Geological Repositories." These regulations were originally issued in 1983 (before EPA's first set of standards had been completed) and revised in 1987 to reflect the NWPA Amendments Act of 1987. NRC's regulation incorporates EPA's generally applicable standards by reference, and includes additional performance requirements for specified individual barriers in the repository system.

YUCCA MOUNTAIN-SPECIFIC REGULATIONS

The Energy Policy Act of 1992 directed EPA to issue an individual dose standard for Yucca Mountain, based upon and consistent with recommendations by the NAS. The process to develop this EPA standard (40 CFR Part 197) and matching NRC implementing regulations (10 CFR Part 63) was complex—it involved an NAS study, multiple lawsuits, and another court remand that required EPA to reconsider certain provisions it had initially proposed. Thus, it was not completed until 2008.

> The EPA Yucca Mountain standard limits doses to members of the public (not total releases of specified radioactive materials) and extends to 1 million years (consistent with a recommendation of the NAS study), with a 15 millirem limit for the first 10,000 years and a 100 millirem limit thereafter. The NRC Yucca Mountain regulations incorporate the new EPA standard and drop the performance standards for individual repository barriers that are contained in the generic regulations (10 CFR Part 60).

Existing regulations in some countries (e.g., Canada, Finland, Sweden, Switzerland, and the UK) include qualitative requirements for the protection of nonhuman organisms and biodiversity; several countries also require that these impacts be explicitly included in future risk and performance assessments. In the United States, EPA standards for the disposal of high-level radioactive waste and TRU waste include a separate groundwater protection standard.

10.2.5. Division of Regulatory Responsibility between EPA and NRC

In the course of the Commission's deliberations, numerous witnesses expressed the view that greater consistency was needed between the EPA and NRC regulatory systems.

Some witnesses also suggested that any effort to rationalize or harmonize the EPA and NRC systems be undertaken before new disposal sites are identified, even for screening purposes, to avoid or at least minimize the perception that standards are being set to ensure that one or more (pre-selected) sites will meet them. This seems particularly important for individual protection requirements, which have been a clear point of contention in the past; however, it is likely to be relevant for many other issues as well. Comments submitted by both the NRC and the EPA in response to the BRC's draft report indicate that coordination between these two regulatory agencies has improved markedly in recent years.

The Commission also received and considered recommendations for a more fundamental redrawing of regulatory roles and responsibilities at the federal level (i.e., transferring all regulatory authority to the NRC or EPA). We concluded that while there are opportunities for improvement in the EPA/NRC regulatory process and in the working relationship between these agencies, the general division of roles and responsibilities that currently exists between EPA (establishing standards) and NRC (licensing and regulating waste management facilities) is appropriate and should be preserved. We return to this point in the next section.

> **Key Questions In Setting A Regulatory Standard For Deep Geological Disposal**
>
> - What should the basis be: a desired level of protection or what is reasonably achievable using today's technology?
> - For how long must compliance be demonstrated?
> - Who is to be protected—individuals or populations?
> - What is the desired level of protection?
> - What is the measure of compliance (e.g., doses to individuals vs. releases to the environment)?
> - How should compliance be demonstrated—primarily through quantitative calculations or through a broader safety analysis that involves multiple lines of qualitative as well as quantitative considerations?

- What level of confidence is required?
- How should the potential for human intrusion be addressed?
- How should retrievability be addressed?
- Can compliance take credit for institutional controls and if so, for how long?
- Should groundwater be separately protected?
- Should there be performance requirements for sub-elements of a disposal facility (e.g., the waste package or the geologic setting)?

10.3. Recommendations for Developing Future Disposal Facility Standards

Without making specific recommendations regarding the standards to be applied to geologic disposal facilities in the future, the Commission recommends a number of general principles or propositions to guide the development of future regulations:

1. The standard and supporting regulatory requirements to license a facility should be generic—that is, applicable to all potential sites.

While there may be advantages to developing standards and requirements that recognize the specific features and characteristics of a particular site, experience with Yucca Mountain indicates that this approach can create suspicions that regulations are being tailored to make a pre-selected site "work." Generally-applicable regulations are more likely to earn public confidence. A generic standard will also support the efficient consideration of multiple sites.

2. Regulatory standards and requirements for compliance demonstrations (including the required level of confidence in the demonstration or "standard of proof") should not go beyond what is scientifically possible and reasonable.

Both the standards themselves and the process used to demonstrate that they have been met must be credible to the scientific community and the public. The Commission has heard the view that some aspects of the current Yucca Mountain regulations lack credibility. A specific concern is the requirement that the compliance demonstration be primarily based on a complex quantitative projection of repository performance for 1 million years. While making calculations over such a long time horizon might be appropriate as a part of establishing a broader safety case, the Commission believes that over-reliance on million-year calculations can reduce credibility rather than enhance it. As the IAEA has warned: "Care needs to be exercised in using the criteria beyond the time where the uncertainties become so large that the criteria may no longer serve as a reasonable basis for decision making."[252]

Whatever the time frame, the standard of proof for compliance should likewise be based on what is scientifically achievable. As discussed above, both existing sets of generic disposal facility and Yucca Mountain-specific regulations emphasize that absolute proof in the normal sense of the word is not possible over long time periods. They therefore stipulate that compliance determinations should be based on a "reasonable expectation" that the standards will be met. This is also the standard of proof defined by EPA[253] and ultimately adopted by the NRC for its Yucca Mountain regulations. This approach has proved workable in the WIPP context and we recommend that it be carried over into new regulations.

3. Rules for demonstrating compliance and for documenting the required level of confidence in the compliance demonstration (i.e., the standard of proof) should be defined at the same time that the performance standards are developed.

Rules for demonstrating compliance (including meeting the standard of proof) are integral to any regulatory standard. These rules should be included when developing the overall standard and should be applied in the way that was expected when the performance standard was adopted. This is particularly important when different agencies are charged with setting the standard (EPA) and implementing the standard (NRC). In these cases, the potential exists for different agencies to apply different regulatory philosophies to the same standard.[254]

4. Standards for a disposal facility should explicitly recognize and facilitate an adaptive, staged approach to development.

Current EPA and NRC regulations were developed before international thinking about repository development shifted in favor of a more staged, adaptive approach (this is also the approach the Commission is recommending for the United States). The NRC, in particular, has a robust and exacting regulatory process for reactor operators and other facility licensees that generally requires very high levels of design specificity and performance assessment at the initial licensing phase. This structure is not necessarily incompatible with a staged, adaptive approach; in fact, the NAS study of staged repository development observed that the "The U.S. licensing process already follows a staged approach" and concluded that "there are no restrictions precluding DOE from implementing Adaptive Staging."[255] However, future disposal facility regulations should be designed to accommodate a process in which decisions about design, construction, and operations might be kept open beyond the initial license application.[256] In general, adaptive staging could make the licensing process more complex by increasing the number of changes made in the course of the process. This in turn would increase the number of regulatory review steps and the potential need for license amendments.[257] Recent NRC planning documents suggest the agency has already recognized that it may need to develop new performance assessment tools that are flexible enough to accommodate different scenarios for the management of spent fuel and HLW (in part to respond to the findings of the BRC in its draft report).[258] More broadly, we believe a revised regulatory structure for future disposal facility development should be designed with express attention to providing the flexibility needed to support an adaptive, staged process.

5. Safety and other performance standards and regulations should be finalized prior to the site-selection process.

If site selection occurs before final performance standards are defined, there are two risks. The first is that time and effort could be spent on a site that should have been ruled out as unsuitable earlier in the process. The second risk is one of perception.

The public and other stakeholders could suspect that standards are being adjusted to fit the site. These considerations argue for setting generic standards that would be applicable to any facility wherever it is located, *before* any particular site is selected for further study. In developing such regulations, however, it will be important to avoid setting excessively detailed and rigid requirements that could prove unworkable when applied to an actual site or that could have the effect of screening out potentially suitable and otherwise promising sites.[259] The Commission believes there is no reason to wait to start the process of developing generic regulations for future geologic disposal facilities. As discussed below, we are not recommending any change in the current allocation of regulatory responsibilities and authorities that would require enabling legislation. Given that we are recommending a flexible process for finding new sites, standards development need not delay early progress on the siting front. Moreover, the fact that the regulatory issues to be resolved have been well defined and extensively analyzed over more than 30 years of EPA and NRC experience in this area, and the fact that some of the key issues have already been tested in court and in the regulatory process, should help expedite the process of developing generic disposal facility safety and performance standards.

6. EPA and NRC should coordinate closely in the development of new disposal regulations.

Problems of coordination between EPA and the NRC in developing repository standards have been widely cited as having contributed to negative perceptions of, and loss of confidence in, the Yucca Mountain project. As we have already noted, the Commission has heard proposals for a fundamental redrawing of regulatory roles and responsibilities for disposal facilities at the federal level (e.g., by consolidating all regulatory authority in the NRC or the EPA). Broadly speaking, however, our examination of the roles of the NRC and EPA, with respect to nuclear waste management under existing law, suggests that while there are opportunities for improvement in the EPA/NRC regulatory process and in the working relationship between these agencies, the general division of roles and responsibilities that currently exists is appropriate and should be preserved.

While we are not recommending a change in the regulatory roles of EPA and NRC, we believe the process of developing EPA standards and NRC regulations for implementing those standards should be carefully coordinated to avoid repeating past problems. For example, the Commission has heard testimony that the processes used to develop standards in the past were confusing and frustrating to the public,[260] and that more coordinated and dedicated efforts are needed in the future to draw not only on the expertise of EPA and NRC but also on input from the knowledgeable public. We have also heard that public disagreements between these agencies over matters of regulatory philosophy can confuse the public and undermine confidence in the regulatory system,[261] and that it is important that such disputes be resolved promptly.[262]

The Commission believes that a coordinated and open process should be used to develop new generic regulations for future disposal facilities, and that that any differences in regulatory philosophy between the two agencies be laid out clearly and resolved as early in the process as possible. We further believe that actions to coordinate the development of new disposal regulations can be undertaken by the Executive Branch without additional action by Congress.

Specifically, we recommend that EPA and NRC begin working together to define an appropriate process (with opportunity for public input) for developing a generic disposal facility safety standard and associated implementing regulations.[263] In addition the two agencies should continue to coordinate efforts during the regulatory development process. This process should be designed to accomplish the following:

- A clear definition of the regulatory issues to be resolved,
- A comprehensive identification of alternative approaches to resolving these issues,
- A thorough and fair analysis of the alternatives,
- A clear explanation of the regulatory choices that are made,
- A shared understanding between the two agencies and with other stakeholders about the compliance demonstration methods and standard of proof that are to be used in implementing the standards.

We also recommend that the administration and Congress ensure that NRC and EPA have sufficient resources to complete this process in a thorough and timely way. The cost of delays in being able to move ahead with finding new sites would certainly be far higher than the cost of a process to establish the necessary standards as soon as possible.

7. The EPA and NRC should also develop a new regulatory framework and standards for deep borehole disposal facilities

As noted in chapter 4 of this report, the Commission has identified deep boreholes as a potentially promising technology for geologic disposal that could increase the flexibility of the overall waste management system and therefore merits further research, development, and demonstration. While a regulatory framework and safety standards for deep boreholes would have much in common with those for mined geologic repositories, the technologies also have key differences. For this reason the Commission recommends that the EPA and NRC develop a new safety standard and regulatory framework for deep boreholes (consistent with the new standard recommended for mined repositories) informed by RD&D efforts aimed at developing a licensed demonstration project.

8. Security and Safeguards

Robust security arrangements are needed at storage and disposal facilities for SNF and HLW, as well as during the transport of these materials, to prevent unauthorized access and acts of sabotage or theft. From a security standpoint, the most sensitive stages at a deep geological repository are when materials are above ground (transported or in a pre-load stage) and during the pre-closure period when materials are emplaced in the disposal facility, but the facility itself is not sealed and could therefore be accessed more easily.[264] As the IAEA has recommended, the regulatory authority (for us, the NRC) will need to provide guidance to the implementing organization concerning the effective application of security measures. Such measures could include physical protection, control and accounting, and verification procedures. Recognizing the importance of international rules, the United States should offer to place all future geologic disposal facilities under IAEA safeguards monitoring.[265]

10.4. Occupational Safety and Health

Another important area of regulation for waste management facilities pertains to the health and safety of facility workers and personnel (as distinct from the protection of the general public). Currently, responsibility for occupational safety and health at nuclear facilities is the shared responsibility of the NRC, OSHA, and (in some cases) the Mine Safety and Health Administration.

On the whole, a white paper commissioned by the BRC finds that the U.S. nuclear industry has had a much better occupational health and safety record than other energy sectors.[266] However, the same report also determined that performance was not uniform across facilities and that further improvement could be achieved by assuring more consistent safety and health performance standards. In terms of occupational safety and health issues specific to the back end of the nuclear fuel cycle, the fact that the United States has not yet opened or operated a deep geologic repository or consolidated storage facility for spent fuel and HLW means there is no direct experience with these types of facilities. However, the United States has had experience with constructing two deep geological facilities (WIPP in the 1980s and the Yucca Mountain Exploratory Studies Facility in the 1990s) and more than a decade of experience operating WIPP. The overall occupational safety and health record for these facilities—and for more than a decade of waste transport operations in connection with WIPP—has been excellent so far. But this record does not argue for complacency. On the contrary, occupational safety and health is an area where continued rigor is warranted and where experience with existing facilities and operations must be looked to for useful insights about how to manage risks to workers at waste management facilities in the future.[267]

For example, constructing facilities deep underground is in and of itself a complex undertaking that poses inherent risks. The major risks to workers at a deep geological repository are the same as those associated with any large-scale underground construction project; they include, principally, traumatic injuries from working around heavy equipment and explosives, lung disease from both dust and diesel exhaust fumes, and noise-induced hearing loss. That said, current construction procedures and technologies make it possible to minimize the risk of traumatic injuries, suppress dust and other respiratory irritants, and protect workers' hearing. Other kinds of facilities could present different risks.

For example, deep boreholes do not involve the construction of underground facilities, but the surface facilities involve occupational hazards similar to those associated with oil and natural gas drilling.

10.5. Waste Classification

NRC regulations and other statutory requirements for the handling of nuclear materials rely on a system for classifying those materials. This section discusses the classification system that is currently in place in the United States for nuclear wastes.

While there have been concerns about aspects of that system for some time, some of the (potentially) most important shortcomings of the current framework are especially pertinent to the wastes that would be generated by fuel cycles that include the reprocessing and recycling of SNF.

Generally speaking, the purpose of waste classification systems is to facilitate the safe and efficient management of waste materials. This goal is best served if the classification system identifies groups or classes of wastes that could be handled and disposed of safely using essentially the same technologies, rather than classifying wastes primarily based on their source of origin.

The most important overarching criticism of the U.S. waste classification system is that it is not sufficiently risk-based. Rather, it is (for the most part) directly or indirectly source-based—that is, based on the type of facility or process that produces the waste rather than on factors related to human health and safety risks.

The legal definitions of SNF, HLW, and tailings are explicitly source-based. The definitions of TRU waste and LLW are indirectly source-based in that these classes of waste are defined by excluding one or more of the source-based waste classes.

A source-based classification system can confound efforts to manage and dispose of wastes based on the risks they pose because wastes in different classes can have essentially the same radionuclide composition and characteristics, while wastes in the same class can have substantially different radionuclide compositions and characteristics.

The Current U.S. Waste Classification System

All classes of nuclear waste defined in federal law apply only to waste that contains radioactive material as defined in the Atomic Energy Act (AEA) of 1954 [AEA, 1954]. The AEA defines three types of radioactive materials: source, special nuclear, and byproduct. Classes of waste have been defined in the AEA and its amendments, or in other federal laws; of necessity these definitions have been used in regulations (which, in some cases, have established sub-classifications). A description of the most important classes of nuclear waste—as defined under the AEA, its amendments or in other federal laws—follows:[268] Spent nuclear fuel (SNF), also called "used nuclear fuel," is fuel irradiated in a nuclear reactor that has not been reprocessed. When declared to be waste, SNF is generally assumed to be destined for disposal in a deep geologic repository. High-level waste (HLW) is the highest-activity primary waste that results from reprocessing SNF. It is ultimately destined for disposal, such as in a deep geologic repository. Transuranic (TRU) waste is waste other than SNF and HLW that contains concentrations of transuranic elements— long-lived alpha-emitting radionuclides created during the irradiation of nuclear fuel (e.g., plutonium)—at levels that are sufficiently high so as to make the waste not generally acceptable for near-surface disposal. TRU waste is generally assumed to be destined for ultimate disposal in a deep geologic repository. There are two sub-classes of TRU waste: remote handled (meaning so radioactive that it must be handled in containers that shield workers from radiation) and contact handled (meaning it does not require shielding). Mill tailings are solid residues from the processing of ores to recover uranium or thorium. Tailings are generally destined for disposal in large, capped piles on the land surface at or near the facilities that produce them. Low-level waste (LLW) is waste other than SNF, HLW, TRU waste, or mill tailings. Most LLW is destined for disposal in near-surface facilities. LLW is further divided by regulation into three subclasses (A, B, and C) that contain increasing radionuclide concentrations. All of these subclasses of LLW are generally acceptable for near-surface disposal.

A fourth subclass of LLW is defined by reference to the concentration limits for Class C wastes: LLW that has radionuclide concentrations in excess of the Class C limits is termed "greater than Class C" or "GTCC" waste.

GTCC wastes are not generally acceptable for near-surface disposal. Disposal in a deep geologic repository and disposal in boreholes at depths up to 1,000 ft are among the alternatives being considered.

For example, the radionuclide content of some of DOE's HLW is similar to that of Class A or B LLW, but because of its source, the HLW must still be managed by disposal in a deep geologic repository.

Generally speaking, it is more often the case that wastes are "over-classified"—in the sense that they are assigned to a more restrictive (and more costly-to-manage) classification than their actual hazard requires—rather than the reverse. Moreover, the requirement that specific disposal sites meet criteria to ensure safety (e.g., dose limits) is designed to prevent any "under-classified" waste from posing an unacceptable risk.

The definition of HLW, in particular, has attracted the most criticism. Much of this criticism focuses on three major issues.

1. HLW is currently defined solely in terms of its source (i.e., wastes from reprocessing), and not in terms of the characteristics that are relevant from a waste management standpoint (e.g., TRU content, radiotoxicity). To the extent that terms such as "highly radioactive," "sufficient concentrations," and "requires permanent isolation" are used to define HLW, they have not been quantified. This is potentially problematic because the liquid waste stream from the front end of a reprocessing plant can have a broad range of characteristics—including characteristics that may be altered by time (decay) or by subsequent processing (which DOE has done with many of its defense wastes). The waste that remains after these changes, while still classified as HLW, may have characteristics similar to TRU waste or LLW. Conversely, some TRU and LLW wastes that do not come from reprocessing can have characteristics similar to HLW.
2. The current system creates obstacles to managing low-concentration HLW as TRU waste or LLW. In 2003, an Idaho district court ruled that any material containing even very small amounts of the radionuclides in HLW had to be disposed of in a deep geologic repository. In response, Congress passed a law designed to allow DOE to close tanks and dispose of material containing HLW (defined as "waste incidental to reprocessing"), provided certain conditions could be met. Applying this exception, however, has proved challenging because of differing views concerning how much radioactive material can be left in tanks and differing views on the performance of vaults and closed tanks.
3. DOE recently decided to classify waste streams bearing radionuclides such as tritium, carbon-14, krypton, and iodine-129 as HLW, even though these radionuclides are not typically part of the HLW stream.[269]

   This approach would have significant ramifications for spent fuel reprocessing because some of these radionuclides (especially tritium and iodine-129) are likely to be distributed in multiple process streams and wastes throughout a plant that uses current reprocessing technology.

As noted in chapter 11, most reprocess/recycle fuel cycles would be expected to generate larger quantities of LLW, compared to the once-through fuel cycle, but would reduce the front-end creation of mill tailings. As with HLW, several concerns have been raised in connection with the current classification system for LLW:

- The current distinction between HLW and LLW has created practical problems in the context of DOE's remediation efforts. A more straightforward approach would be to use a quantitative boundary—such as concentration limits on shorter- and longer-lived radionuclides, similar to the LLW Class C limits—to make this distinction. This would allow a particular waste to fall into either class depending on its characteristics; it would also allow the effects of waste processing to be taken into account.
- Currently LLW is subdivided into classes—Class A, B, C, or GTCC—according to a list of specified radionuclides and concentrations. If a waste contains only radionuclides other than those on the list it is automatically categorized as Class A, regardless of its radionuclide concentration or the level of hazard it poses. The NRC developed the current list of radionuclides in 1982 by anticipating the types of LLW that might be produced in the future. However, the NRC's foresight was not perfect and wastes now exist, or have the potential to exist, that contain radionuclides not on the list. The most important current example is uranium. More than 500,000 metric tons of concentrated depleted uranium currently exist in the United States, much of which is destined for disposal.

  It is considered Class A LLW but NRC staff analyses indicate that near-surface disposal of this material is not likely to meet performance objectives at some sites. There are many other examples that can be found. For example, closed fuel cycles would release gaseous krypton-85 from spent fuel and current regulations would require that a substantial fraction of such releases would need to be captured and disposed of as waste.

  DOE has also considered separating the radioactive fission product cesium from spent fuel and allowing the cesium-137 it contains (which has a 30-year half-life) to decay to innocuous levels in an engineered surface storage facility, after which it would be disposed of as LLW. However, cesium-137 is accompanied by long-lived cesium-135 and, since the latter is also absent from the 10 CFR 61 list, this leaves the appropriate classification of the decayed cesium open to question. Some of these wastes may be determined to be GTCC and, as a consequence, may become "orphans" for reasons discussed below.
- Because the definition of LLW is implicitly source-based and thus not risk-informed, LLW, TRU waste, and HLW can all contain similar radionuclide concentrations but would nevertheless be managed differently. Similarly, the sub-classification of civilian LLW into Classes A, B, C, and GTCC under 10 CFR 61 is not fully risk-informed. In addition, no disposal pathway has been specified for GTCC waste, which is currently orphaned with nowhere to go. DOE is currently developing an environmental impact statement (see endnote 17) aimed at identifying such a pathway and closing this gap (as noted previously, deep geologic disposal in either a repository or boreholes is being considered for this category of waste).

- Compared to the once-through fuel cycle, future fuel cycles that involve reprocessing would produce considerably more GTCC waste under the current classification system. Reprocessing would leave metal cladding hulls along with a wide range of waste forms generated in the process of recovering, purifying, and fabricating plutonium or other transuranic elements. These would be considered TRU wastes under certain EPA regulations and GTCC waste under NRC regulations. In both cases, these wastes would be orphaned because they are not considered SNF or HLW destined for a repository. Moreover, because the material is not of defense origin, it cannot be accepted at WIPP under current law.

In light of these shortcomings, a number of alternative classification systems, or changes to the current system, have been proposed—both by the NRC and the IAEA —over the years. As yet, however, no comprehensive reforms have been implemented in the United States. Instead, the NRC has used case-by-case exemptions to address issues with particular wastes as they arise.

Adopting the most recent IAEA waste classification proposal would represent a major departure from current U.S. practice and would need to be carefully evaluated.

Though many stakeholders believe the time has come for an overhaul of the U.S. waste classification system, there is also considerable concern that changes could have unintended consequences—especially considering the complex web of laws and regulations that rely on the current system. If changes are made, it will be important to assure the public and other stakeholders that wastes are not being inappropriately re-assigned into lower classes and that protections for human health and safety and the environment remain rigorous.

The fact is that the current approach to classification—for nuclear waste generally, and for LLW in particular—appears to be working to provide adequate public protection, despite its shortcomings and complexities.

Nevertheless, the decision to pursue alternative fuel cycles—especially if they include reprocess and recycle elements—seems likely to strengthen the case for a more comprehensive reconfiguring of the current waste classification system.

Recent developments suggest that the NRC may consider revising the LLW classification system.

In 2009 the NRC directed its staff to include in a future budget request a proposal "for a comprehensive revision to risk-inform the 10 CFR Part 61 waste classification framework ..." In 2010, an NRC official reported that the staff was working on the issue and is considering a range of options including specific changes to the current structure (e.g. along the lines of recent recommendations by the International Commission on Radiological Protection), a new classification system (e.g., one based on site-specific analysis), and other changes to 10 CFR Part 61 beyond the classification system.[270]

Additionally, the NRC staff is planning to identify a number of options for changing the definition of HLW as part of developing a framework for licensing fuel reprocessing plants and plans to send a paper on the framework to the NRC Commissioners by the end of fiscal year 2011.

The Commission endorses and encourages efforts underway at the NRC to review and potentially revise the U.S. waste classification system (while recognizing that a comprehensive statutory revision to establish a risk-based system will eventually be required).

11. Advanced Reactor and Fuel Cycle Technologies

All of the commercial nuclear power reactors operating in the United States today were built based on reactor designs that are at least 30 years old and in some cases even older.

Technology has advanced in the several decades since, so while reactors in operation today are required to meet all relevant NRC safety standards, new plants are expected to achieve still higher levels of safety, efficiency, and reliability than their predecessors. Clearly, any comprehensive and forward-looking strategy for managing the back end of the nuclear fuel cycle in the United States needs to consider the potential impact not only of current technology but of further technology advances in the decades ahead.

More importantly, we cannot be sure today of the national and global context that will determine which nuclear fuel cycle technologies and systems will be considered for use in the future.

Concerns over global climate change and greenhouse gas emissions, the cost and sustainability of alternatives to nuclear power, and any number of other factors may appear very different to future generations than they do to us today. The integrated and flexible strategy that we propose for nuclear waste management puts a premium on creating and preserving options that could be employed by future generations to respond to the particular circumstances they face.

RD&D is a key to maximizing those options. To that end, the charter of the BRC asks the Commission to evaluate existing fuel cycle technologies and R&D programs in terms of specific criteria (those listed in the charter include "cost, safety, resource utilization and sustainability, and the promotion of nuclear nonproliferation and counter-terrorism goals"). Or, as Energy Secretary Steven Chu expressed the charge to the Commission in opening remarks at the Commission's first meeting in March 2011, our task was "to look at all the science and technology and all the other things that would influence how we deal with the back end of the fuel cycle."

This chapter discusses the Commission's conclusions concerning the impact of new reactor and fuel cycle technology developments, both for the nature and magnitude of the immediate and longer-term nuclear waste management challenges this country faces and in terms of the potential to provide options for advancing broader nuclear-energy-related policy objectives in the decades ahead.

Additional information concerning the current DOE nuclear R&D program and nuclear RD&D infrastructure needs is available in separate Commission documents (at www.brc.gov).

Finally, this chapter includes a short discussion of nuclear workforce needs. While clearly a cross-cutting issue we chose to cover it here because of its links to science and technology and to the kinds of entities (universities, national laboratories, reactor vendors, etc.) that would be involved in nuclear energy RD&D.

An appropriately educated and trained workforce is obviously needed, not only to conduct RD&D work but to design, build, and operate all the facilities involved in the nuclear fuel cycle, from mines and enrichment facilities to commercial power plants and storage and disposal facilities.

11.1. Advanced Technologies and the Nature of the Nuclear Waste Management Challenge

All of the commercial nuclear reactors operating in the United States today and the vast majority of reactors operating worldwide are light-water reactors operating on the once-through fuel cycle. This means ordinary water serves as the reactor coolant and enriched uranium is used only once in the reactor core and is then stored pending final disposition (as opposed to undergoing reprocessing to separate still usable constituents for re-use as reactor fuel).

Technologies exist today or are under development that would allow spent fuel to be at least partly re-used; systems have also been proposed that could—in theory and at some point in the future—possibly allow for the continuous recycle of reactor fuel, thereby fully "closing" the fuel cycle. Substantial uncertainties exist, however, about the cost and commercial viability of the more advanced of these technologies; in addition, significant concerns have been raised about their impacts on weapons proliferation risks and other aspects of the fuel cycle (e.g., the production of LLW) even if they could be successfully deployed. Without getting into these debates, the central point for purposes of this discussion is that expanded deployment of reprocess and recycle technologies would clearly affect the quantity and composition of nuclear material slated for final disposition and in this way have implications for managing the back end of the fuel cycle.

At the same time, technological advances also hold promise for improving the efficiency and resource utilization of the once-through fuel cycle. To the extent that these improvements make it possible to increase the quantity of electricity produced for every unit of reactor fuel used, this will also have an impact on the overall quantity of spent fuel generated to meet a given level of nuclear power demand.[271] Thus, a central question for the Commission was whether any currently available reactor and fuel cycle technologies, or any not-yet commercial technologies that are now under development, have the potential to change either the fundamental nature of the nuclear waste management challenge this nation confronts over the next several decades or the approach the United States should take to implement a strategy for the storage and ultimate disposition of SNF and high-level radioactive waste.

To answer this question the Commission reviewed the most authoritative available information on advanced reactor and fuel cycle technologies, including the potential to improve existing light-water reactor technology and the once-through fuel cycle, as well as options for partially or fully closing the nuclear fuel cycle by reprocessing and recycling SNF.

We concluded that while new reactor and fuel cycle technologies may hold promise for achieving substantial benefits in terms of broadly held safety, economic, environmental, and energy security goals and therefore merit continued public and private R&D investment, *no currently available or reasonably foreseeable reactor and fuel cycle technology developments—including advances in reprocessing and recycling technologies—have the potential to fundamentally alter the waste management challenge this nation confronts over at least the next several decades, if not longer.*[272] Put another way, we do not believe that today's recycle technologies or new technology developments in the next three to four decades will change the underlying need for an integrated strategy that combines safe storage of SNF with expeditious progress toward siting and licensing a disposal facility or facilities. This is particularly true of defense HLW and some forms of government-owned spent fuel that can and should be prioritized for direct disposal at an appropriate repository.

The above conclusion rests on several practical observations. First, the United States has a large existing inventory (on the order of 65,000 metric tons) of spent fuel and will continue to accumulate more spent fuel as long as its commercial nuclear reactor fleet continues to operate. In addition, the U.S. inventory includes materials with a very low probability of re-use under any scenario, including high-level radioactive waste from past nuclear weapons programs and some forms of government-owned spent fuel. Second, the timeframes involved in developing and deploying *either* breakthrough reactor and fuel-cycle technologies *or* waste disposal facilities are long: on the order of multiple decades even in a best-case scenario. Given the high degree of uncertainty surrounding prospects for successfully commercializing advanced reactor and fuel cycle concepts that are, for the most part, still in the early R&D phases of development it would be imprudent to delay progress on developing disposal capability—especially since that capability will be needed under any circumstances to deal with at least a portion of the existing HLW inventory. The final and most important point, which further strengthens this conclusion, is that all nuclear energy systems generate waste streams that require long-term isolation from the environment: nuclear fission creates radioactive fission products.

Our conclusion concerning the need for geologic disposal capacity stands independently of any position one might take about the desirability of closing the nuclear fuel cycle in the United States. The Commission could not reach consensus on that question. *As a group we concluded that it is premature at this point for the United States to commit irreversibly to any particular fuel cycle as a matter of government policy given the large uncertainties that exist about the merits and commercial viability of different fuel cycles and technology options. Rather, in the face of an uncertain future, there is a benefit to preserving and developing options so that the nuclear waste management program and the larger nuclear energy system can adapt effectively to changing conditions*. Future evaluations of potential alternative fuel cycles must account for linkages among all elements of the fuel cycle (including waste transportation, storage, and disposal) and for broader safety, security, and non-proliferation concerns.

To preserve and develop those options, we believe RD&D should continue on a range of reactor and fuel cycle technologies, described in this report, that have the potential to deliver societal benefits at different times in the future. If and when technology advances change the balance of market and policy considerations to favor a shift away from the once-through fuel cycle, that shift will be driven by a combination of factors, including—but hardly limited to—its waste management impacts. *In fact, safety, economics, and energy security are likely to be more important drivers of future fuel cycle decisions than waste management concerns per se*. We also note that other elements of our proposed approach to managing the back end of the fuel cycle—including, notably, our recommendations concerning the need to move forward with consolidated storage capacity—will provide the flexibility needed to take full advantage of advanced technologies if and when these technologies materialize.

The remainder of this chapter summarizes the Commission's findings with respect to the potential benefits and trade-offs associated with different broad categories of advanced nuclear reactor and fuel cycle combinations. We also briefly discuss the need for continued public and private investment in nuclear energy R&D, the status of DOE's nuclear energy R&D program, and the adequacy of existing regulatory and legal frameworks to accommodate new types of technologies and facilities.

11.2. Results of a High-Level Comparison of Reactor and Fuel Cycle Alternatives

As directed by our charter, the Commission undertook to evaluate existing fuel cycle technologies and R&D programs in terms of a set of broad criteria that will have a critical influence on the nuclear energy industry's prospects going forward (e.g., safety, cost, security, etc.). In doing so, we relied on the numerous studies that have been undertaken in the last decade to assess and compare various reactor and fuel cycle options.[273] It is important to emphasize that the Commission could not and did not attempt to draw definitive or quantitative conclusions about the relative merits of different technology combinations. This is because the numerous studies we considered—although collectively they analyze a wide array of strategies and technologies—use often very different underlying parameters and assumptions. As a result, the quantitative results of these studies are not comparable. Additionally, many of the potential technologies require considerable development before a defensible comparison could be made. Thus, it is impossible at this time to distill quantitative comparisons across alternative nuclear energy systems and then draw definitive conclusions based on those comparisons.

We approached the task of comparing advanced nuclear energy systems by first identifying three representative alternatives to the once-through light water reactor (LWR) strategy. One of these alternatives is already in use; the other two are substantively different from the once-through cycle and have received extensive previous study. We then focused on the major qualitative differences between these alternatives and the existing once-through LWR fuel cycle, based on the findings contained in the literature available to the

Commission. The results, which are summarized in table 5, indicated a wide range of trade-offs in terms of safety, cost, resource utilization and sustainability, waste management, and the promotion of nuclear non-proliferation and counter-terrorism goals. These trade-offs complicate any effort to compare the relative merits of different nuclear energy systems, particularly given uncertainty about future technological developments and social conditions.

Moreover (and as we have already noted) the conclusions reached by different technology assessments and comparative analyses are heavily influenced by input assumptions and by the relative weight given to different policy objectives (e.g., reducing waste vs. minimizing proliferation risk vs. maximizing resource utilization)—making it difficult to compare results across studies.

The four systems (one baseline plus three alternatives) considered in our qualitative comparison are characterized as follows:

- *Once-through fuel cycle with light-water reactor technology*: We chose this system as the baseline because it is the dominant fuel-cycle and technology combination currently in use in the United States and in the majority of the world's nuclear nations. That said, future technology advances can be used to improve on this system; an example might include the ability to achieve higher fuel burnup using improved cladding and improved safety features.
- *Modified open cycle using mixed-oxide (MOX) fuel with light-water reactor technology:* This system was selected for comparison chiefly because it is the only alternative fuel cycle strategy that is currently being utilized on a commercial scale.[274] Used in France since the 1970s, MOX fuel is also used in reactors in Germany, Switzerland, Belgium and Japan. The United States is currently building a MOX fuel fabrication facility in South Carolina to utilize excess defense plutonium, and the United Kingdom, China, and Russia are also in various stages of operation or planning for the use of MOX fuel.
- *Closed fuel cycle system with fast reactors*: This system was considered because it has the theoretical potential to maximize the use of uranium resources and therefore to be sustainable for centuries while simultaneously reducing the amount of long-lived radionuclides in resulting waste streams. Lower amounts of radionuclides would allow a larger quantity of waste to be placed in a given repository; in addition, this fuel cycle would greatly reduce uranium mining requirements and eventually eliminate the need for uranium enrichment.
- *Once-through fuel cycle with high-temperature reactors*: The defining feature of this fourth system is a high-temperature reactor that can achieve temperatures greater than 600°C (light water reactor outlet temperatures are about 300°C).[275] It was selected because it has the potential to displace the use of fossil fuel across all energy sectors, not just electricity production. Examples of energy-intensive industries where high-temperature nuclear process heat could be used include cement and steel manufacturing, and petroleum refining. High-temperature nuclear process heat could also be used to produce hydrogen for transportation fuels by directly decomposing water instead of using electrolysis or decomposing natural gas, and the high power conversion efficiency can also make dry cooling and thermal desalination of seawater practical.

Many additional system options exist that have received varying levels of study. For example, nuclear energy systems that involve a fast-spectrum reactor capable of achieving very high temperatures by using a molten salt or gas coolant, or a thermal-spectrum, high-temperature molten-salt reactor using thorium have also been proposed.

Such systems could potentially offer many of the combined benefits of the alternatives listed. However, these systems have not received systematic study and the component technologies for these types of systems are less well developed. Other concepts, such as fusion energy, are even further from being successfully demonstrated—but if they ever prove feasible they may have even larger impacts on fuel cycles and nuclear waste generation.

The results of this comparison for the baseline strategy and the three nuclear energy systems selected for comparison are shown in table 4. The entries in this table generally refer to a steady-state condition. The Commission recognizes that in some cases a long transition time is necessary to reach a steady state. Each of the four nuclear energy systems is assumed to produce the same amount of electric power and the outcomes are stated in relative terms in relation to the baseline strategy.

Table 4. A comparison of the existing once-through, conventional light-water reactor fuel cycle with representative advanced nuclear energy systems in the long term[276]

Criterion	Once-through lwR	Once-through with High-temperature Reactor
Nuclear energy Description	Clad uranium oxide fuels irradiated in lwRs with evolutionary improvement	High-temperature reactors (such as those using graphite based Fuels) capable of temperatures over 600°C operating On a once-through fuel cycle. Being pursued in Doe's next Generation nuclear Plant project
		Safety
Reactor and fuel cycle safety277	Baseline, with potential for further improvement	Potential for improvement; all must meet similar regulatory Requirements
		Cost
Capital and operating costs	Baseline	Test reactors have operated well, but demo (Fort st. Vrain) was Unreliable. Fuel costs are uncertain and may be high. RD&D Is needed on to provide a basis for design, licensing, and Evaluating long-term economic viability.

Criterion	Once-through lwR	Once-through with High-temperature Reactor
		Sustainability
uranium utilization[279]	Baseline	similar uranium requirements although can vary by design
Climate change impacts	Baseline	Potential for major reduction in carbon dioxide by using nuclear process heat in fossil-energy-intensive industries and to produce hydrogen for non-carbon-based transportation fuels
energy security	Baseline	Potentially large benefit in reducing petroleum imports now used to fuel non-electricity sectors
	Non-proliferation and counter-terrorism	
non-proliferation	Baseline	Reference designs require similar enrichment capacity capable of producing 8%-20% uranium enrichment. Fuel is more difficult to reprocess than lwR fuel.
Counter-terrorism	Baseline	similar to baseline
	Waste management	
Disposal safety: toxicity and longevity of waste	Baseline	Repository: similar to baseline Fuel Cycle: similar public and occupational risk from mining and milling
Volume of waste280	Baseline	~10X increase in snF volume going to repository. About the same non-mill tailings llw as baseline.
Repository space requirements	Baseline	~25% reduction due to higher reactor efficiency.tae 4. A comparison of the existing once-through, conventional Light-water reactor Fuel cycle with representative advanced Nuclear energy systems in the long term[276]

lwR modified open Cycle	Fast-spectrum Reactor with Closed Fuel Cycle
Clad uranium- and mixed-oxide fuels irradiated in lwRs with evolutionary improvements. moX fuel is irradiated once and then sent to repository.	Fast-spectrum liquid-metal-cooled reactors capable of continuous recycle of actinides
	Safety
Potential for improvement; all must meet similar regulatory requirements	Potential for improvement; all must meet similar regulatory requirements
	Cost
Capital cost increased because of need to build reprocessing and moX fuel fabrication plants. Operating costs also increased due to the high cost of fabricating fuels containing Pu. Cost of electricity increased a few to several percent. technology is relatively mature with evolutionary improvements largely in the hands of industry.	Previously built reactors (mostly prototype/demo) were often unreliable and not economic. significant capital cost for recycle facilities. RD&D is needed to provide a basis for design, licensing, and evaluating long-term economic viability.[278] operating costs relative to baseline largely depend on the future price of uranium, fuel fabrication cost, and operational reliability.
	Sustainability
~19% reduction in uranium requirements	~95% reduction in uranium requirements
About the same as the baseline	About the same as baseline
About the same as the baseline	modest benefit from potential for long term reliance on indigenous uranium resources
	Non-proliferation and counter-terrorism

Table 4. (Continued)

Criterion	Once-through lwR	Once-through with High-temperature Reactor
involves use of reprocessing, enrichment, and moX fuel fabrication technology, and deployment of facilities for same. increased proliferation risk from substantial normalized inventory of Pu or Pu-plus other actinides in reactors and the fuel cycle.		involves use of reprocessing and plutonium-bearing fuel fabrication technology, and deployment of facilities for same. Enrichment technology needed during transition to fast reactors. Increased proliferation risk from substantial normalized inventory of Pu or Pu-plus other actinides in reactors and the fuel cycle.
involves production and inventory of co-processed nuclear materials (u/np/Pu) and 5%-10% enriched uranium, and fuels containing same. Increased security risk due to separated materials and additional facilities and transportation.		involves production and inventory of co-processed nuclear materials (u/np/Pu) and fuels containing same. Increased security risk due to separated materials and additional facilities and transportation.
		Waste management
Repository: noticeable reduction in the amount of tRu in wastes. tailored waste form for ~90% of the Hlw Fuel Cycle: 15%–20% reduction in fuel cycle public and occupational risk from reduced mining and milling. Although there is an increase in emissions from reprocessing, overall risk is reduced as a result of reduced risks on the front end.		Repository: tailored waste form for fission products; potential for reduction in long-term repository dose from tRu elements if recycle is sustained for decades to centuries Fuel Cycle: ~85% reduction in fuel cycle public and occupational risk from reduced mining and milling, increase from emissions from reprocessing
similar repository waste volume: less snF/Hlw, more secondary waste. ~20% decrease in near-surface wastes, esp. mill tailings and depleted uranium. Besides mill tailings and depleted uranium, about the same amount of llw as baseline.		~40% increase in repository waste volume: less Hlw, more secondary waste. ~95% decrease in near-surface wastes, primarily due to mill tailings and depleted uranium. ~40% decrease in non-mill tailings llw due to greatly reduced throughput in the front end of the fuel cycle.
similar to baseline, with some reduction in long-term decay heat generation.		~75% decrease in repository space required when tRu are recovered and recycle is sustained over many decades to a couple of centuries.

The fact that there are no clear winners among the main alternative fuel cycles summarized in the table and others considered by the Commission suggests that the United States should pursue a policy of keeping multiple options open. That said, certain fuel cycle strategies and technologies are clearly better developed than others—research in some areas has been underway for decades and it is possible that more mature technologies could be implemented more quickly, perhaps within a few decades. Other concepts are barely at the proof-of-principle stage and would require substantial investments of time and funding (and in some cases a number of revolutionary technical developments) to bring them to a level of maturity sufficient to evaluate their suitability for further development and potential implementation. Consequently, the level and duration of R&D effort needed to advance these concepts varies widely. Ironically, near-term funding needs for technologies that are relatively more developed can be greater than for technologies still in an earlier phase of the RD&D process—particularly in the case of technologies that are ready to be demonstrated. At that point, large investments may be needed to provide the demonstration facilities required to make further progress. In the next section, we explore U.S. nuclear energy R&D plans and programs and offer suggestions for addressing the challenges facing those programs.

11.3. The Case for Continued Public and Private Investment in Nuclear Energy Rd&D and the Status of the Current Doe Program

The results of our qualitative assessment suggest that while it is too early to select "winners," advanced nuclear energy systems could offer a range of benefits in terms of broadly held policy goals with respect to safety, cost, security, etc. In a world facing rising energy demand and significant resource and environmental concerns, including the threat of climate change, preserving an improved nuclear energy option could be extremely valuable. Therefore, the Commission concludes that the United States should continue to pursue a program of nuclear energy RD&D both to improve the safety and performance of existing nuclear energy technologies and to develop new technologies that could offer significant advantages in terms of the multiple evaluation criteria identified in our charter (i.e., safety, cost, resource utilization and sustainability, waste management, and non-proliferation and counterterrorism). We believe a well-designed federal RD&D program is critical to enabling the U.S. to regain its role as the global leader of nuclear technology innovation and should be attentive to opportunities in two distinct realms:

1. Near-term improvements in the safety and performance of existing light-water reactor technology as currently deployed in the United States and elsewhere as part of a once-through fuel cycle, and in the technologies for storing and disposing of SNF and HLW.
2. Longer-term efforts to advance potential "game-changing" nuclear technologies and systems that could achieve very large benefits across multiple evaluation criteria compared to current technologies and systems. Examples might include fast-spectrum reactors demonstrating passive safety characteristics that are capable of continuous actinide recycling and that use uranium more efficiently, or reactors that—by using molten salt or gas coolants—achieve very high temperatures and can thereby supply process heat for hydrogen production or other purposes, or small modular reactors with novel designs for improved safety characteristics and the potential to change the capital cost and financing structure for new reactors.

At the same time the federal government and nuclear industry should ensure that necessary efforts and investments are made to maintain valuable existing capabilities and assets, including critical infrastructure and facilities such as are present at a number of DOE sites, as well as vital human expertise (in the form of highly trained nuclear scientists, engineers, and workers, etc.).

In making these recommendations, the Commission is mindful that federal RD&D funding of all kinds will be under enormous budget pressure in the years ahead. It will therefore be especially important to focus scarce public resources on addressing key gaps or needs in the U.S. nuclear RD&D infrastructure and to leverage effectively the full range of resources that exist in industry, the national laboratories, and the academic community. This could include funding well-designed, multipurpose test facilities that can be used to advance knowledge in several areas of inquiry. Such facilities would be available to scientists from different institutions around the country (an example is the Advanced Test Reactor National Scientific User Facility at Idaho National Laboratory) and exemplify the kind of RD&D infrastructure that could yield particularly high returns on public investment. Furthermore, while this Commission is charged with making recommendations to the government, we also want to clearly emphasize the importance and value of industry RD&D efforts, such as those of the Electric Power Research Institute, and the importance of continuing and stable industry RD&D investment in reactor and fuel cycle technologies.

In recent years, DOE's budget for nuclear energy R&D has totaled approximately $500 million per year. The Commission is not making a specific recommendation with respect to funding levels in future years, recognizing that this is a decision that will have to be made in the context of larger energy policy considerations and increasingly difficult federal budget constraints.

Generally speaking, however, the Commission concurs with recent findings issued by the President's Council of Advisors on Science and Technology concerning the need for better coordination of energy policies and programs across the federal government; for a substantial increase in federal support of energy-related research, development, demonstration, and deployment; and for efforts to explore new revenue options to provide this support.[281] Meanwhile, with federal discretionary budgets under increasing pressure, the ability to articulate a clear direction or agenda for the U.S. nuclear energy R&D program, to prioritize elements of that agenda, and to set performance objectives and evaluate the effectiveness of related activities on an ongoing basis will obviously be critical.

To that end, the Commission believes that DOE's nuclear energy R&D Roadmap provides a good, science-based step toward the development of an effective, long-term RD&D program. The Roadmap should be periodically updated in the future (we recommend once every four years) and in the process should be informed by broader strategic planning efforts, such as DOE's recently launched Quadrennial Technology Review and Quadrennial Energy Review process.

In addition, it should explicitly apply the evaluation criteria noted in the BRC's charter and it should build in the flexibility needed to respond to unexpected technology developments and changing societal concerns and preferences. (The recent and still-unfolding events at the Fukushima Daiichi nuclear power plant are just one example of the type of development that should be reflected in future updates of the roadmap.) Finally, we urge DOE to support future versions of the Roadmap with more detailed, frequently updated, and transparent research and implementation plans.

Additional principles or objectives that should guide DOE's approach to nuclear energy RD&D in the future include the following:

- System assessments and evaluations must account for the interconnections among the various elements of the nuclear fuel cycle (including transportation, storage, and disposal) and for broader safety, security, and nonproliferation concerns. For example, adding facilities to one phase or section of the nuclear fuel cycle could change overall system costs or otherwise affect the performance of the system as a whole. RD&D investment and technology choices can be made most effectively only if the interconnections between and among the elements of the fuel cycle system are well understood.
- Nuclear energy RD&D going forward will continue to involve a broad range of participants including universities, industry, and national laboratories in cooperation with international research partners. Integrating the efforts of these disparate participants will require a concerted effort and is essential if DOE is to maximize the

value of the RD&D it supports. DOE should undertake efforts to strengthen coordination and organizational and mission alignment across laboratories, energy hubs, innovation centers, and other entities.

- Federal cost sharing with industry to license new reactor designs has been extremely successful and should be pursued where practical. Indeed, federal support has bolstered U.S. technical leadership in the nuclear energy arena generally and played a role in developing the state-of-the-art AP-1000 and ESBWR[282] reactor designs specifically. These designs employ the most advanced passive safety systems developed to date.
- Safety concerns, along with nuclear weapons proliferation and nuclear material safeguards and security (discussed in the following chapter), deserve special attention in the R&D roadmap and in plans for demonstration facilities. Integrating safety, security and safeguards considerations in future evaluations of advanced nuclear energy systems and technologies will allow the United States to maintain consistency between its technology development agenda, its commercial interests, and its international policy agenda.
- As a result of the focus on repository design issues specific to the Yucca Mountain site, generic R&D on deep geologic disposal for the last few decades has been assigned a lesser priority within DOE's R&D portfolio. The move by DOE to absorb the R&D responsibilities of the Office of Civilian Radioactive Waste Management into the Office of Nuclear Energy presents an opportunity to better integrate waste management considerations into the DOE nuclear energy research agenda.
- Going forward, the nuclear energy R&D program should include an emphasis on the development of disposal and waste form alternatives that are optimized to work with potential natural and engineered barriers in the disposal system. If alternative nuclear energy systems are deployed in the future, however, they will likely generate a greater variety of waste streams than the current system. Efforts to manage these wastes will benefit from an improved understanding of different combinations of geologic disposal environments, engineered barriers and waste forms.

Finally, one area outside the DOE RD&D program where the Commission has identified a specific need for increased funding involves ongoing work by the NRC to develop a regulatory framework for novel components of advanced nuclear energy systems. This is a priority because a regulatory framework can help guide the design of new systems and lower barriers to commercial investment by providing greater confidence that new systems can be successfully licensed.

In its draft report to the full Commission, the Reactor and Fuel Cycle Technology Subcommittee recommended that 5 to 10 percent of total federal funding for reactor and fuel cycle technology RD&D be directed to the NRC to support its work in this area; the Commission agrees that adequate funding for this activity should be provided. We also support the NRC's current risk-informed, performance-based approach to developing regulations for advanced nuclear energy systems.

11.4. Workforce Development

The effective conduct of the U.S. nuclear enterprise (whether that enterprise is expanded, maintained at the current level, or diminished in the future) will require a properly trained workforce, including scientists and engineers in many disciplines as well as skilled workers for site evaluation, construction, operation, decommissioning and closing nuclear facilities—including facilities in the nuclear waste management program. At the professional level there has been healthy growth in the number of students pursuing a nuclear engineering career over the last decade. Several factors account for this, including the availability of federal funding and a recent increase in the number of new plants proposed or under construction in the United States and around the world. In addition, there has been noteworthy progress in developing programs to prepare skilled labor from many different building trades and crafts for the entire spectrum of work at nuclear facilities. Finally, the training available to first responders in other sectors has resulted in improved capabilities for responding to transportation accidents as well as incidents at fixed facilities involving hazardous materials, including radioactive wastes.

Nevertheless, workforce needs in the nuclear industry and in other high-tech sectors of the U.S. economy are expected to grow in the coming years. According to a 2008 report prepared by the Directors of the National Laboratories: "

A recent industry study pointed out that over the next five years, half of the nation's nuclear utility workforce will need to be replaced." The Directors called for both government and industry actions to support the development of the future nuclear workforce.[283] Based on testimony presented to the BRC Subcommittee on Reactor and Fuel Cycle Technology, the Commission concurs with this general finding. We recommend expanded federal, joint labor-management, and university-based support for advanced science, technology, engineering, and mathematics training to develop the skilled workforce needed to support an effective waste management program as well as a viable domestic nuclear industry.

12. International Issues

> The United States has long been a global leader in the development of nuclear technologies and policies and in international efforts to address issues of nuclear security and safety.

Throughout its deliberations the Commission has been acutely aware of the international implications of future U.S. actions with respect to innovation in nuclear technologies and the management of the back end of the nuclear fuel cycle.

In our view, international collaborations and considerations are especially important in three areas: safety, non-proliferation, and nuclear security (counterterrorism). This chapter highlights our main conclusions and recommendations in each of these areas.

Unfortunately, our failure to develop a broadly-accepted domestic spent fuel storage and disposal strategy has limited our non-proliferation policy choices in the context of nuclear fuel cycles. In addition to supporting non-proliferation objectives, our international nuclear safety goals will also be served by establishing and implementing effective waste management strategies.

Overall, we believe the United States must continue to strengthen its leadership role on the world stage to assure the safe, secure, and responsible application of nuclear technology, particularly if rising resource demands coupled with climate change concerns prompt a significant global expansion of nuclear capacity in coming decades.

12.1. International Nuclear Safety

Recent events in Japan have reinforced the importance of a focus on nuclear safety. Although the radiological releases in Japan will have no direct impacts of significance on the United States, the events at Fukushima are certain to affect attitudes toward nuclear technology here and abroad. Even if the health consequences of the Fukushima accident prove to be small compared to the direct impacts of the earthquake and tsunami, economic ramifications—including the permanent loss of contaminated land and six costly reactors—and the potential danger of a nuclear accident remain abiding public concerns. These concerns must be directly and forthrightly addressed.

At a minimum, events in Japan will have to be carefully scrutinized to see what can be learned from them and to identify any needed changes in the U.S. regulatory system. Insights gained from Fukushima should also have an influence on the direction of research and development efforts and on the design of advanced nuclear energy systems.

Events in Japan also reinforce the need for expanded international efforts to promote the safe operation of existing and planned nuclear installations, including facilities for spent fuel storage and disposal. A significant expansion of nuclear power is planned in the years ahead in countries such as China, Russia, India, and Korea. Over 60 countries that do not currently have nuclear power plants have approached the IAEA to explore the possibility of acquiring one and the IAEA anticipates that about 15 of these emerging nuclear nations will proceed over the next decade or two. Several of these "new-entrant" countries have already committed to construction. All will have to provide for safe storage and disposal of their nuclear fuel as part of a larger commitment to ensure the safety of all nuclear facility operations.

Safety is an inescapable, continuing, expensive, and technologically sophisticated demand that all new entrants to commercial nuclear power will have to confront over the full lifecycle of these systems—from preparing for construction through decommissioning.

The nature and scope of the safety challenges involved might not be fully apparent to new entrants. Managing these challenges requires that robust institutional, organizational and technical arrangements be in place at the very early stages of a nuclear program. Also needed are sufficient technical knowledge and experience, strong management, continued peer-review and training, and an enduring commitment to excellence and a robust safety culture. Many countries will not initially be able to obtain the needed level of expertise and experience on their own. Thus, relevant international organizations and industry groups should expand the assistance available to such countries as they tackle the planning, design, construction, operation and regulation of nuclear energy systems.

All nations that have or plan to construct nuclear reactor facilities will, of course, also face the paramount task of providing for the safe storage and ultimate disposition of spent fuel. Here again, international efforts are needed to help new-entrant countries successfully manage these challenges.

The capacity to pursue nuclear technology in the United States will depend to a large extent on other countries' success in achieving a high level of safety performance. Many of these countries have not yet demonstrated that they have the infrastructure or the commitment to a safety culture that provides confidence they will succeed. Moreover, as the events at Fukushima have shown, even a nation that has successfully operated nuclear reactors can have difficulties preparing for and responding to major accidents. Since Fukushima, the international community has worked to strengthen global nuclear safety, particularly through the work of the IAEA and its focus on enhanced international safety standards and expanded safety services. However, the Commission does not believe the task of enhancing global reactor safety should fall to the IAEA alone. In that spirit, we strongly support recent actions undertaken by the World Association of Nuclear Operators (WANO), which shortly after the Fukushima accident asked every nuclear power plant operator in the world to take "specific actions to verify their ability to deal with a station blackout or a beyond-design-basis event like fire or flood."[284] In addition, the WANO Governing Board established a high level commission tasked with recommending changes to WANO programs and organization as a result of lessons learned from Fukushima. That commission has recommended that WANO better define its roles and responsibilities in an emergency situation and develop a worldwide integrated event response strategy. It has also urged WANO to add emergency preparedness as a core review area to each and every WANO peer review, to conduct peer reviews for all initial reactor start ups, and to look more closely at fuel storage—including fuel pools and dry cask storage.[285] In addition to the work being done by the operators of nuclear power plants, efforts by international nuclear vendors to embrace additional and voluntary self-governance regimes have also gained traction. Several "principles of conduct"[286] recently adopted by a number of the world's leading civilian nuclear technology vendors describe a commitment to undertake good faith efforts in several areas: safety, security, environmental protection, compensation for nuclear damage, nonproliferation, and ethics. Participating vendors express their intention to follow these principles in designing and exporting nuclear power plant technologies. These principles are based upon practices de rived from the experience of power plant vendors and operators and IAEA standards.[287] *The Commission supports these efforts and encourages industry to continue pushing forward with the implementation of global best practices across a range of issues related to the safe and secure operation of nuclear energy facilities.*

In sum, the United States should work with the IAEA and other interested nations to launch a major international effort, encompassing international organizations, regulators, vendors, operators, and technical support organizations, to enable the safe application of nuclear energy systems and the safe management of nuclear wastes in all countries that pursue this technology. The United States should also participate in other new and ongoing IAEA initiatives to address safety challenges. Finally, we believe DOE and NRC should be explicitly directed and funded to offer nuclear safety assistance and guidance to new-entrant countries who request it.

12.2. Non-Proliferation Considerations

Because enrichment, reprocessing and recycled fuel fabrication facilities typically produce or utilize large amounts of separated materials (including enriched uranium and plutonium) as part of their operations, they present higher proliferation risks and are therefore considered particularly sensitive elements of the fuel cycle. The technologies used in these facilities can not only serve nuclear power needs, but can give countries the technical and physical capacity to obtain the direct-use nuclear materials required for a weapons program. Proliferation risks are varied: they may include the potential for countries to secretly divert materials from civilian nuclear facilities that they have declared to the IAEA under the NPT, or the potential for countries to apply know-how and equipment from declared programs to the construction of clandestine production facilities (e.g., clandestine enrichment plants). Finally, there is the risk that under some circumstances countries might withdraw from the NPT and then overtly misuse materials and facilities.

A number of institutional and technical approaches exist under the NPT and other international and bilateral agreements to address these risks. These include the application of IAEA safeguards to detect the diversion of nuclear materials in a timely manner and to verify peaceful uses of declared civil nuclear energy infrastructure; the IAEA's ability to verify the absence of clandestine production facilities in countries that have ratified the IAEA Additional Protocol; international agreements by nuclear supplier nations to apply export controls to detect and prevent transfers of dual-use equipment to clandestine production facilities; the use of national technical means and human intelligence to detect clandestine production efforts; and initiatives aimed at developing international fuel cycle facilities as a way to provide emerging nuclear energy nations with reliable and affordable access to fuel enrichment and reprocessing services without the need to develop indigenous capacity; and the international system of bilateral and multilateral security and mutual defense agreements that reduce regional security concerns that could otherwise lead some countries to seek nuclear weapons capability.

None of these measures offers a perfect solution to the problem of nuclear proliferation, but together they can help reduce proliferation risks to a manageable level. The ability to identify and isolate non-compliant programs by itself can help ensure that problem countries do not come to be viewed as role models by other emerging nuclear energy nations. In the sections that follow we review the main elements of the current international non-proliferation regime and offer recommendations for improving and strengthening these elements through further U.S. investments and policy leadership.

12.2.1. The Treaty on the Non-Proliferation of Nuclear Weapons (NPT)

The NPT provides the foundation of the international nuclear non-proliferation regime. Opened for signature in 1968, the Treaty entered into force in 1970. It currently has 189 signatories,[288] divided between nuclear weapon states (NWS) and non-nuclear weapon states (NNWS). Virtually all states in the international system have signed and ratified the treaty: only Israel, India, and Pakistan have declined to sign, and North Korea is the only state that has joined the treaty but later exercised its right to withdraw.

The NPT is designed to promote three main objectives: to limit the spread of nuclear weapons, to encourage eventual nuclear disarmament, and to provide a framework and enable widespread access to peaceful uses of nuclear energy. The key provisions of the NPT therefore outline rights and responsibilities for state parties in the area of nuclear non-proliferation, nuclear energy, and disarmament.[289] Article I states that no NWS may "transfer, assist, encourage or induce" any NNWS to "manufacture or otherwise acquire nuclear weapons." Article II requires NNWS parties not to "receive, manufacture or otherwise acquire" nuclear weapons and "not to seek or receive any assistance in the manufacture of nuclear weapons." Article IV protects the right of all states to peaceful nuclear energy, conditional on their being in compliance with their Article II commitment: "Nothing in this Treaty shall be interpreted as affecting the inalienable right of all the Parties to the Treaty to develop research, production and use of nuclear energy for peaceful purposes without discrimination and in conformity with Articles I and II of this Treaty." Article VI of the NPT calls for all parties to work towards nuclear disarmament: "Each of the Parties to the Treaty undertakes to pursue negotiations in good faith on effective measures relating to cessation of the nuclear arms race at an early date and to nuclear disarmament." As noted above, Article VI is often treated as exclusively applicable to NWS, though it clearly states that *each of the parties* to the treaty must pursue "negotiations in good faith"[290] in pursuit of nuclear disarmament.

Although the NPT provides a legal framework for the global non-proliferation regime, the workhorse of the regime has been the IAEA safeguards system. This system is used to verify NPT compliance and to affirm that governments are not using civil nuclear energy programs for nuclear-weapons purposes.

All signatories to the NPT are required to have a comprehensive safeguards agreement (CSA) in place. These CSAs cover "all source or special fissionable material in all peaceful nuclear activities within the territory of a State, under its jurisdiction, or carried out under its control anywhere."[291] Because IAEA safeguards depend on correct and complete declarations of countries' nuclear materials and activities, CSAs play an important role in verifying country reports. Typically, they rely heavily on nuclear material accounting measures, complemented by containment and surveillance techniques such as tamperproof seals and cameras that the IAEA installs at facilities. Verification measures include on-site inspections, visits, and ongoing monitoring.

Unfortunately, some events of the last several decades have challenged the efficacy and credibility of CSAs. In particular, Iraq's engagement in a clandestine nuclear weapons program from the mid-1980s to the early 1990s violated its safeguards obligations under the NPT. In response, the IAEA broadened the scope of materials and facilities covered by the safeguards and strengthened safeguards techniques.[292]

In 1992, the IAEA Board of Governors reaffirmed the agency's authority to conduct "special inspections" of suspected undeclared sites in NPT non-nuclear weapon states and in 1997 the IAEA Board of Governors adopted a new safeguards model. Known as the "Additional Protocol" or AP, the protocol gave IAEA inspectors increased access to all aspects of a non-nuclear weapon state's nuclear program, even where nuclear material is not involved; required states to provide more detailed information on their nuclear program; allowed for the use of improved verification technologies (i.e., environmental sampling); and required more extensive inspections at declared nuclear sites.[293] There are currently 104 countries with Additional Protocol agreements in force but some key countries, like Iran, have refused to ratify the AP.[294]

Even with the Additional Protocol in place, plans to expand global nuclear energy production and concerns over the spread of sensitive nuclear technologies are placing increased strain on international safeguards. To the extent that this expansion involves new reprocessing and enrichment facilities, one of the most vexing challenges for the safeguards system will be verifying physical materials at these facilities. In large bulk-handling facilities with high volume throughputs (hundreds to thousands of metric tons) and complicated equipment schematics, material unaccounted for or "MUF" can represent a substantial proliferation challenge. Even as a small percentage of facility throughput, the quantity of material unaccounted for can be significant. Over the last 15 years, material accounting efforts have failed on multiple occasions to detect and resolve anomalies in a timely fashion. These lapses have involved large amounts of MUF that remained unresolved for months, years, or even decades.[295]

The Commission endorses R&D efforts on modern safeguards technologies and urges continued U.S. government support for IAEA's work in this area. The National Nuclear Security Administration is the principal federal sponsor of nuclear non-proliferation-related research and development and is currently (in conjunction with the national laboratories) supporting work on safeguards systems analysis and enhancements, safeguards-by-design, nuclear material control and accountability (MC&A) improvements, modern inventory controls, software and hardware development, collaborative information technology tools, and real-time process monitoring and data integration systems. *Support for the development of novel safeguards technologies is not only imperative because of the fundamentally important nature of the threat, but because of compounding issues related to the development of these technologies.* The IAEA finds itself constrained financially, lacking the resources to perform research and development on the necessary technologies, while tasked with ever-increasing responsibilities.[296] In addition, the size of the "safeguards market" simply doesn't support the cost-effective production of units or major R&D investments by commercial players. As a result, the IAEA remains reliant on the R&D efforts of national governments.

12.2.2. Multilateral/Multi-National Fuel Cycle Services Options

Proposals for "multi-nationalizing" certain fuel cycle facilities or operations as a way to provide access to sensitive parts of the nuclear fuel cycle are not new and have been discussed in multiple forms since the 1946 Acheson-Lilienthal report and Eisenhower's 1953 Atoms for Peace speech.[297] Striking a balance between the reliable provision of fuel supply services on the one hand and guaranteeing adherence to non-proliferation norms on the other hand is difficult, to say the least. In concert with the IAEA, several countries, including the United States, have proposed an array of strategies to provide countries with credible, cost-efficient options for an assured nuclear fuel supply, including the development of backup supplies or "fuel banks" of enriched uranium, multi-national fuel cycle centers, and government-to-government agreements.

Today, as shown in figure 20, the majority of nuclear energy programs worldwide are small, with less than 10 GWe of capacity (fewer than 10 reactors). Furthermore, while some new uncertainty has been introduced by the Fukushima accident, the number of countries with small nuclear energy programs is still widely expected to grow further. In 2011, Iran's first power reactor reached criticality at Bushehr, adding another country to the list shown in figure 20. In addition, 65 more countries are currently participating in IAEA technical cooperation projects related to the introduction of nuclear power. Because most national nuclear energy programs are small, the combined installed nuclear capacity in these countries accounts for less than 15 percent of total global nuclear generation capacity. Given this feature of the current global nuclear energy market, there are compelling practical and economic reasons for countries to make use of regional or multi-national fuel cycle facilities and services, rather than developing their own nuclear fuel cycle capabilities.

In 2004, the Director General of the IAEA appointed an international expert group to consider options for possible multilateral approaches to developing facilities on the front and back ends of the nuclear fuel cycle. Their report, *Multilateral Approaches to the Nuclear Fuel Cycle*, was released in February of 2005;[299] it categorized the options for offering assured fuel supply into three major and distinct categories: assurances of services not involving the ownership of fuel cycle facilities, conversion of existing facilities to multinational facilities, and the construction of new jointly-owned facilities.

Within the first option, it is generally assumed that a functional market exists for whatever fuel service is required, either through state-owned enterprises or commercial enterprises. Of course, market options currently vary across the fuel cycle (i.e., more commercial options exist for enrichment than they do for reprocessing, and none exist for spent fuel and HLW disposal). While a diversity of supply options alone does not necessary reflect the health of a market and its ability to answer demand, it can affect countries' confidence that their ability to access supplies is really "assured." In some cases, the ability to access fuel supplies via existing and perfectly healthy markets is not sufficient for a country to forgo developing its own indigenous fuel cycle development, ostensibly the case in Iran. Assurance of sufficient supply, beyond that available through the normal market, can be strengthened through additional agreements, including by supplier and government consortia and through the IAEA.

As the 2005 IAEA report noted, the advantages and disadvantages of either converting a national facility to an international facility or to building a new multi-nationally managed facility will vary depending on the type of facility being discussed (enrichment, reprocessing, etc.). The advantages of converting an existing facility to multi-national ownership include lower capital investment required, no further dissemination of facility construction know-how, strengthened proliferation resistance due to multi-national management and operating teams, and pooled expertise and resources. Disadvantages include the potential need for additional facilities in politically diverse countries to provide adequate assurance that fuel supplies will not be withheld for ideological reasons, the need to balance existing property rights, potential proliferation risks due to an increased number of international partners, added international management demands, and the potential need to back-fit safeguards depending on the host nation's prior approach.

The advantages to building a new fuel cycle facility under multi-national controls include the ability to incorporate safeguards during construction instead of back-fitting these controls, the ability to pool expertise and resources, the ability to size the facility economically, and the opportunity to strengthen proliferation safeguards. The disadvantages of building new facilities include potentially higher proliferation risks due to broader access to know-how (depending on the management model chosen), uncertain commercial competitiveness, and potential for breakout and retention of fissile materials.

Regardless of the advantages and disadvantages of each of these options, it is clear that cross-cutting technical, legal, cultural, political and financial factors will affect perceptions about their relative feasibility and desirability. These factors may be decisive in any future multilateral or unilateral efforts to develop multi-national fuel cycle facilities.

Longer term, the United States should support the use of multi-national fuel-cycle facilities, under comprehensive IAEA safeguards, as a way to give more countries reliable access to the benefits of nuclear power while simultaneously reducing proliferation risks. We note that the term "multinational fuel cycle facility" is commonly understood to encompass facilities associated with all aspects of the nuclear fuel cycle.

The Commission wishes to stress that our support for multi-national management of such facilities should not be interpreted as support for additional countries becoming involved in enrichment or reprocessing facilities, but rather reflects our view that if these capabilities were to spread it would be far preferable—from a security and non-proliferation standpoint—if they did so under multi-national ownership, management, safeguards, and controls.

The Urenco uranium enrichment facilities—which are owned by the UK, Germany, and the Netherlands—are a long-standing example of facilities under multi-national ownership. Other examples of multi-national approaches to providing fuel supply "assurance" include the IAEA's $150 million fund for uranium purchases,[300] Russia's creation of an International Uranium Enrichment Center,[301] a 120 MT LEU Fuel Bank[302] in Angarsk, and the UK's Nuclear Fuel Assurance Plan.[303] The latter is basically a bilateral agreement that is supposed to serve as a model for government-to-government arrangements between supplier and recipient states, where the supply of low enriched uranium is not disrupted for non-commercial (political) reasons.

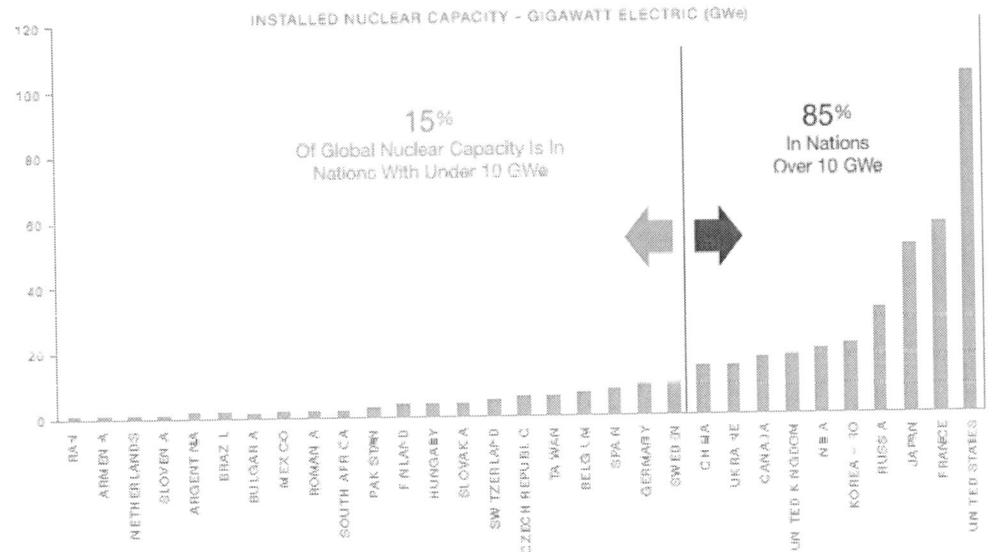

Figure 20. Worldwide distribution of Civil nuclear energy generation capacity in 2010.[298]

Although discussions of multi-national facilities and fuel services typically focus on securing enrichment and reprocessing facilities, the same concepts can be applied to the disposal of spent fuel and HLW. All countries with nuclear power will have to store SNF and HLW for some period of time and ultimately provide for disposal (internally or multi-nationally) of the spent fuel or of the high-level radioactive waste components that remain if the spent fuel is reprocessed.

Spent fuel contains approximately 1 percent plutonium and the self-protecting nature of the radioactivity will diminish over time making the plutonium more accessible. Thus, it is in the best interests of the United States and the international community to have spent fuel under effective and transparent control and to assure that in the coming century no spent fuel becomes "orphaned" anywhere in the world with inadequate safeguards and security.

Fuel "take-away" arrangements[304] would allow countries, particularly those with relatively small national programs, to avoid the very costly and politically difficult step of providing for waste disposal on their soil and to reduce associated safety and security risks. Fuel take-away could also provide a strong incentive for emerging nuclear nations to take key actions, such as ratifying the IAEA Additional Protocol, that can strengthen the non-proliferation regime and further isolate the currently small number of problem states.

The United States has implemented a relatively small but successful initiative to ship spent foreign research reactor fuel to U.S. facilities for storage and disposal.

This program has demonstrated meaningful non-proliferation and security benefits. *A similar capability to accept spent fuel from foreign commercial reactors, in cases where the President would choose to authorize such imports for reasons of U.S. national security, would be desirable within a larger policy framework that creates a clear path for the safe and permanent disposition of U.S. spent fuel.* Unfortunately, the failure to develop broadly-accepted domestic spent fuel storage and disposal strategies thus far limits U.S. non-proliferation policy choices.

The Commission believes the availability of spent fuel take-away would provide substantially greater incentives for some emerging nuclear nations to forgo the indigenous development of sensitive fuel-cycle facilities in return for access to regional or international facilities. In that context, government support for limited fuel supply and take-away initiatives to advance U.S. national security interests can be part of a comprehensive strategy for maintaining the nuclear energy option while simultaneously addressing proliferation and security concerns.

12.3. Security and Counterterrorism

As stated in a communiqué issued by the Washington Nuclear Summit on April 13, 2010, "Nuclear terrorism is one of the most challenging threats to international security, and strong nuclear security measures are the most effective means to prevent terrorists, criminals, or other unauthorized actors from acquiring nuclear materials...Success will require responsible national actions and effective international cooperation."[305] To date, the United States has worked to enhance global capacity to prevent, detect, and respond to nuclear terrorism by conducting multilateral activities aimed at strengthening the operations, plans, policies, procedures, and interoperability of partner nations through a variety of activities. Most recently, these activities have included the 2010 Nuclear Summit, the Nunn-Lugar Cooperative Threat Reduction Program, the Global Threat Reduction Initiative, and the Global Initiative to Combat Nuclear Terrorism.

Held in April 2010 and attended by 47 nations, the U.S.- hosted 2010 Nuclear Security Summit was launched with the goal of securing all vulnerable nuclear material worldwide within four years. Other efforts since that time have included signing a plutonium disposition protocol with Russia,[306] returning Russian origin high-enriched uranium (HEU) back to Russia,[307] converting the Kyoto University research reactor in Japan from HEU to low-enriched uranium (LEU), and pursuing ratification to an amendment of the Convention on Physical Protection of Nuclear Materials that would extend and strengthen the Convention's coverage of peaceful nuclear material in storage or use at domestic nuclear facilities, rather than merely in international transit. In preparation for the next summit, some U.S. experts are proposing the development of an international "nuclear material security framework agreement [that] would identify the threats to humankind from vulnerable fissile and radiological materials...and list actions and commitments required to mitigate them."[308]

The domestic division of regulatory responsibility for nuclear security and counter-terrorism is discussed in chapters 5 and 9 of this report; those chapters also describe security measures implemented at U.S. reactor sites following 9/11. As the United States continues to improve its ability to secure and protect nuclear facilities and materials, the Commission urges continued U.S. leadership to improve nuclear security and strengthen nuclear safety standards worldwide. Reviews conducted post-Fukushima will undoubtedly examine the safety and security benefits that could be achieved by improving instrumentation to measure key plant safety parameters including pool water levels under conditions of station blackout and severe damage, reviewing and strengthening procedures for connecting portable pumps and power supplies, and potentially accelerating the transfer of SNF out of reactor pools and into dry storage. The Commission urges that these reviews be completed expeditiously and that unclassified results be widely shared with regulators and other appropriate entities around the world.

Finally, the Commission finds that is important for the U.S. government to continue to support the IAEA's physical protection programs as well as efforts by the WINS to promote global best practices regarding nuclear security. Overall, the physical protection of nuclear material and facilities to deter terrorist activity remains a very high priority in today's security environment as the potential theft and sabotage of nuclear materials and facilities continues to be a real threat.[309]

Furthermore, the theft of weapons-usable material or any act of nuclear-related sabotage or terrorism anywhere in the world could create real consequences here in the United States, particularly if it leads to a detonation or large release of radioactivity. For this reason, the United States has a direct interest in encouraging and enabling all nations to uphold their national and international obligations for the security and safety of nuclear materials and facilities.

13. Near-Term Actions

The Commission recognizes that it will take time, commitment to action, and new authorizing legislation to implement our most important recommendations, particularly the recommendation to establish a new waste management organization.

Given uncertainty about how long that might take and the fact that under current law DOE remains responsible for the nuclear waste management activities of the federal government, it is important that those steps that do not require the new organization to be in place be initiated as soon as possible.

Specifically, the Commission urges near-term action in the areas described below. To ensure continued progress, we also urge the Secretary of Energy to task a senior official with sufficient authority to coordinate all of the DOE elements involved in the implementation of the Commission's recommendations.

Financing the Waste Program

- DOE should offer to enter into negotiations with contract holders to revise current contracts to provide a new fee payment option in which payments to the Waste Fund each year would be based on actual appropriations from the Waste Fund, with the remainder of the nuclear waste fee being placed in a third-party escrow account by the contract holder until needed.
- The Administration should work with the appropriate congressional committees and the Congressional Budget Office to reclassify receipts from the nuclear waste fee as discretionary offsetting collections and allow them to be used to offset appropriations for the waste program.
- The Administration, DOE, and DOJ should also work with nuclear utilities and other stakeholders toward a fair resolution of outstanding litigation and damage claims.

Establishment of a New Organization

- The appropriate congressional committees should begin hearings on establishment of an independent waste management organization as soon as practicable. The Commission recognizes that there are many details that need to be worked out in creating a new institution, and believes that the sooner the process of obtaining the views of interested parties and developing a detailed legislative proposal can begin, the better.

Storage

- Using existing authority in the NWPA, DOE should begin laying the groundwork for implementing consolidated storage and for improving the overall integration of storage as a planned part of the waste management system without further delay. Specific steps that DOE could take in the near term include:

- Performing the systems analyses and design studies needed to develop a conceptual design for a highly flexible, initial federal spent fuel storage facility.
- Preparing to respond to requests for information from communities, states, or tribes that might be interested in learning more about hosting a consolidated storage facility.
- Working with nuclear utilities, the nuclear industry, and other stakeholders to promote the better integration of storage into the waste management system, including standardization of dry cask storage systems. This effort should include development of the systems analyses needed to provide quantitative estimates of the system benefits of utility actions such as the use of standardized storage systems or agreements to deliver fuel outside the current OFF priority ranking. (These analyses would be needed to support the provision of incentives to utilities to undertake actions such as using standardized storage systems or renegotiating fuel acceptance contracts.)

- The Administration should request, and Congress should provide funding for, the National Academy of Sciences to conduct an independent investigation of the events at Fukushima and their implications for safety and security requirements at SNF and HLW storage sites in the United States.
- DOE, NRC and industry should continue a vigorous research and regulatory oversight effort in areas such as spent fuel and storage system degradation phenomena, vulnerability to sabotage and terrorism, and others.

Transportation

- DOE should complete the development of procedures and regulations for providing technical assistance and funds (pursuant to section 180 (c) of the NWPA), and begin providing funding, for working with states and regional state-government groups and training local and tribal officials in areas likely to be traversed by spent fuel shipments, in preparation for movement of spent fuel from shutdown reactor sites to consolidated storage.
- DOE and other federal agencies should reexamine and address those recommendations from the 2006 NAS *Going the Distance* study that have not yet been implemented. As a part of this reexamination, the NRC should reassess its plans for the Package Performance Study without regard to the status of the Yucca Mountain project, and if it is found to have independent value, funding should be provided from the Nuclear Waste Fund so that the NRC can update these plans and proceed with those tests.

Disposal

- DOE should keep a repository program moving forward through valuable, non-site specific activities, including R&D on geological media, work to design improved engineered barriers, and work on the disposal requirements for advanced fuel cycles. The work of the Used Fuel Disposition Campaign of DOE's Office of Used Nuclear Fuel Disposition Research and Development in this area should be continued.
- DOE should develop an RD&D plan and roadmap for taking the borehole disposal concept to the point of a licensed demonstration.

Facility Siting

- To ensure that future siting efforts are informed by past experience, DOE should build a data base of the experience that has been gained and relevant documentation produced in efforts to site nuclear waste facilities, in the United States and abroad. This would include storage facility and repository siting efforts under the NWPA by both DOE and the Nuclear Waste Negotiator.

Regulatory Actions

- EPA and the NRC should work together to define an appropriate process (with opportunity for public input) for developing a generic safety standard for geologic disposal sites. The implementation of this standard-setting process should be coordinated with the aim of developing draft regulations for mined repositories and deep borehole facilities.
- The NRC should continue efforts to review and potentially revise the existing waste classification system.
- A portion of federal nuclear energy RD&D resources should be directed to the NRC to accelerate the development of a regulatory framework and to support anticipatory research for novel components of advanced nuclear energy systems. An increased degree of confidence that new systems can be successfully licensed is important for lowering barriers to commercial investment.

Occupational Safety and Health for the Nuclear Workforce

- The jurisdictions of safety and health agencies should be clarified and aligned.
  New site-independent safety standards should be developed by the safety and health agencies responsible for protecting nuclear workers through a coordinated joint process that actively engages and solicits input from all relevant constituencies. Efforts to support uniform levels of safety and health in the nuclear industry should be undertaken with federal, industry, and joint labor–management leadership. Safety

and health practices in the nuclear construction industry should provide a model for other activities in the nuclear industry.

Nuclear Workforce Development

- DOE, in cooperation with the U.S. Department of Labor and the Bureau of Labor Statistics, should lead a public–private initiative to develop ongoing labor demand projections and forecast capacity for the nuclear workforce, in areas including science, technology, engineering and mathematics; crafts; and emergency response and hazardous material (HAZMAT).
  This capacity will help inform expanded federal, joint labor–management, and university-based support for critical high-skill, high-performance nuclear workforce development needs, including special attention to the expansion of the emergency response and HAZMAT-trained workforce.

International

- DOE should identify any legislative changes needed to authorize and direct the U.S. waste management program to support countries that pursue nuclear technologies in developing capacity for the safe management of the associated radioactive wastes and to encourage broad adherence to strengthened international norms for safety, security, and non-proliferation for all nuclear infrastructure and materials.

APPENDIX A: BRC COMMISSIONERS, STAFF, SENIOR CONSULTANTS AND CHARTER

The members of the Blue Ribbon Commission on America's Nuclear Future are:

Lee H. Hamilton, Co-Chair - Director of The Center on Congress at Indiana University; former Member, U.S. House of Representatives (D-IN).

Brent Scowcroft, Co-Chair - President of the Scowcroft Group; former National Security Advisor to Presidents Gerald Ford and George H.W. Bush.

Mark H. Ayers, President, Building and Construction Trades Department, AFL-CIO.

Vicky A. Bailey, Former Commissioner, Federal Energy Regulatory Commission; former Indiana PUC Commissioner; former DOE Assistant Secretary for Policy and International Affairs.

Albert Carnesale, Chancellor Emeritus and Professor, University of California, Los Angeles.

Pete V. Domenici, Senior Fellow, Bipartisan Policy Center; former U.S. Senator (R-NM).

Susan Eisenhower, President, Eisenhower Group, Inc.

Chuck Hagel, Distinguished Professor at Georgetown University and the University of Nebraska at Omaha; former U.S. Senator (R-NE).

Jonathan Lash, President, Hampshire College; former President, World Resources Institute.

Allison M. Macfarlane, Associate Professor of Environmental Science and Policy, George Mason University.

Richard A. Meserve, President, Carnegie Institution for Science and Senior Counsel, Covington & Burling LLP; former Chairman, U.S. Nuclear Regulatory Commission.

Ernest J. Moniz, Professor of Physics and Cecil & Ida Green Distinguished Professor, Massachusetts Institute of Technology.

Per Peterson, Professor and Chair, Dept. of Nuclear Engineering, Univ. of California – Berkeley.

John Rowe, Chairman and Chief Executive Officer, Exelon Corporation.

Phil Sharp, President, Resources for the Future; former Member, U.S. House of Representatives (D-IN).

BRC Staff

John Kotek – Staff Director

Matthew Milazzo – Deputy Director, Staff Liaison to the Reactor and Fuel Cycle Technology Subcommittee

Tom Isaacs – Lead Advisor

Alex Thrower – Counsel, Staff Liaison to the Transportation and Storage Subcommittee

Natalia Saraeva – Research Associate, Staff Liaison to the Disposal Subcommittee

Marika Tatsutani – Lead Writer and Editor

Mary Woollen – Government and Community Liaison Irie Harris – Office Manager

BRC Senior Consultants

Dr. Tom Cotton – *Senior Technical Advisor* Allen Croff – *Senior Technical Advisor*

Dr. Glenn Paulson – *Senior Technical Advisor*

DEPARTMENT OF ENERGY, WASHINGTON, DC 20585

Blue Ribbon Commission on America's Nuclear Future
U.S. Department of Energy, Advisory Committee Charter

1. *Committee's Official Designation.* Blue Ribbon Commission on America's Nuclear Future (the Commission).

2. *Authority.* The Commission is being established in accordance with the provisions of the Federal Advisory Committee Act (FACA), as amended, 5 USC. App. 2, and as directed by the President's Memorandum for the Secretary of Energy dated January 20, 2010: Blue Ribbon Commission on America's Nuclear Future. This charter establishes the Commission under the authority of the U.S. Department of Energy (DOE).
3. *Objectives and Scope of Activities.* The Secretary of Energy, acting at the direction of the President, is establishing the Commission to conduct a comprehensive review of policies for managing the back end of the nuclear fuel cycle, including all alternatives for the storage, processing, and disposal of civilian and defense used nuclear fuel, high-level waste, and materials derived from nuclear activities. Specifically, the Commission will provide advice, evaluate alternatives, and make recommendations for a new plan to address these issues, including:

   a) Evaluation of existing fuel cycle technologies and R&D programs. Criteria for evaluation should include cost, safety, resource utilization and sustainability, and the promotion of nuclear nonproliferation and counter-terrorism goals.
   b) Options for safe storage of used nuclear fuel while final disposition pathways are selected and deployed;
   c) Options for permanent disposal of used fuel and/or high-level nuclear waste, including deep geological disposal;
   d) Options to make legal and commercial arrangements for the management of used nuclear fuel and nuclear waste in a manner that takes the current and potential full fuel cycles into account;
   e) Options for decision-making processes for management and disposal that are flexible, adaptive, and responsive;
   f) Options to ensure that decisions on management of used nuclear fuel and nuclear waste are open and transparent, with broad participation;
   g) The possible need for additional legislation or amendments to existing laws, including the Nuclear Waste Policy Act of 1982, as amended; and
   h) Any such additional matters as the Secretary determines to be appropriate for consideration.

The Commission will produce a draft report to the Secretary and a final report within the time frames contained in paragraph 4.

4. Description of Duties. The duties of the Commission are solely advisory and are as stated in Paragraph 3 above.

A draft report shall be submitted within 18 months of the date of the Presidential memorandum directing establishment of this Commission; a final report shall be submitted within 24 months of the date of that memorandum. The reports shall include:

a) Consideration of a wide range of technological and policy alternatives, and should analyze the scientific, environmental, budgetary, financial, and management issues, among others, surrounding each alternative it considers. The reports will also include a set of recommendations regarding policy and management, and any advisable changes in law.

b) Recommendations on the fees currently being charged to nuclear energy ratepayers and the recommended disposition of the available balances consistent with the recommendations of the Commission regarding the management of used nuclear fuel; and

c) Such other matters as the Secretary determines to be appropriate.

5. Official to Whom the Committee Reports. The Commission reports to the Secretary of Energy.
6. Agency Responsible for Providing the Necessary Support. DOE will be responsible for financial and administrative support. Within DOE, this support will be provided by the Office of the Assistant Secretary for Nuclear Energy or other Departmental element as required. The Commission will draw on the expertise of other federal agencies as appropriate.
7. Estimated Annual Operating Cost and Staff Years. The estimated annual operating cost of direct support to, including travel of, the Commission and its subcommittees is $5,000,000 and requires approximately 8.0 full-time employees.
8. Designated Federal Officer. A full-time DOE employee, appointed in accordance with agency procedures, will serve as the Designated Federal Officer (DFO). The DFO will approve or call all of the Commission and subcommittee meetings, approve all meeting agendas, attend all Commission and subcommittee meetings, adjourn any meeting when the DFO determines adjournment to be in the public interest. Subcommittee directors who are full-time Department of Energy employees, as appointed by the DFO, may serve as DFOs for subcommittee meetings.
9. Estimated Number and Frequency of Meetings. The Commission is expected to meet as frequently as needed and approved by the DFO, but not less than twice a year.

The Commission will hold open meetings unless the Secretary of Energy, or his designee, determines that a meeting or a portion of a meeting may be closed to the public as permitted by law. Interested persons may attend meetings of, and file comments with, the Commission, and, within time constraints and Commission procedures, may appear before the Commission.

Members of the Commission serve without compensation. However, each appointed non-Federal member may be reimbursed for per diem and travel expenses incurred while attending Commission meetings in accordance with the Federal Travel Regulations.

10. Duration and Termination. The Commission is subject to biennial review and will terminate 24 months from the date of the Presidential memorandum discussed above, unless, prior to that time, the charter is renewed in accordance with Section 14 of the FACA.
11. Membership and Designation. Commission members shall be experts in their respective fields and appointed as special Government employees based on their knowledge and expertise of the topics expected to be addressed by the

Commission, or representatives of entities including, among others, research facilities, academic and policy-centered institutions, industry, labor organizations, environmental organizations, and others, should the Commission's task require such representation. Members shall be appointed by the Secretary of Energy. The approximate number of Commission members will be 15 persons. The Chair or Co-Chairs shall be appointed by the Secretary of Energy.

12. Subcommittees.

    a) To facilitate functioning of the Commission, both standing and ad hoc subcommittees may be formed.
    b) The objectives of the subcommittees are to undertake fact-finding and analysis on specific topics and to provide appropriate information and recommendations to the Commission.
    c) The Secretary or his designee, in consultation with the Chair or Co-Chairs, will appoint members of subcommittees. Members from outside the Commission may be appointed to any subcommittee to assure the expertise necessary to conduct subcommittee business.
    d) The Secretary or his designee, in consultation with the Chair or co-Chairs will appoint Subcommittees.
    e) The DOE Committee Management Officer (CMO) will be notified upon establishment of each subcommittee.

13. Recordkeeping. The records of the Commission and any subcommittee shall be handled in accordance with General Records Schedule 26, Item 2 and approved agency records disposition schedule. These records shall be available for public inspection and copying, subject to the Freedom of Information Act, 5 USC. 552.
14. Filing Date.

Date filed with Congress: ____ March 1, 2010

Signed

Carol A. Matthews
Committee Management Officer

Appendix B: Full Commission, Subcommittee and Regional Comment Meetings

March 25 & 26, 2010 – Washington DC – Full Commission Meeting
May 25 & 26, 2010 – Washington, DC – Full Commission Meeting
July 7, 2010 – Washington, DC – Disposal Subcommittee Meeting
July 12 & 13, 2010 – Idaho Falls, ID – Reactor & Fuel Cycle Technologies Subcommittee Meeting

July 14 & 15, 2010 – Hanford Site/Kennewick, WA – Full Commission Meeting
August 10, 2010 – Maine Yankee Site/Wiscasset, ME – Transportation & Storage Subcommittee Meeting
August 19, 2010 – Washington, DC – Transportation & Storage Subcommittee Meeting
August 30 & 31, 2010 – Washington, DC – Reactor & Fuel Cycle Technologies Subcommittee Meeting
September 1, 2010 – Washington, DC – Disposal Subcommittee Meeting
September 21 & 22, 2010 – Washington, DC – Full Commission Meeting
September 23, 2010 – Washington, DC – Transportation & Storage Subcommittee Meeting
October 12, 2010 – Washington, DC – Reactor & Fuel Cycle Technologies Subcommittee Meeting
October 21 & 22, 2010 – Finland – Disposal Subcommittee Site Visits and Meetings
October 23-26, 2010 – Sweden – Disposal Subcommittee Site Visits and Meetings
November 2, 2010 – Chicago, IL – Transportation & Storage Subcommittee Meeting
November 4, 2010 – Washington, DC – Disposal Subcommittee Meeting
November 15 & 16, 2010 – Washington, DC – Full Commission Meeting
January 6 & 7, 2011 – Aiken, SC and Augusta, GA – Savannah River Site Visit and Meeting
January 26, 27 & 28, 2011 – Carlsbad and Albuquerque, NM – Waste Isolation Pilot Plant Site Visit and Meetings
February 1 & 2, 2011 – Washington, DC – Full Commission Meeting
February 3, 2011 – Washington, DC – Classified (Closed) Meeting
February 8-11, 2011 – Japan – Site Visits and Meetings
February 17 & 18, 2011 – Russia – Meetings
February 20, 21 & 22, 2011 – France – Site Visits and Meetings
May 13, 2011 – Washington, DC – Full Commission Meeting
June 21-28, 2011 – United Kingdom and France – Site Visits and Meetings
September 12, 2011 – Denver, CO – Regional Public Meeting
October 12, 2012 – Boston, MA – Regional Public Meeting
October 18, 2011 – Atlanta, GA – Regional Public Meeting
October 20, 2011 – Washington, DC – Regional Public Meeting
October 28, 2011 – Minneapolis, MN – Regional Public Meeting
December 2, 2011 – Washington, DC – Full Commission Meeting

Appendix C: Status of Nuclear Waste Management Programs in Other Countries

[NOTE: Much of the information herein was derived from a 2009 report of the U.S. Nuclear Waste Technical Review Board (NWTRB) on the status of international repository programs. This appendix may therefore not reflect more recent developments, although we have attempted to update the information below based on information provided to the Commission. For nations covered herein but not covered in the NWTRB report, the information was gathered by BRC staff].

Canada: Canada currently has 18 operating nuclear power plants, which together account for nearly 15% of the country's total electricity production. Responsibility for managing nuclear waste rests with the Nuclear Waste Management Organization (NWMO), a private corporation formed by nuclear plant owners. The Organization's key policies and decisions must be approved by the government, which regulates nuclear waste management activities through the Canadian Nuclear Safety Commission. Similar to the approach taken in the United States, owners of nuclear power plants pay into a Nuclear Fuel Waste Act Trust Fund. Canada does not reprocess commercial used nuclear fuel. The Organization's "Adaptive Phased Management" (APM) plan is guided by five fundamental values: Integrity, Excellence, Engagement, Accountability, and Transparency. It consists of a technical method and a management system. The technical method envisions disposal in an appropriate geologic formation with the option of shallow underground storage at the disposal site. It includes the potential for retrievability, continuous monitoring, flexible design, and an ongoing technical and social research program. The management system calls for collaborative and phased decision making; continuous learning; open, inclusive and transparent engagement; and pursuit of a willing and informed host community in one of the four nuclear provinces. Currently NWMO is in the early stages of the siting phase and a number of communities have expressed interest in learning more about the program in order to inform a decision about their interest in volunteering to host such a facility. The proposed process for selecting a deep geologic repository site does not contain a firm schedule for completing this process or an anticipated start date for repository operations.Canada does not have an independent, centralized interim-storage facility for used nuclear fuel.

Finland: Finland currently has four operating nuclear power plants, which together account for nearly 30% of the country's total electricity production. Responsibility for waste management rests with Posiva Oy, a joint company created by Finland's two nuclear utilities in 1995. The government's Radiation and Nuclear Safety Authority serves as independent regulator. Nuclear power generators pay into a nuclear waste management fund; their annual obligation depends on the gap between estimated waste disposal and plant decommissioning costs and the level of the fund at that point in time. Finland does not reprocess commercial used nuclear fuel. In 2000, the government approved Olkiluoto, a migmatite site in the municipality of Eurajoki, for a deep geologic repository. (Two of Finland's four existing nuclear reactors and a new reactor that is currently under construction are also located at Olkiluoto.) The site was subsequently approved by Finland's Parliament (in 2001) and is currently being characterized at depth using an underground research tunnel known as Onkalo (construction of the tunnel began in 2004). Selection of the Olkiluoto site has the support of the host community, which could have exercised veto power over the government's decision (instead, the Eurajoki Municipal Council approved a positive statement about the site). The community had negotiated a benefits package with Posiva Oy in 1999. Key decisions concerning long-term health and safety requirements, the design of engineered barrier systems, and the methodology to be used for demonstrating compliance with post-closure standards have been taken; details are available from the NWTRB report and other sources. Earlier regulations stipulated that waste emplaced at the site be retrievable in the future; this requirement was lifted in 2008 but Posiva is still obliged to present a plan and cost estimate for waste retrieval when it applies for a license to construct the Olkiluoto repository. The anticipated start date for repository operations is 2020. Finland does not have an independent, centralized interim-storage facility.

France: France has 58 operating nuclear plants, which together account for 76% of the country's total electricity production. A new 1.6 GW plant is currently under construction.

Responsibility for managing and disposing of nuclear waste falls to the National Agency for Radioactive Waste Management, a government-owned public service agency which reports to the government's Ministries of Environment, Industry, and Research. France's Nuclear Safety Authority is the independent regulator. In France, all owners of a nuclear license are responsible individually for assessing costs of decommissioning their plants and for the long-term management of the waste and spent fuel— which is not considered "waste" in France. They are also responsible for establishing the necessary financial provisions and for ear-marking the necessary assets for the exclusive coverage of those costs. They are individually responsible for managing those assets, which will be disbursed when the relevant decommissioning and long-term management activities start on an industrial basis. The necessary R&D for long-term waste management is financed through an additional tax on nuclear installations, which is transferred to a fund that goes to the National Waste Agency. France requires reprocessing in the fuel cycle; accordingly, only high-level waste and long-lived intermediate-level waste are authorized for disposal in a deep geologic repository. In 1999, construction began on an underground research facility in argillite rock at a location near the village of Bure in the Meuse area located at the boundary of the Haute-Marne region; the area was subsequently approved for a long-term repository site in 2006. The National Agency has recently identified a 30 square kilometer area to locate the repository. The selection of this area was carried out in consultation with the mayors and authorities from both the Meuse and Haute Marne region. Consultations continue on where to locate surface facilities and the lay-out of the underground facility and its access. Local government in the Meuse and Haute-Marne area have been associated with the site-characterization program in various ways, and both can expect to benefit from a series of measures designed to support their development, funded through a dedicated tax on basic nuclear installations. France has established health and safety requirements for a deep repository site, identified a methodology for demonstrating compliance with post-closure standards, and decided on the design of engineered barrier systems at the site (the plan is to place vitrified waste in stainless steel packages). Current policy stipulates that the repository must be designed to be "reversible" for at least 100 years, a concept that implies technical retrievability. Specific conditions for meeting this requirement will be prescribed by the French Parliament after a license application has been submitted. France currently expects its repository to become operational in 2025. All commercial high level waste is slated for disposal and is stored in a special facility within the spent fuel reprocessing complex at La Hague.

Japan: Japan has 53 nuclear power plants, which prior total electricity production. In addition, three new nuclear power plants (totaling 3.7 GW) are under construction. The Nuclear Waste Management Organization, a private, non-profit entity formed by nuclear power plant owners, is responsible for waste management. The Nuclear and Industrial Safety Agency, a unit within Japan's Ministry of Economy, Trade, and Industry, is the independent regulator. The Ministry maintains two funds to cover costs associated with radioactive waste management: nuclear power plant owners pay into a High-Level Waste Fund; owners of reprocessing plants and mixed-oxide fuel fabrication plants pay into a TRU Waste Fund. Commercial spent nuclear fuel from Japan has been reprocessed in France and the United Kingdom; in addition, reprocessing takes place in Japan at a small facility in Tokai. A large new reprocessing facility at Rokkasho Village is expected to open in the next few years pending the results of pre-service testing. Two underground research laboratories to investigate deep geologic disposal (in granite and sedimentary rock) are under construction, but no decision has been reached in terms of selecting a site for a long-term repository. Requirements for such a repository (with regard to health and safety, retrievability, design of engineered barriers, etc.) have also not been established. The Nuclear Waste Management Organization has adopted a transparent, voluntary approach to identifying potential sites—thus, both the mayor of the host community and the governor of the prefecture must agree to participate. Localities that agree to be included in an initial survey can receive up to $18 million; if they subsequently agree to participate in surface-based site investigations they can receive up to $65 million. One town (Toyo-cho) initially agreed to participate but later withdrew. The national government has since indicated that it may play a more proactive role in the site selection process going forward. Japan had been constructing an independent, centralized interim-storage facility at Mutsu in Aomori Prefecture but those plans have been put on hold in the aftermath of the March 2011 earthquake and tsunami. Japan has not projected a date for opening a permanent repository.

Russia: Russia currently has 33 nuclear reactors in operation (including a 600 MWe fast breeder reactor) which together account for nearly 16% of the country's total electricity production. Another 9 reactors are under construction (including a 800 MWe fast breeder reactor). Radioactive waste management and spent fuel waste management are divided into two different programs. Radioactive waste management is the responsibility of the newly created federal state enterprise "RosRAO" within the structure of the federal corporation Rosatom (which runs the country's nuclear power complex) and Rosatom itself. The new Federal Law on Radioactive Waste Management (came into force in 2011) establishes a legal framework for radioactive waste management in Russia and requires creation of a unified state system for radioactive waste management. Among other provisions, the law authorizes a single-purpose organization (so-called "national operator" – currently Rosatom) to conduct main activities related to waste management activities (e.g. receiving, storing, securing and disposing of radioactive waste); decision-making regarding siting, construction, commissioning etc. of waste-management facilities remains the responsibility of the Federal Government. The law also establishes a framework for a new funding mechanism (analogous to the Nuclear Waste Fund in the United States). Some federal budget resources have also been allocated for the program (the total for 2016 to 2020 is $13 billion in U.S. dollars). Meanwhile, a system for managing spent nuclear fuel is being developed by Rosatom. It is not clear whether implementing this system will be the responsibility of Rosatom or one of its subsidiaries. The pending Federal Law on Spent Nuclear Fuel Management will provide the legal framework for the national program. As work continues on drafting this legislation, Rosatom has gone ahead with developing plans for the construction and commissioning of an underground rock laboratory by 2015 and a final repository by 2021. Several sites have been proposed as candidates for such a facility, including a granite site on the Kola Peninsula (in the Murmansk region), Krasnokamenks in Chita (4,300 miles east of Moscow), and the Nignekamensk Rock Mass in the Krasnoyarsk Territory of Siberia. Site selection efforts are currently underway on the Kola Peninsula. Russia plans to close its fuel cycle as much as possible and use plutonium in MOX fuel in fast breeder reactors. However, current reprocessing capacities are limited to about 100 metric tons per year. A new reprocessing plant in the city of Zheleznogorsk (in the Krasnoyarsk Territory) is being redesigned from a previous version and is expected to commence operations in the 2025–2030 timeframe.

Although most of Russia's spent nuclear fuel is being stored at reactor sties, there is a centralized interim wet (pool) storage facility located in Zheleznogorsk. Its storage capacity of 7,200 metric tons was recently expanded to 8,600 metric tons, allowing it to safely store/accept spent VVER-1000 (PWR-1000) fuel through 2025. In addition, a dry storage facility for spent RBMK (BWR) fuel with a total capacity of 8,600 metric tons was commissioned in late 2011, also in Zheleznogorsk, with the first SNF is scheduled to arrive in early 2012.

Low-level radioactive wastes and some intermediate-level wastes are processed and stored at 16 sites in Russia (within the structure of the federal state enterprise RosRAO).

Russia currently has a program to "take-back" spent fuel of Russian origin for reprocessing from commercial and research reactors abroad. However, due to limits on available reprocessing capabilities, the spent fuel that has been accepted under this program is being held in wet (pool) storage.

Spain: Spain has eight operating nuclear power plants, which together account for 18% of the country's total electricity production. Management of nuclear waste is the responsibility of the Spanish National Company for Radioactive Waste, a government-owned corporation. The Nuclear Safety Council is the independent regulator, although the Ministry of Industry, Tourism, and Trade is required by law to make a final decision concerning the disposition of used nuclear fuel. Operators of nuclear power plants pay into a nuclear decommissioning fund that was established to cover the costs of both decommissioning plants and managing radioactive waste. Some used nuclear fuel from Spanish reactors has been reprocessed in the past at the La Hague and Sellafield facilities, but current national policy does not contemplate any further reprocessing. No decision has been made regarding a deep geologic repository for high-level waste and used nuclear fuel, but in 2006 Spain initiated a process to site a centralized temporary facility. That process required voluntary participation by potential host communities and resulted in recent selection of a site for a consolidated storage facility in the town of Villar de Cañas, located in the "autonomous community" (the level of government in Spain roughly equivalent to a state) of Castilla la Mancha.

Sweden: Sweden currently has 10 operating nuclear power plants, which together account for 42% of the country's total electricity production. The Swedish Nuclear Fuel and Waste Management Company, a private corporation formed by nuclear power plant owners, is responsible for waste management. The Radiation Safety Authority within Sweden's Ministry of the Environment is the independent regulator. Owners of nuclear power plants pay fees into a nuclear waste fund. The fees vary from year to year and from plant to plant, depending on the estimated costs of disposing of used nuclear fuel and the level of the fund. Small amounts of used nuclear fuel from Swedish reactors have been reprocessed in the past at facilities in France and England (none of the resulting high-level waste was returned to Sweden), but Sweden's current plans do not include reprocessing. In 2001, the government approved a proposal by the Swedish Nuclear Fuel and Waste Management Company to investigate three potential sites for a long-term geologic repository—at Östhammer, Oskarshamn (Oskarshamn was also the site of an underground research laboratory constructed in the early 1990s) and area in the northern part of Tierp. Later, municipal councils in Östhammer and Oskarshamn consented to further investigation, while Tierp opted-out. The site at Östhammer was selected for a repository in 2009. The value of benefits for communities was estimated as $300 million. The local community at Östhammer, which could have vetoed its selection as a geologic repository site, will receive 25% of the benefits. In addition, the community at Oskarshamn, which was *not* selected, will receive about 75% of the benefits for participating in the siting process. A license application for the Östhammer repository was submitted to the Radiation Safety Authority for review in March 2011. Concurrently, Sweden's Environmental Court will rule on the application. Based on the findings of the Safety Authority and the Court, the Swedish government will decide whether to approve the license application. Regular operation of the repository is expected to begin after several years of trial operation. Current plans call for transporting waste to the site using a specially designed ship and for placing used nuclear fuel in a copper canister that has a cast-iron insert for support and is surrounded by bentonite clay. Details concerning safety standards, post-closure compliance demonstration, and other requirements applicable to the Östhammer repository are available from the NWTRB report and other sources. Sweden currently expects to start repository operations in 2023. Sweden also has an independent, centralized interim-storage facility for used nuclear fuel: the CLAB facility, also located in Oskarshamn, was commissioned in 1985.

United Kingdom: The United Kingdom currently has 19 nuclear reactors that together account for one-sixth of the country's electricity generation. In October 2010, the UK government approved the construction of up to eight new nuclear power stations. All nuclear installations in the UK are subject to regulation by the Office for Nuclear Regulation and by environmental authorities. Responsibility for designing and developing a geological disposal facility for higher activity wastes rests with the Nuclear Decommissioning Authority (NDA). The NDA has a baseline disposal plan that envisions first emplacement of legacy intermediate level waste in 2040, emplacement of legacy high level waste and spent fuels in 2075, emplacement of spent fuel from new reactors in 2130, and commencement of facility closure in 2175. The UK has accumulated a substantial legacy of radioactive waste from a variety of different nuclear programs, both civil and defense-related. For decades, the UK struggled to find a solution to the problem of long-term radioactive waste management. The nearest the UK came was a planning application for a "Rock Characterisation Facility" as the first step towards geological disposal in Cumbria in 1994. The application went to a public inquiry and was rejected in 1997, largely on the basis of the site selection process used and scientific and technical uncertainties at the time. Recognizing that the existing approach was unworkable, the government undertook a more fundamental review of options for managing radioactive wastes in the long term. In 2001, the UK government initiated the "Managing Radioactive Waste Safely" (MRWS) program, which provided for public consultation on the waste management issue with the goal of finding a practicable solution for the UK's higher activity wastes. The process was designed to work in an open and transparent way that inspired public confidence, was based on sound science, and ensured the effective use of public monies. Having collected feedback from a public consultation process, an independent body, the Committee on Radioactive Waste Management (CoRWM) was set up to recommend specific program options. In July 2006, CoRWM announced an integrated package of recommendations for pursuing geological disposal, coupled with safe and secure interim storage and a program of ongoing research and development. Following publication of a white paper in 2008, the UK government launched a search for an engineered, underground site to serve as a permanent disposal facility for high-level radioactive wastes. The government invited communities across the country to learn more about what it would mean to potentially host this facility. To date, only a group of communities in West Cumbria, near the Sellafield nuclear site in northwest England, have sought to examine the possibility further. They formed a West Cumbria MRWS Partnership, including a range of local stakeholders, to examine the proposal and make recommendations to the local decision-making bodies on whether to proceed further. A comprehensive local consultation is currently underway to gauge public view prior to submission of the final recommendation whether to proceed or not. The U.K. has taken a noteworthy approach to providing benefits to potential host communities. One element is an "Engagement Package" which Government agrees upon each year to support the running costs of the MWRS partnership, including all the research, project management, consultants, travel expenses, staff time and public engagement work. In 2011 the support costs were approximately 1.2 million pounds. This kind of Engagement Package is anticipated to continue throughout the whole siting process, and be extended to individual host communities as they enter the process actively, to cover their own costs. Note, however, that the definition of Engagement Package does not cover any 'incentive' type payments - only reimbursing actual costs incurred. A "Community Benefits Package" will only be paid when a host community has passed the time at which it can withdraw from the process (i.e. when a final planning application is submitted for the actual facility to be built). The Community Benefits Package would, however, be agreed upon well before that point. A site has not yet been selected so there are no specific agreements to date regarding what amount of money or investment any community would receive for hosting the facility, only a promise in the Government's policy that these kinds of benefits be available to the community that finally agrees to host a repository. Recognizing this, but wanting reassurance at the same time, the current partnership has developed a number of principles for community benefit that have been agreed to by the Government, so that the community's understanding of the type and scale of benefits meets their expectation. These principles will form the basis for any future negotiation.

APPENDIX D: BRC COMMISSIONED PAPERS

1. Joe T. Carter et al., U.S. Radioactive Waste Inventory and Characteristics Related to Potential Future Nuclear Energy Systems, Rev. 2, May 2011
2. Peter C. Chestnut et al., The Role of Indian Tribes in America's Nuclear Future, April 29, 2011
3. Rodney C. Ewing, Standards & Regulations for the Geologic Disposal of Spent Nuclear Fuel and High-Level Waste, March 4, 2011
4. Cliff W. Hamal et al., Spent Nuclear Fuel Management: How Centralized Interim Storage Can Expand Options and Reduce Costs, May 16, 2011
5. Elizabeth Helvey, Overview of the Section 180(c) Program: History, Lessons Learned and Potential Next Steps, April 18, 2011.
6. Joseph S. Hezir, Budget and Financial Management Improvements to the Nuclear Waste Fund, May 2011
7. Judith A. Holm, Innovative Stakeholder Involvement Processes in Department of Energy Programs: A Selective Accounting, April 2011
8. Hank Jenkins-Smith, Public Beliefs, Concerns and Preferences Regarding the Management of Used Nuclear Fuel and High-Level Radioactive Waste, February 2011
9. Andrew C. Klein, Nuclear Energy R&D Infrastructure Report for The Blue Ribbon Commission on America's Nuclear Future, March 16, 2011

10. Jim Lieberman et al., Overview of the Nuclear Regulatory Commission and Its Regulatory Process for the Nuclear Fuel Cycle for Light Water Reactors, February 25, 2011
11. Richard C. Moore, Enhancing the Role of State and Local Governments in America's Nuclear Future: An Idea Whose Time Has Come, May 2011
12. Michael O'Hare et al., Nuclear Waste Facility Siting and Local Opposition, January 2011
13. Scott D. Sagan, The International Security Implications of U.S. Domestic Nuclear Power Decisions, April 18, 2011
14. Stoneturn Consultants, From Three Mile Island to the Future: Improving Worker Safety and Health In the U.S. Nuclear Power Industry, March 14, 2011
15. Eileen M. Supko and Michael H. Schwartz, Overview of High-Level Nuclear Waste Materials Transportation: Processes, Regulations, Experience and Outlook in the U.S., January 2011.
16. Seth P. Tuler and Roger E. Kasperson, Social Distrust: Implications and Recommendation for Spent Nuclear Fuel and High-Level Radioactive Waste Management, January 29, 2011
17. Van Ness Feldman, P.C., Legal Memorandum, Co-Mingled and Defense-Only Repositories, updated Dec. 9, 2011
18. Van Ness Feldman, P.C., Federal Commitments Regarding Used Fuel and High-Level Wastes, August 13, 2010 (revised November 12, 2010)
19. Gary Vine, Abridged History of Reactor and Fuel Cycle Technologies Development, March 15, 2011
20. Thomas Webler et al., Options for Developing Public and Stakeholder Engagement for the Storage and Management of Spent Nuclear Fuel (SNF) and High Level Waste (HLW) in the United States (Updated), June 6, 2011
21. Chris Whipple, Disposal of Spent Nuclear Fuel and High-Level Radioactive Waste, September 10, 2010
22. Paul P.H. Wilson, Comparing Nuclear Fuel Cycle Options: Observations and Challenges, June 25, 2011
23. R.G. Wymer and A.G. Croff, Current U.S. Department of Energy Nuclear Energy RD&D Programs and Plans, January 2011
24. Van Ness Feldman, P.C., Legal Analysis of Commission Recommendations for Near-Term Actions, July 29, 2011 (revised October 11, 2011).
25. Van Ness Feldman, P.C., Legal Background and Questions Concerning the Federal Government's Contractual Obligations Under the "Standard Contracts" with "Utilities," December 20, 2010

These papers are available at www.brc.gov.

End Notes

[1]. The resulting crisis at the Fukushima Daiichi plant was by no means the most tragic aspect of the disaster that occurred in Japan on March 11, 2011: on the contrary, more than 23,000 people were lost and immense damage was caused by the immediate impacts of the earthquake and tsunami.

[2] "Spent fuel" is sometimes also referred to as "used fuel." The difference in terminology in fact reflects a profound policy issue as to whether the material should be seen as a waste or a resource. We use the term "spent fuel" in this report, but, as discussed in chapter 11, we believe it is premature to resolve that policy debate.

[3]. The Nuclear Power Plant Exporters' "Principles of Conduct" are available at http:// nuclear principles.org/the-principles/.

[4]. The inter-temporal, inter-generational dimensions of this ethical obligation have long been recognized in the U.S. context and internationally. The 1996 IAEA "Joint Convention on the Safety of Spent Fuel and Radioactive Waste Management," for example, speaks of the need to avoid "compromising the ability of future generations to meet their needs and aspirations." (See http://www.iaea.org/ Publications/Documents/Conventions/ jointconv. html.) Put another way, plans for geologic disposal must not impose reasonably predictable impacts on future generations that are greater than those permitted for the current generation.

[5]. The "Open Government Directive" was sent by Peter Orszag to the heads of executive departments and agencies on December 8, 2009. See: http://www.whitehouse.gov/sites/ default/files/omb/assets/memoranda_2010/m10-06.pdf.

[6]. Atoms of a given element, such as uranium, can exist in different forms or "isotopes," depending on the number of neutrons present in the nucleus of the atom. Different elements are distinguished by their unique atomic number, which reflects the number of protons in the atomic nucleus. Uranium has an atomic number of 92, which means that all uranium atoms have 92 protons. A U-235 atom differs from a U-238 atom in that its nucleus holds three fewer neutrons—143 neutrons instead of 146—in combination with 92 protons.

[7]. Some reactors – such as the CANDU reactors employed in Canada and elsewhere – can use natural uranium as fuel. There have been around 3,300 truck and rail shipments of spent nuclear fuel in the United States since the mid-1960s (firm numbers before that are not available).

[8]. There have been a very few barge shipments--most notably from Brookhaven National Laboratory and the Shoreham plant, both on Long Island--but they were multimodal shipments that used truck or rail for most of the distance. Information on shipments to date is taken from National Research Council, *Going the Distance? The Safe Transport of Spent Nuclear Fuel and High-Level Radioactive Waste in the United States*, Washington, DC: The National Academies Press, 2006. See Table 3.2 on p. 188.

[9]. The amount of fission products is roughly proportional to the amount of electricity generated, no matter what fuel cycle is used. As a result, any nuclear energy system produces wastes that will require disposal.

[10]. The figure shows the composition of a standard PWR fuel assembly which achieved a burnup of 33 gigawatts per ton after ten years of cooling. Adapted from "Fuel Cycles for Sustainable Development and Waste Minimization," presentation by M. Salvatories, CAE/DEN Cadarache, France at World Nuclear University Summer Institute 2007 (available at: http://www.world-nuclear-university. org/html/summer_institute/2007/ 2007SI-lecture%20 Materials/0725/0725_Massimo%20SALVATORES_1/0725- 1-Massimo%20Salvatores .pdf).

[11]. People are routinely exposed to low levels of radiation in everyday life. These low-level exposures can come from natural sources (e.g., cosmic rays, certain minerals) and from man-made sources (e.g., building materials, medical procedures such as x-rays, certain cancer treatments, etc.).

[12]. Riley E. Dunlap, Michael E. Kraft and Eugene A. Rosa, eds., *Public Reactions to Nuclear Waste: Citizens' Views of Repository Siting*, Duke University Press, 1993.

[13] Half-life is the time required for half of the initial atoms of a given amount of a radionuclide to decay.

[14]. Weight (e.g., metric tons) is not the best measure of the nuclear waste challenge. However, it is commonly used, including in federal law, so we adopt the same practice for purposes of this report.

[15] Sources for the information shown in the figure include the following: U.S. Environmental Protection Agency, "Radiation Health Effects," available at http://www. epa.gov/radiation /understand/health_effects.html and World Nuclear Association, "Nuclear Radiation and Health Effects," available at http://www.world-nuclear.org/info/inf05.html. See also http:// www.naturalnews.com/032136_radiation_exposure_chart.htmland http://www. Information isbeautiful. net/visualizations/radiation-dosage-chart/.

[16] U.S. Department of Energy, "Supplemental Analysis for the U.S. Disposition of Gap Material – Spent Nuclear Fuel," (DOE/EIS-0218-SA-4), January 2009.

[17] U.S. Department of Energy, "Draft Environmental Impact Statement (EIS) for the Disposal of Greater-ThanClass C (GTCC) Low-Level Radioactive Waste (LLRW) and GTCC-like Waste" (DOE/EIA-0375D), 2011. More information available at http://www.gtcceis.anl. gov/.

[18] John McKenzie, Presentation to the Blue Ribbon Commission on America's Nuclear Future, March 25, 2010, available at http://www.brc.gov/pdfFiles/NR_ Briefing_100325.pdf.

[19] The Navy has also designed and built large shipping casks that each hold one loaded spent fuel canister and are designed for shipment by rail. The canisters were intended to be transferred at Yucca Mountain into disposal overpacks and then be directly disposed in the repository.

[20] More details on the 1995 Agreement and the 2008 modifications to it can be found in a white paper prepared for the BRC by Van Ness Feldman, P.C.: "Federal Commitments Regarding Used Fuel And High-Level Wastes," August 31, 2010 (Revised November 12, 2010). Available at www.brc.gov.

[21] There is a similar agreement with the State of Colorado under which Navy SNF has to be moved out of that state by January 1, 2035. The penalty for noncompliance amounts to $15,000 for each day of delay beyond the deadline.

[22] United States Government Accountability Office, *DOE NUCLEAR WASTE: Better Information Needed on Waste Storage at DOE Sites as a Result of Yucca Mountain Shutdown*, GAO-11-230, March 2011.

[23] See presentation of Susan Burke, Idaho National Laboratory Oversight Coordinator for the State of Idaho, "Idaho's Perspective on the Blue Ribbon Commission Report," available at http://brc.gov/sites/ default/files/ meetings/presentations/brc_comments_idahos_perspective.pdfIdaho

[24] http://www.brc.gov/sites/default/files/comments attachments/brc_nrcomentletter.pdf

[25] The Atomic Energy Commission was the nation's first overarching nuclear regulatory authority. It was established by the Atomic Energy Act of 1946.

[26] The National Academy of Sciences (NAS) was established by an Act of Congress in 1863 to provide independent advice to the U.S. government on matters related to science and technology. The National Research Council, which was created in 1916 to extend the scope of the NAS in its advisory role, conducts much of the research and technical analysis, and issues many of the reports released under NAS auspices. Today, the National Research Council is operated collaboratively by the NAS and by the National Academy of Engineering as the working arm of both organizations. (The NAS, the National Academy of Engineering, and the National Research Council—along with the Institutes of Medicine— together comprise the "National Academies.") For simplicity in this report, we use the acronym "NAS" to encompass work that may be undertaken by, or studies that have been issued by, the National Research Council.

[27] In 1957, the NAS published *The Disposal of Radioactive Waste on Land* (Washington, DC: The National Academies Press). This report recommended geological disposal and specifically recommended disposal in cavities mined in salt beds or domes.

[28] ERDA, along with the newly formed NRC, took the place of the AEC in 1975. Soon after, in 1977, the functions and responsibilities of ERDA were assumed by the newly formed DOE.

[29] A statement by Representative Morris Udall of Arizona, on the floor of the House of Representatives in 1987, summed up the general mood of dismay. Referring to the site selection process in the original NWPA, Representative Udall said, "We created a principled process for finding the safest, most sensible place to bury these dangerous wastes. Today, just 5 years later, this great program is in ruins. Potential host states no longer trust the technical integrity of the Department of Energy's siting decisions."

[30] By 1989, DOE was relying on the Negotiator to find an MRS site, with linkages to the repository removed. According to a DOE report to Congress in 1989 concerning the schedule for an MRS facility: "[T]he reference schedule for the MRS facility assumes that (1) a site will be obtained through the efforts of the Nuclear Waste Negotiator and (2) the statutory linkages specified in the Nuclear Waste Policy Amendments Act between the MRS facility and the repository (see Section 4) are modified" (U.S. Department of Energy, *Reassessment of the Civilian Radioactive Waste Management Program: Report to the Congress by the Secretary of Energy*, November 29, 1989, DOE-RW-0247).

[31] Keith Schneider, "Grants Open Doors for Nuclear Wastes," *New York Times*, January 9, 1992, available at http:// www.nytimes.com/1992/01/09/us/grants-open-doors-fornuclear-waste.html?scp= 2&sq=Office +of+the+ nuclear+waste +negotiator&st=cse&pagewanted =print.

[32] Several organizations that commented on the BRC Disposal Subcommittee's draft report have pointed out that the Yucca Mountain site was ranked first among candidate sites in the DOE assessments that led up to the 1987 Amendments.

[33] See MSNBC, "Store Nuclear Waste on Reservation? Tribe Split," June 26, 2006, accessible at http://www.msnbc. msn.com/id/13458867/.

[34] Governor Gary R. Herbert, State of Utah, letter to the Blue Ribbon Commission on America's Nuclear Future, September 13, 2011. Available at www.brc.gov.

[35] Richard B. Stewart, "Solving the U.S. Nuclear Waste Dilemma," *Environmental Law and Policy Review* (forthcoming, 2011). See http://www.brc.gov/sites/default/ files/meetings/ attachments/stewart_ elpar_ article.pdf.

[36] Massachusetts Institute of Technology, *The Future of the Nuclear Fuel Cycle: An Interdisciplinary MIT Study*, Cambridge, MA, 2011, p. 4.

[37] See description in Luther J. Carter, *Nuclear Imperatives and Public Trust: Dealing with Radioactive Waste*, Washington, D.C.: Resources for the Future, 1987, pp. 84-89.

[38] E. Michael Blake, "Where new reactors can (and can't) be built," *Nuclear News*, November 2006, pp. 23-25. Other states, such as Minnesota, have adopted moratoria on new nuclear reactors, but these moratoria are not necessarily tied to the waste issue.

[39] U.S. Nuclear Regulatory Commission, 42 FR 34391, July 5, 1977.

[40] "Waste Confidence and Waste Challenges: Managing Radioactive Materials," Remarks Prepared for NRC Chairman Dale E. Klein, Waste Management Symposium, Phoenix, Arizona, February 25, 2008.

[41] At an August 22, 2007 briefing to the Nuclear Regulatory Commission on new reactor issues ((the briefing was attended by all four NRC commissioners), Marvin Fertel of NEI called for the NRC to reaffirm the waste confidence decision: "[W]e believe that it would be prudent and reasonable for NRC to consider reaffirming their waste confidence position that they currently have in rulemaking...The thing that we think would be harmful to decision-making at the companies and then to the licensing process themselves is to have this [uncertainty about when and whether Yucca Mountain would be licensed] become an issue in individual proceedings. We think it would delay proceedings. We think the potential for that could actually impact decision-making by corporate boards... So our recommendation would be for the Commission to look at going forward to update the rulemaking and to have that behind us as soon as possible as this licensing process begins and particularly as the companies make decisions. [F]irm decisions [about moving ahead with new reactors] are still being discussed and evaluated at the Board level. So anything we can do from our standpoint to relieve what people perceive as risks, we think is important and that's one that we do perceive as a risk." In a September 7, 2007 follow-up memo (SRM M070822) on the meeting to the Executive Director for Operations and the General Counsel, the secretary of the Commission reported that the Commission agreed: " The Commission agreed with the nuclear industry view that it was appropriate to update the NRC's waste confidence findings in the near term. Accordingly, the staff should include waste confidence in its proposal to the Commission regarding potential rulemaking to resolve issues that are generic to COL applications, as required by the Staff Requirements Memorandum to COMDEK-07-0001/COMJSM-07-0001." See http:// pbadupws.nrc.gov/docs/ML0724/ML072400432.pdf,

[42] U.S. Nuclear Regulatory Commission news release No. 10-162, September 15, 2010.

[43] Testimony of Jack Spencer, The Heritage Foundation, before the Subcommittee on Energy and Power, Committee on Energy and Commerce, United States House of Representatives, June 3, 2011, http://www.heritage.org/ research/testimony/2011/06/the-american-energy-initiative.

[44] The NWTRB report presents all of its country-specific information in tables, using alphabetical groupings of three countries at a time.

[45] Most ratepayers are, of course, also taxpayers (and vice versa). For clarity, we refer to taxpayers and ratepayers as distinct groups here and in the main body of the report.

[46] Spent nuclear fuel and other high-level radioactive waste often also contains toxic or hazardous chemicals, but these are not primary drivers of the disposal concerns and issues that are the subject of the Blue Ribbon Commission's work.

[47] In the past, a number of concepts have been advanced periodically in hopes of eliminating the need for long-term nuclear waste disposal options (including permanent repositories). One program at Los Alamos National Laboratory, for example, focused on accelerator-driven systems for transmuting waste; it eventually evolved into a more comprehensive effort known as the Advanced Fuel Cycle Initiative. Advanced fuel cycle technologies are discussed in chapter 11 of this report.

[48] An international review of options for disposal of high-level waste and spent fuel conducted by the National Academy of Sciences specifically examined technologies for separating out and transmuting long-lived radionuclides to produce wastes that have shorter half-lives and that therefore pose less of a challenge for long term disposal. The NAS study concluded that "this option should be considered a supplement to, but not a substitute for, continued surface storage or geological disposition." It also concluded that "Geological disposition followed by closing the repository (geological disposal) is nevertheless the only permanent and final solution to the waste problem." National Research Council, *Disposition of High-Level Waste and Spent Nuclear Fuel: The Continuing Societal and Technical Challenges*, Washington DC: The National Academies Press, 2001, Chapter 7.

[49] For example, see endnotes 52, 53 and 56

[50] According to a report issued by the OECD's Nuclear Energy Agency (NEA) in 2008 and titled *Moving Forward with Geological Disposal of Radioactive Waste*: "The overwhelming scientific consensus world-wide is that geological disposal is technically feasible. This is supported by the extensive experimental data accumulated for different geological formations and engineered materials from surface investigations, underground research facilities and demonstration equipment and facilities; by the current state-of-the-art in modeling techniques; by the experience in operating underground repositories for other classes of waste; and by the advances in best practice for performing safety assessments of potential disposal systems." See p. 7 of report available at: http://www.oecd-nea.org/rwm/reports/2008/nea6433- statement.pdf and http://www.oecd-nea.org/rwm/documents/FSC_moving_flyer_A4.pdf.

[51] On July 19, 2011 the European Commission adopted the "radioactive waste and spent fuel management directive" that had been proposed by the Commission for the European Union on November 3, 2010. That directive supports disposal as the necessary long-term end point for radioactive waste: "Temporary storage is an important stage in the overall management of radioactive waste, in particular for spent fuel and HLW, allowing effective cooling and radiation levels to decrease thereby making handling safer. However, there is also a broad consensus that storage of spent fuel and radioactive waste, including long-term storage, is only an interim solution requiring active and permanent institutional controls. In the longer term, only disposal with its inherent passive safety characteristics can guarantee protection against all potential hazards." See http://eur-lex.europa.eu/LexUriServ/LexUriServ.do?u ri=CELEX:52010PC0618:EN:HTML:NOT.

[52] Massachusetts Institute of Technology, *The Future of the Nuclear Fuel Cycle: An Interdisciplinary MIT Study*, Cambridge, MA, 2011, p. 59.

[53] The description in this paragraph is adapted from U.S. Department of Energy, *Final Environmental Impact Statement: Management of Commercially Generated Radioactive Waste*, Volume 1, October 1980, DOE/EIS-0046F Volume 1 of 3 UC-70.

[54] International Atomic Energy Agency, *Scientific and Technical Basis for Geological Disposal of Radioactive Wastes*, Vienna, 2003.

[55] Nuclear Waste Technical Review Board, *Survey of National Programs for Managing High-Level Radioactive Waste and Spent Nuclear Fuel: A Report to Congress and the Secretary of Energy*, Washington DC, October 2009, available at http://www.nwtrb.gov/reports/ reports.html.

[56] A similar conclusion is reached in several submissions made to the BRC (e.g. Hansen, et. al, *Geologic Disposal Options in the USA*, SAND2010-7975C).

[57] For a description of different borehole disposal concepts, see Fergus Gibb, "Deep borehole disposal (DBD) methods," *Nuclear Engineering International*, March 25, 2010, at http://www.neimagazine.com/story.asp?storyCode=2055862. See also: Patrick V. Brady, Bill W. Arnold, Geoff A. Freeze, Peter N. Swift, Stephen J. Bauer, Joseph L. Kanney, Robert P. Rechard, Joshua S. Stein, *Deep Borehole Disposal of High-Level Radioactive Waste*, SAND2009-4401, August 2009, at http://www. mkg.se/uploads/Bil_2_Deep_Borehole_Disposal_HighLevel_Radioactive_Waste_-_Sandia_Report_2009-4401_ August_2009.pdf. In addition, the Commission received a number of public comments about deep boreholes.

[58] Chris Whipple, "Disposal of Spent Nuclear Fuel and High-level Radioactive Waste," ENVIRON International Corporation, September 10, 2010, at http://www.brc.gov/ sites/default/files/documents/disposal_of_spent_nuclear_fuel_and_high_level_radio active_waste_rev4.pdf.

[59] EPA's existing disposal standards (40 CFR Part 191) apply to any disposal method for SNF and HLW.

[60] It is important to recognize that retrievability is not an absolute or binary characteristic—rather it is a relative one. The question is how easy (or difficult) would it be to retrieve materials from a geologic disposal facility and over what time frame. Wastes that were disposed of geologically could always, if absolutely necessary, be recovered somehow—although different methods of disposal could make it more or less expensive to do so.

[61] The OECD/NEA's International Project on Reversibility and Retrievability defines retrievability as "the possibility to reverse the step of waste emplacement" and reversibility as a term that "implies a disposal programme that is implemented in stages and that keeps options open at each stage, and provides the capability to manage the repository with flexibility over time." See: *International Understanding of Reversibility of Decisions and Retrievability of Waste in Geological Disposal* at http://www.oecd-nea.org/rwm/rr/ documents/RR_Leaflet.pdf.

[62] Specifically, current regulations stipulate that the option of waste retrieval must be preserved throughout the period of waste emplacement and thereafter until the completion of a performance confirmation program and subsequent NRC review.

[63] Bill W. Arnold, Peter N. Swift, et al, "Into the Deep," *Nuclear Engineering International*, March 25, 2010. http://www.neimagazine.com/story.asp?storyCode=2055856.

[64] U.S. Department of Energy, "Analysis of the Total System Life Cycle Cost of the Civilian Radioactive Waste Management Program, Fiscal Year 2007," (DOE/RW0591),Washington DC, July 2008.

[65] U.S. Department of Energy, "Fiscal Year 2007 Civilian Radioactive Waste Management Fee Adequacy Assessment Report," (DOE/RW-0593), Washington DC, July 2008.

[66] Electric Power Research Institute, *Industry Spent Fuel Storage Handbook*, July 2010, available at http://my.epri.com/ portal/server.pt?Abstract_id=000000000001021048.

[67] The figure is from a presentation to the Blue Ribbon Commission by Dr. John Kessler of EPRI. In his presentation, Dr. Kessler predicted that utilities "will continue with on-site storage on a plant-by-plant basis—barring clear, compelling national guidance."

[68] It is worth noting that the NRC's 2010 Waste Confidence finding is being challenged in court by three states; it is also worth noting that the NRC took care to reaffirm as part of this finding its view that final disposal in a mined repository will still be necessary.

[69] See study conducted for the BRC by Cliff W. Hamal, Julie M. Carey and Christopher L. Ring and titled "Spent Nuclear Fuel Management: How centralized interim storage can expand options and reduce costs," May 16, 2011, available at http://brc.gov/sites/default/ files/documents/centralized_interim_storage_of_snf.pdf. See also Massachusetts Institute of Technology, *The Future of the Nuclear Fuel Cycle: An Interdisciplinary MIT* Study, Cambridge, MA, 2011, p. 50.

[70] Derived from U.S. Department of Energy, *Report to Congress on the Demonstration of the Interim Storage of Spent Nuclear Fuel from Decommissioned Nuclear Power Reactor Sites* (DOE/RW-0596), December 2008. This table contains updated information from reactor site managers as of January 2012.

[71] Nevada Nuclear Waste Task Force, "Comments on the: Transportation and Storage Subcommittee Draft Report to the Full Commission," June 9, 2011.

[72] "Because the NWPA required DOE to enter into contracts with the owners and generators of SNF, rather than merely create a statutory program, changes in the program, even if directed by a statutory change, can potentially cause further breaches of contract, which then can create additional monetary liability." Statement of Michael F. Hertz, Deputy Assistant Attorney General Civil Division, before the Blue Ribbon Commission on America's Nuclear Future, February 2, 2011, pp. 10-11.

[73] Van Ness Feldman P.C., "Legal Background and Questions Concerning the Federal Government's Contractual Obligations Under the 'Standard Contracts' with 'Utilities,'" Memorandum prepared for the BRC, revised Dec. 20, 2010, at pp. 22-23.

[74] In March, 2011, the Department of Justice offered all remaining SNF litigants a New Framework settlement agreement that would be applicable to the 78 nuclear power reactors covered by a Standard Contract that had not previously entered into a settlement agreement. Under this framework, which uses a higher acceptance rate for determining the government's performance liability than has been used in previous settlements, there is no "crossover point" at which the government will have caught up with its acceptance obligations. Rather, the government's liability for spent nuclear fuel storage costs will continue until the spent fuel is removed from reactor sites covered by a New Framework settlement agreement. As of October 6, 2011, eight utilities have accepted the New Framework settlement agreement. See memorandum provided by David K. Zabransky, Director, Office of Standard Contract Management Office of General Counsel, to Steve Isakowitz, Chief Financial Officer, DOE on October 26, 2011 and available at http://www.brc.gov/index. php?q= generalcomment/response-brc-request-information.

[75] NWPA, Sections 144-149.

[76] The regulations cover multiple types of dry cask technologies as well as dry vaults. While no ISFSIs using pools have been proposed, there is little doubt that pools – the storage technology for which there is most experience – would not raise any new technical issues.

[77] This would be consistent with common practice in Sweden and France, where fuel is removed from reactor pools within a year after discharge and moved to central pool storage pending later disposition. (In Sweden, the fuel is stored for disposal; in France it is stored for reprocessing.)

[78] It is worth noting that nearly 60 percent of the fuel discharged from the six reactors at Fukushima prior to the earthquake and tsunami had already been transferred into a shared pool, leaving relatively small inventories of spent fuel (compared to U.S. practice) in the reactor pools. This shared pool appears to have survived the disaster relatively unscathed.

[79] The MRS Review Commission concluded that "in view of the continuing delay in the building of a repository... it would be in the national interest to have available a safety net of storage capacity for emergency purposes, such as an accident at a nuclear power plant, which would make it advantageous to have the plant's spent fuel pool available for decontamination of affected parts of reactors and for storage of debris." The Review

Commission recommended construction of a Federal Emergency Storage (FES) facility with a capacity limit of 2,000 metric tons. See U.S. Monitored Retrievable Storage Review Commission, *Is there a need for federal interim storage? Report of the Monitored Retrievable Storage Commission*, University of Michigan Library, 1989. Accessible at http://www.brc.gov/sites/default/ files/documents/is_there_a_need_for_interim_ sto rage_s.pdf

[80] U.S. Department of Energy, Office of Civilian Radioactive Waste Management, *1987 OCRWM Mission Plan Amendment* (DOE/RW-0128), June 1987, p. 116.

[81] It is worth noting that the opportunity to host an R&D facility of this type might itself be among the inducements for a community interested in being considered for a consolidated storage facility. A national center for ongoing research on all aspects of the storage of spent fuel could be a significant ancillary benefit for a community willing to host a storage facility.

[82] "If standardization is not mandated by the Federal government, then an MRS facility that accepts waste early could promote standardization by reducing the variety of spent fuel forms and packages to be handled and limiting the number of reactors providing storage for other than intact, unpackaged spent fuel." U.S. Monitored Retrievable Storage Review Commission, *Is there a need for federal interim storage? Report of the Monitored Retrievable Storage Commission*, University of Michigan Library, 1989, p. 97. More generally, actions that could be taken now or in the near term to specify standardized requirements for SNF package designs could be of substantial value to the waste management program, and to utilities and cask vendors, in the future. Standardization would help ensure that SNF and HLW would not need to be re-packaged before being shipped and emplaced at a storage or disposal facility. See for example, Nuclear Waste Technical Review Board, *Technical Advancements and Issues Associated with the Permanent Disposal of High-Activity Wastes: Lessons Learned from Yucca Mountain and Other Programs*, Washington DC, June 2011, p. 44.

[83] The MRS Review Commission evaluated occupational doses to workers in the no-MRS, linked MRS, and unlinked MRS systems and concluded that the unlinked MRS system would result in the lowest doses because of "greater reliance on remote operations and remote handling facilities" at the MRS. U.S. Monitored Retrievable Storage Review Commission, *Is there a need for federal interim storage? Report of the Monitored Retrievable Storage Commission*, University of Michigan Library, 1989, p. 13.

[84] The MRS Review Commission explicitly evaluated the argument that a system using dual-purpose storage/ transportation casks for storage at reactors would provide as much flexibility as a system including a centralized MRS facility and concluded that it would not because they could not be certain "that a dual-purpose cask could be developed that could be used for prolonged storage and then transported without having to be returned to a spent fuel pool or opened." Ibid., p. 95.

[85] The recent MIT fuel cycle study (titled *The Future of the Nuclear Fuel Cycle*; full cite given in previous endnotes and available at http://web.mit.edu/mitei/docs/spotlights/ nuclear-fuel-cycle.pdf) refers to storage on the order of a century. The NRC is evaluating the implications of storage for a period of up to 300 years. As noted in the main text, the fact that storage periods of this length are being evaluated should not be taken as an indication that storage over multi-century timeframes would be desirable or even defensible from either a cost or risk management standpoint.

[86] A.C. Kadak and K. Yost, *Key Issues Associated with Interim Storage of Used Nuclear Fuel*, MIT, 2010, pp. 27-28.

[87] Staged development of a centralized storage facility is discussed in a paper developed for the BRC by Cliff W. Hamal, Julie M. Carey and Christopher L. Ring and titled "Spent Nuclear Fuel Management: How centralized interim storage can expand options and reduce costs," *May 16, 2011*, pp. 48-50. Available at http://brc.gov/sites/default/files/ documents/ centralized_interim_storage_of_snf.pdf.

[88] American Physical Society, *Consolidated Interim Storage of Commercial Spent Nuclear Fuel: A Technical and Programmatic Assessment*, American Physical Society Panel on Public Affairs, February 2007

[89] It is possible that the contractual obligation for waste acceptance would remain with DOE for some time and perhaps indefinitely—even after a new waste management organization is established.

[90] Eileen M. Supko and Michael H. Schwartz, *Overview of High-Level Nuclear Waste Materials Transportation: Processes, Regulations, Experience and Outlook in the U.S.*, Energy Resources International, Inc., ERI-2030-1101, January 2011, p. 74.

[91] While the Standard Contract allows DOE to give priority to fuel at shutdown sites, the Department has declined to consider this option in the past because of concerns about equity impacts on contract holders. U.S. Department of Energy, *Report to Congress on the Demonstration of the Interim Storage of Spent Nuclear Fuel from Decommissioned Nuclear Power Reactor Sites* (DOE/RW0596), Dec. 2008, p. 3.

[92] This chart uses GAO's estimate of $4.5 million/year M&O costs for stranded fuel at a shutdown site. While the figure indicates a repository, the estimates would apply to any facility capable of accepting spent fuel in 2030.

[93] Some of the data used in preparing this chart were derived from the following report: Government Accountability Office, *Nuclear Waste Management; Key Attributes, Challenges, and Costs for the Yucca Mountain Repository and Two Potential Alternatives,* GAO-10-48, November 2010.

[94] This cost analysis is based on capacity for either interim storage or disposal being available for spent fuel.

[95] U.S. Nuclear Regulatory Commission SECY-11-0137, "Prioritization of Recommended Actions To Be Taken in Response to Fukushima Lessons Learned," Oct. 3, 2011 (available at http://www.nrc.gov/reading-rm/doc-collections/ commission/secys/2011/2011-0137scy.pdf).

[96] In the course of the BRC's deliberations, Commission members with appropriate clearances were briefed by officials from DOE, NRC, and other agencies regarding issues of fuel storage and transportation safety and security. These briefings also covered related research efforts and the additional security measures that have been implemented at some sites. We are confident that the NRC's current analytical and regulatory processes are adequate to make needed assessments, and to adapt as appropriate.

[97] Over time, spent fuel "cools" thermally and radioactively and requires less shielding to be handled directly. In this way it loses some of the characteristics that would make it difficult to remove and transport for unauthorized purposes. Depending on burnup, spent fuel may no longer be self-protecting after a century or so of storage.

[98] Material for this section was developed from presentations to the BRC Transportation and Storage Subcommittee by Mr. Philip Brochman, NRC Office of Nuclear Security and Incident Response, Sept.23, 2010 (available at http://www.brc.gov/index.php?q =meeting /openmeeting-4.).

[99] Electronic mail from Dr. Brittain Hill, NRC, to Alex Thrower, BRC staff, Feb. 23, 2011 (available at http://www.brc.gov/sites/default/files/comments/attachments/post_9- 11steps_ b_hill.pdf).

[100] Additional background about NRC's security programs is available at http://www.nrc.gov /reading-rm/doc-collections/fact-sheets/security-enhancements.pdf).

[101] U.S. Nuclear Regulatory Commission Staff Requirements Memorandum dated Aug. 26, 2010 (SECY10-0014 Enclosure 1, found at http://wba.nrc.gov:8080/ves/ view_contents.jsp).

[102] 102 U.S. Nuclear Regulatory Commission paper dated Aug. 26, 2010 (SECY-10-0114, Enclosure 1), found at https://adamsxt.nrc.gov/WorkplaceXT/getContent ?id=release&vsId =%7B1214CFFE-E9C0-4742-B109-74DCD84A1B84%7D&objectStoreName=Main.__.Library&object Type= document.

[103] Robert Alvarez et al., "Reducing the Hazards from Stored Spent Fuel Power-Reactor Fuel in the United States," *Science and Global Security* 11: 1-51, 2003.

[104] Allan S. Benjamin et al., *Spent Fuel Heatup Following Loss of Water During Storage*, Sandia National Laboratory (NUREG/CR-0649, SAND77-1371), 1979.

[105] Robert Alvarez et al., "Reducing the Hazards from Stored Spent Fuel Power-Reactor Fuel in the United States," *Science and Global Security* 11: 1-51, 2003, p. 21.

[106] U.S. Nuclear Regulatory Commission, "Reducing the Hazards from Stored Spent Power-Reactor Fuel in the United States," *Science and Global Security*, Vol. 11, pp. 203–211.

[107] National Research Council, Committee on the Safety and Security of Commercial Spent Nuclear Fuel in Storage, *Safety and Security of Commercial Spent Nuclear Fuel Storage*, Washington DC: The National Academies Press, 2006, p. 5 (accessible at http://www.nap. edu/catalog. php?record_id=11263).

[108] Ibid.

[109] Ibid at p. 3

[110] The term "hardened on-site storage" is not currently defined in regulations, and is not commonly used by the industry.

[111] Michelle Boyd, "Principles for Safeguarding Nuclear Waste at Reactors," submission to the BRC, May 11, 2010 (the principles may be found at http://www.citizen.org/ documents/PrinciplesSafeguarding2009.pdf).

[112] Charles W. Pennington, "Storage and Transportation of Spent Fuel: Does Storage/Transport System Hardening Enhance Safety and Security," submission to the BRC Transportation and Storage Subcommittee, Sept. 2010 (accessible at http://www.brc.gov/sites/default/ files/meetings/ presentations/c_pennington_presentation_ final.pdf). Mr. Pennington subsequently submitted a detailed critique of the HOSS proposal as presented by Mr. David Kraft at the subcommittee meeting in Chicago, IL on Nov. 2, 2010. Mr. Kraft's submittal can be found at http://www.brc.gov/sites/ default/files/meetings/presentations/panel_ present_to_brc_11- 2-10.pdf. Mr. Pennington's critique was submitted to the BRC on January 20, 2011 and is available at http://brc.gov/ sites/default/files/comments/ attachments/recapitulating_and_ expanding_upon_safety_of_dry_storage_-_final.pdf.

[113] Nuclear Energy Institute [NEI 06-12, Revision 2), "B.5.b Phase 2 & 3 Submittal Guideline." This document was initially designated for Official Use Only – Security Related Information, and so is unavailable to the public. However, it was made publicly available on May 9, 2011 and can be found on NRC's ADAMS system at http://www. nrc.gov/reading-rm/adams.html with accession number ML070090060.

[114] See remarks in the comments of Bill Borchardt, Executive Director for Operations of the Nuclear Regulatory Commission, and Anthony Pietrangelo, Senior Vice President and Chief Nuclear Officer of the Nuclear Energy Institute, in the transcript of the March 29, 2011 meeting of the Senate Committee on Energy and Natural Resources on the accident at the Fukushima Daiichi reactor complex, accessible at http://dpwsa. powergenworldwide. com/index/display/wire-news-display/1389933775.html.

[115] Michelle Boyd, "Principles for Safeguarding Nuclear Waste at Reactors," submission to the BRC, May 11, 2010 (the principles may be found at http://www.citizen.org/ documents/PrinciplesSafeguarding2009.pdf).

[116] Another country that has grappled with the siting issue is Germany, which in the late 1990s commissioned an expert committee (not unlike the BRC) to look at the problem of nuclear waste. The German committee developed a relatively straightforward plan in which the siting organization was to do an initial screening of the entire country for geologically suitable sites, based on a short set of criteria. From the subset of potentially suitable sites, weighted criteria were to be used to reduce the number of potential locations to five. At that point, the five affected municipalities were to be asked whether they wished to go forward with a more detailed evaluation. The hope was that at least two sites would survive this next cut, and assuming approval could be obtained from the local communities, the plan was to build two underground facilities for further technical analysis in preparation for a final decision. However, because of a change of government, the German plan was never implemented.

[117] According to a March 2010 document issued by the NEA's Forum on Stakeholder Confidence: "History shows that the search for sites for radioactive waste management facilities has been marred by conflicts and delays. Affected communities have often objected that their concerns and interests were not addressed. In response, institutions have progressively turned away from the traditional "decide, announce and defend" model, and are learning to "engage, interact and co-operate." This shift has fostered the emergence of partnerships between the proponent of the facility and the potential host community, as shown in a recent NEA study. Working in partnership with potential host communities enables pertinent issues and concerns to be raised and addressed, and creates an opportunity for developing a relationship of mutual understanding and mutual learning, as well as for developing solutions that will add value to the host community and region. Key elements of the partnership approach are being incorporated into waste management strategies, leading increasingly to positive outcomes." See: http://www.oecd-nea.org/rwm/fsc/docs/ FSC_ partnership_flyer_U.S._letter.pdf.

[118] Under Finland's Nuclear Energy Act of 1987, the consent of the host municipality is required for any major nuclear installation (including reactors as well as repositories). Thus, local acceptance was a necessary prerequisite for any decision in principle to approve the Olkiluoto repository. Interestingly, when a proposal for the Olkiluoto repository first came up for a vote by the local town council, it was vetoed. See http://www.finlex.fi/fi/laki/ kaannokset/1987/en19870990.pdf.

[119] Like the U.S. program, the Finnish program included a siting schedule. However, that schedule allowed considerably more time than in the U.S. case. The schedule set by Finnish government in 1983 called for repository construction to begin in 2010, and targeted 2020 as the date when spent fuel would begin to be accepted for final disposal. See http://www.worldenergy.org/documents/ p000915.pdf.

[120] The Swedish Act on the Management of Natural Resources gives municipalities a veto over siting permits. While the government has the right, under certain circumstances, to disregard such vetoes, neither SKB nor the Swedish Parliament favored siting a repository without the consent of the selected municipality. The government's choice not to exercise its override authority, in other words, represents a discretionary policy decision. See SKB RD&D Programme 1998, p. 30 (accessible at http://www.skb.se/ upload/publications/pdf/ RD&D98webb.pdf) and Rolf Lidskog & AnnCatrin Andersson, "The management of radioactive waste: A description of ten countries" (accessible at http://www.edram.info/ en/edram-home/joint-activities/ status-report-skb-report/ index.php), p. 71.

[121] Claes Thegerström (CEO of the SKB), "Comment on Disposal Subcommittee Draft," submission to the full Commission on June 29, 2011 (available at: http://www.brc. gov/sites/default/files/comments/attachments/brb-text_5_ sweden.pdf).

[122] See "Next phase for French geological disposal," January 5, 2012, at http://www.world-nuclear-news.org/WR-Next_phase_for_French_geological_disposal-0501127.html.

[123] In a presentation before the Commission, Liz Dowdeswell, former President of the NWMO, summarized the organization's perspective this way: "We believed that fundamentally the selection of an approach for long-

term management was really about developing a contract between science and society, a contract that would allow all of us to continue to benefit from technology, but also would mitigate risk and, most importantly, would respect the values of our citizens."

124 This siting process was described to the Blue Ribbon Commission at its September 20, 2010 meeting. See http:// www.brc.gov/sites/default/files/meetings/presentations/ alvaro_ rodriguez_usa_21-09-10.pdf and http://www. brc.gov /sites/default/files/meetings/ attach ments/alvaro_atc_articulo_para_la_ens.pdf

125 National Research Council, *Disposition of High-Level Waste and Spent Nuclear Fuel: The Continuing Societal and Technical Challenges, Summary*, Washington DC: The National Academies Press, 2001. An even earlier National Academies study, issued in 1990 and titled *Rethinking High-Level Radioactive Waste Disposal*, likewise called for an adaptive approach. According to the abstract: "This alternative approach emphasized flexibility; time to assess performance and a willingness to respond to problems as they are found, remediation if things do not turn out as planned, and revision of the design and regulations if they are found to impede progress toward the health goal already defined as save disposal. To succeed, however, this alternative approach will require significant changes in laws and regulations, as well as in program management."

126 In follow-on study sponsored by DOE, the National Academies elaborated on this central conclusion by describing two approaches to staging: (1) "Linear staging, involving a single, predetermined path to a well-defined end point, with stages viewed as milestones at which cost and schedules are reviewed and modified as needed" (this is the approach that in the Academies' view characterized the current U.S. program); and (2) "adaptive staging, which emphasizes deliberate continued learning and improvement and in which the ultimate path to success and the end points themselves are determined by knowledge and experience gathered along the way." The report, which was issued in 2003, concluded by recommending that adaptive staging should be the approach used in geologic repository development. See: National Research Council, *One Step at a Time: The Staged Development of Geologic Repositories for High-Level Radioactive Waste*, Washington DC: The National Academies Press, 2003.

127 The *One Step at a Time* report argues strongly for the use of a periodically-revised safety case as a central feature of adaptive staging: "Two primary roles of the safety case are: (1) to guide the work of the implementer while adapting the program at each stage, and (2) to provide the implementer with a vehicle for making the safety arguments understandable by a wide audience."

128 The concept of a broad "safety case" that integrates all available lines of evidence supporting the safety of a repository, including institutional as well as technical factors, has become increasingly prominent internationally over the last decade. See International Atomic Energy Agency And The OECD Nuclear Energy Agency, *Geological Disposal Of Radioactive Waste: Safety Requirements*, No. Ws-R4, International Atomic Energy Agency, Vienna, 2006, sections 3.40-3.53 (available at www-pub.iaea.org/MTCD/ publications/ PDF/Pub1231_web.pdf).

129 The Commission recognizes that many arguments for the safety of the repository are distributed throughout the regulatory process in the regulations and the justification documents supporting them, in requirements for reporting, and elsewhere. See EPRI comments on the BRC Disposal Subcommittee's Draft Report to the Full Commission, July 1, 2011 available at http://www.brc.gov/sites/default/ files/comments/attachments/ epri_ comments_on_the_brc_ disposal_subcommittee_draft_report_-_final_0.pdf.

130 "Complex performance assessments may be necessary for licensing, but the public and policy-makers may not find them easy to understand. The probabilistic performance assessment methods and results developed for Yucca Mountain are a state-of-the-art achievement and very valuable. They verge on being incomprehensible because of their complexity, however, except perhaps to organizations able to afford a large cadre of experts. Future repository programs still may have to produce complex performance assessments for compliance purposes, but they also must produce more-realistic, less-complex performance assessments for non-regulators. As experience with the Yucca Mountain program clearly shows, not only the regulators decide the fate of a repository program. The audience for the safety case is much broader." See Nuclear Waste Technical Review Board, *Technical Advancements And Issues Associated With The Permanent Disposal Of High-Activity Wastes: Lessons Learned from Yucca Mountain and Other Programs*, Washington DC, June 2011.

131 "If adopted, Adaptive Staging would lead DOE to ...Focus more strongly on achieving the degree of technical and societal consensus needed to begin waste emplacement, rather than on the emplacement of all waste."National Research Council, *One Step at a Time: The Staged Development of Geologic Repositories for High-Level Radioactive Waste*, Washington DC: The National Academies Press, 2003, pp. 7-8.

132 Ibid., pp. 22-23.

133 Unless the provisions of an agreement would require additional legislative authorizations not already provided in the law establishing the waste management organization.

[134] This is very well demonstrated in Sweden's repository program which began by establishing an underground rock laboratory.

[135] A similar approach has just been adopted in a recent directive of the European Commission requiring members of the European Community to develop programs and schedules for developing disposal facilities: "Member States will have to draw up national programmes and notify them to the Commission by 2015 at the latest. The Commission will examine them and can require changes. National programmes have to include plans with a concrete timetable for the construction of disposal facilities, as well as a description of the activities needed for the implementation of disposal solutions, costs assessments and a description of the financing schemes. They will have to be updated regularly." Source: "Nuclear waste: Commission welcomes adoption of radioactive waste directive," Brussels, 19 July 2011, accessible at http://ec.europa.eu/energy/nuclear/waste_management/waste_ management _en.htm.

[136] This contention is supported by a 2008 report of the National Academy of Sciences, which concluded: "When done well, public participation improves the quality and legitimacy of a decision and builds the capacity of all involved to engage in the policy process. It can lead to better results in terms of environmental quality and other social objectives. It also can enhance trust and understanding among parties. Achieving these results depends on using practices that address difficulties that specific aspects of the context can present." See National Research Council, *Public Participation in Environmental Assessment and Decision Making*, Washington, DC: The National Academies Press, 2008.

[137] For this reason, the BRC provided funding for key NGO and community stakeholders to travel to its deliberative meetings.

[138] For example, the waste management organization could provide funding for independent monitoring and testing on a candidate repository site, provided that these activities do not interfere with other site development activities or compromise the site's integrity. In fact, Section 116 of the NWPA provides for grants to states and affected units of local governments for a number of purposes, including "any monitoring, testing, or evaluation activities with respect to site characterization programs with regard to such site," while Section 117 adds the proviso "except that such monitoring and testing shall not unreasonably interfere with or delay onsite activities." Funding for monitoring, testing, or evaluation activities is also provided for affected tribes. Under these provisions, over $4 million was provided to Inyo County, CA for the Inyo Regional Ground Water Monitoring Program, and over $31 million was provided to Nye County, NV for a Science & Verification Program that included the Nye County Early Warning Drilling Program, which provided data used in the Yucca Mountain project (Office of Civilian Radioactive Waste Management Office of Business Management, "Summary of Program Financial & Budget Information as of January 31, 2010").

[139] For example, a report from 1980 on the subject pointed out that states have a "constitutional responsibility to ensure the health and safety of their citizens," as well as "jurisdiction over local authorities and land use," and that states therefore believed "it is both undesirable and impartial for disposal procedures to be wholly federally determined." See Pat Choate and John Bowman, *Radioactive Waste Management: State Concerns, A Report to the Office of Technology Assessment from the Academy for Contemporary Problems*, 1980, p. 3.

[140] Ibid p. 11.

[141] An absolute state veto had been opposed by the State Planning Council established by President Carter to provide advice on intergovernmental relations, as well as by others. See U.S. Congress Office of Technology Assessment, *Managing the Nation's Commercial High-Level Radioactive Waste*, (OTA-O-171), March, 1985, p. 180.

[142] The state of Nevada's strong opposition to the proposed Yucca Mountain repository is well known, but other examples abound. In Utah, efforts to site a private centralized storage facility were blocked when the Utah delegation successfully pushed for congressional designation of a wilderness area that prevented access to the proposed site. Utah took this action despite its tradition of hostility toward past federal efforts to designate wilderness lands and national monuments within the state.

[143] The Commission recognizes that more than one community, state, or tribe might be affected by a proposed repository. The waste management organization should therefore be directed to consult with any state, affected unit of local government, or Indian tribe that it determines may be so affected and to include any reasonable and appropriate provisions relating to their interests in negotiated agreements, as the Nuclear Waste Negotiator was directed and empowered to do under Section 403(b) of the NWPA.

[144] In one notable instance—the NRC Agreement States program—regulatory authorities of the federal government under the Atomic Energy Act have been delegated to states. Although the current Agreement States program does not cover licensing of a repository, it does suggest a similar enforcement model might give a host state or tribe sufficient regulatory oversight to assure a meaningful public safety role. Alternative approaches through

memoranda of understanding or other binding agreements may also be acceptable and should be explored as part of the negotiating process.

145 For instance, the Western Governors' Association (WGA) adopted a definition of consent-based siting for spent fuel storage facilities as part of its Resolution 11-3. The resolution requires the written consent of the governor of the hosting state. Members of the WGA's WIEB High-Level Waste Committee also believe that written consent should be required for all future disposal and storage decisions. For details see WGA's comments, dated September 13, 2011, on the BRC draft report.

146 Mixed waste is waste that contains, in addition to radioactive materials, materials that are defined as hazardous under RCRA (e.g., a chemical such as toluene).

147 Current federal law—including aspects of the Atomic Energy Act, the Commerce Clause, and the doctrine of intergovernmental immunity on federal reservations—has the effect of preempting almost all forms of state regulation over a high-level waste facility.

148 Elements that were essential to the success of the Environmental Evaluation Group have been summarized by the two scientists who served as director of the organization. See R.H. Neill and M.K. Silva, "EEG's Independent Technical Oversight on WIPP, a TRU Waste Geologic Repository" in the conference proceedings of the 9th International High Level Radioactive Waste Management Conference, Session T-1, Las Vegas, NV, April 29–May 3, 2001.

149 The article quotes Sullivan as stating that "the same problems that existed 20 years ago still exist today. Among them is the lack of trust that western states have of the federal government to either follow through on a long-term policy or to actually work in a state's own interest." See: http://wyofile. com/2011/02/sullivan-i-was-right-to-veto

150 Peter C. Chestnut et al., *The Role Of Indian Tribes In America's Nuclear Future*, BRC Commissioned paper April 29, 2011, available at: http://www.brc.gov/sites/default/files/ documents/the_role_of_indian_tribes_in_americas_nuclear_ future-2011-04-29_final.pdf

151 See, for example, Executive Order 13175 "Consultation and Coordination with Indian Tribal Governments" available at: http://www.epa.gov/fedreg/eo/ eo13175.htm.

152 Peter C. Chestnut et al., *The Role of Indian Tribes in America's Nuclear Future*, BRC commissioned paper April 29, 2011, accessible at http://www.brc.gov/sites/default/files/ documents/the_role_of_indian_tribes_in_americas_nuclear_ future-2011-04-29_final.pdf. According to this paper, "It is critical to remember that any entity created by the federal government...must have a formal working relationship directly with all potentially affected Indian Tribes. An example of a formal working relationship with Indian tribes is the Memorandum of Agreement between the Tennessee Valley Authority (TVA) and the Tennessee State Historic Preservation Officer of Knox County, TN. This Agreement calls for consultation with federally recognized Indian tribes that are participants or invited signatories to the Agreement."

153 In France, direct financial benefits for the region surrounding the proposed repository are spelled out in law. In addition, a range of other programs to promote development are being provided. While the particular government—utility mechanism that is used for this purpose may be unique to the French situation, the concept of promoting regional development through activities that go beyond financial benefits and waste-management-related employment is worthy of careful examination.

154 For example, the authorization of federal funds to build a new road so that shipments of transuranic waste could bypass Santa Fe (known as the Santa Fe bypass) was important in the WIPP context.

155 In the past, DOE often did not make the most of these opportunities. For example, WIPP was managed for years by DOE personnel located in Albuquerque rather than at an office in Carlsbad near the facility. It was only late in the process that DOE relocated its top WIPP management to Carlsbad. Likewise, the TRANSCOM tracking system used in the transportation program was originally based out of Oak Ridge, Tennessee. It was later relocated to Albuquerque and finally moved to Carlsbad in 2005.

156 Benefits provided by the current NWPA include cash payments of up to $20 million per year (Section 171) and special consideration for selection for DOE research projects (Section 174).

157 Provisions for evaluating and providing compensation are contained in Sections 116 and 118 of the NWPA.

158 Throughout this report we use the term "management" to refer to these three activities (i.e., transportation, storage, and disposal).

159 Outside of the United States and Germany, the implementing organizations are all dedicated public or private entities rather than a ministry or department of the national government.

160 In 2010, Senator Voinovich introduced the "United States Nuclear Fuel Management Corporation Establishment Act of 2010" (S. 3322), co-sponsored by Senators Alexander and Murkowski, and Representative Upton

introduced a companion bill (H.R. 5979) in the House. There was no legislative activity on these bills in the 111th Congress.

161 *Report of the Advisory Panel on Alternative Means of Financing and Managing Radioactive Waste Management Facilities (AMFM Panel): A Report to the Secretary of Energy*, December 1984 (transmitted to Congress in April 1985), in the BRC library at http://www.brc.gov/sites/default/files/ documents/amfm_1984_s.pdf.

162 Belgium, France, Japan, Spain, and United Kingdom have established public companies to implement high level waste management programs. In Canada, Finland, Sweden, and Switzerland, waste producers have set up implementing bodies to undertake these tasks. Only the United States and Germany have assigned the job to a government department. See International Association for Environmentally Safe Disposal of Radioactive Materials (EDRAM), *Report on Radioactive Waste Ownership and Management of Long-Term Liabilities in EDRAM Member Countries*, June 2005, available at http://www. edram.info/fileadmin/edram/pdf/ EDRAM WGonWOwnershipFinal_271005.pdf.

163 The TVA board provides an example of how the need for expertise and stakeholder representation might be balanced. It has nine members appointed by the President and confirmed by the Senate. Key qualifications specified in law include "management expertise relative to a large for-profit or nonprofit corporate, government, or academic structure" and "support for the objectives and missions, of the Corporation, including being a national leader in technological innovation, low-cost power, and environmental stewardship." That is, Board members must be both capable of and invested in ensuring that the Corporation achieves its mission. In appointing members of the Board, the President must consider recommendations from governors of states in the service area; individual citizens; business, industrial, labor, electric power distribution, environmental, civic, and service organizations; and the congressional delegations of the states in the service area. Furthermore, the President must "seek qualified members from among persons who reflect the diversity, including the geographical diversity, and needs of the service area of the Corporation."

164 Section 302(d) of the NWPA limits use of the Waste Fund to "non-generic research, development, and demonstration activities under this Act." An example of such non-generic research is the OCRWM Science and Technology program initiated by OCRWM in 2002 to improve existing technologies and develop new technologies so as to achieve efficiencies and life-cycle cost savings in the waste management system (transportation, waste handling, and disposal) and to increase confidence in repository performance. Robert J. Budnitz, "Status of OCRWM's New Science and Technology Program," presentation to the National Research Council's Board on Radioactive Waste Management, December 12, 2002.

165 This could include addressing the need for complex adjustments to the nuclear waste fee schedule if spent nuclear fuel becomes a feedstock.

166 In our proposal, responsibility for the treatment and storage of defense waste would remain with DOE.

167 Based on an evaluation conducted by DOE pursuant to the NWPA and titled *An Evaluation of Commercial Repository Capacity for the Disposal of Defense High-Level Waste*, (DOE/DP/0020/1), Washington DC, 1985.

168 This general approach, in which government and not the implementing organization defines the policy framework that will guide future waste management activities is common to most countries with a significant waste management program. A review of 11 countries that are members of the International Association for Environmentally Safe Disposal of Radioactive Materials (EDRAM) shows that in all cases general waste management policy is set by government, rather than the implementing organization. International Association for Environmentally Safe Disposal of Radioactive Materials, *Report on Radioactive Waste Ownership and Management of Long-Term Liabilities in EDRAM Member Countries*, June 2005.

169 For example, "the economic and social well-being of the people living in [the Tennessee] river basin" is one of the general purposes identified in the legislation that established TVA [48 Stat. 69, 16 USC. sec. 831v]; consequently, TVA sees economic development of the region as a key part of its mission and has an economic development program for that purpose (http://www.tva.com/econdev/index.htm). Similarly, ENRESA, which is Spain's national corporation for radioactive waste management, has established the ENRESA Foundation to promote social welfare and socio-economic development, the environment, education, and culture in areas that host ENRESA facilities.

170 The NWPA already requires annual audits of the activities of OCRWM by GAO, a comprehensive annual report by OCRWM on its activities and expenditures, and an annual report to Congress from the Secretary of the Treasury (after consultation with the Secretary of Energy) on the financial condition and operations of the Waste Fund.

171 Spain, for example, may offer a useful model: the government provides policy direction to the waste management organization, ENRESA, through ministerial review and approval of a General Radioactive Waste Plan that is revised and resubmitted every four years.

[172] The CRA requires federal agencies that promulgate rules to submit certain information to each House of Congress and the General Comptroller about the rule. Generally, major rules may not become effective until 60 days after submission to Congress. During those 60 days, Congress could pass a joint resolution to disapprove the major rule. The President could veto a congressional joint resolution of disapproval. In that case Congress would have 30 days to override the President's veto. If Congress does not override the veto, the rule becomes effective. In legislation establishing the waste management organization and setting nuclear waste policy direction, Congress could provide itself CRA-like authority to review the organization's Mission Plan update.

[173] These ten countries are Belgium, Canada, France, Germany, Japan, Spain, Sweden, Switzerland, Taiwan, and the United Kingdom.

[174] The AMFM Panel recommended that a "Waste Fund Oversight Commission" be established for the specific purpose of ensuring that NWF fees are being used cost-effectively and to approve or disapprove proposed changes to the level of the fee. In its 2001 update of the AMFM study, DOE instead recommended that the Federal Energy Regulatory Commission (FERC) serve this purpose.

[175] The National Academies' *One Step at a Time* report also recommended a stakeholder advisory board.

[176] Waste management facilities include disposal and interim storage facilities as well as any new transportation infrastructure required to construct, operate or decommission a geologic repository or interim storage facility.

[177] The NWPA does provide (in a separate section) for local government representation on a review panel that would advise DOE in the context of a negotiated "benefits agreement" between the federal government and a state or tribe hosting a repository or MRS facility. However, local interests account for only a small part of the representation on this panel.

[178] Clinch River MRS Task Force, "Position on the Proposed Monitored Retrievable Storage Facility," October 10, 1985.

[179] The Voinovich/Upton bill deals with this issue by providing that contracts and settlements remain the liability of DOE until 10 years after termination of the license of the reactor involved. The new federal corporation would take liability under the existing contracts no later than 10 years after license termination; it would also be liable for all new contracts and for any negotiated transfer of liability between DOE and the corporation.

[180] For more details see "R&D Activities for Used Nuclear Fuel Disposition Storage, Transportation & Disposal," by William Boyle, Director, Office of Used Nuclear Fuel Disposition Research & Development, DOE Office of Nuclear Energy, presented to the NWTRB winter meeting, February 16, 2011 and accessible at http://www.nwtrb.gov/ meetings/2011/feb/boyle.pdf.

[181] The "polluter pays" principle for high-level waste disposal was first established by the AEC in 1970 when it established rules for the solidification and disposal of high-level wastes from reprocessing. However, the waste generators were going to pay when they actually delivered the waste for disposal, leaving the federal government to come up with the funds needed to develop a disposal system before the government could be reimbursed for this expense by the waste generators. In the NWPA, Congress departed from this approach and opted for an up-front fee to generate the revenues to build the system without having to rely on taxpayer funds, to ensure that adequate funds were available as needed.

[182] U.S. Congress Office of Technology Assessment, *Managing the Nation's Commercial High-Level Radioactive Waste*, OTA-O-171, March, 1985, p. 93, pp. 106-107.

[183] Opening Statement of Senator J. Bennett Johnston, Chairman, at a hearing before the Senate Committee on Energy and Natural Resources, March 1, 1994.

[184] Data Source: U.S. DOE Office of Civilian Radioactive Waste Management, Office of Business Management "Summary of the Program Financial & Budget Information," as of January 31, 2010.

[185] Belgium, Canada, Finland, France, Germany, Japan, Spain, Sweden, Switzerland, UK, and United States.

[186] International Association for Environmentally Safe Disposal of Radioactive Materials (EDRAM), *Report on Radioactive Waste Ownership and Management of Long-Term Liabilities in EDRAM Member Countries*, June 2005, Tables 7.4 and 7.5, accessible at http://www.edram.info/fileadmin/ edram/pdf/EDRAMWGonWOwnershipFinal_271005.pdf.

[187] See extended discussion in Joseph S. Hezir, "Budget and Financial Management Improvements to the Nuclear Waste Fund (NWF): Background Report to the Blue Ribbon Commission on America's Nuclear Future," May 2011. Accessible at http://brc. gov/ sites/default/files/documents/ brc_hezir_nwfbudget_051511.pdf.

[188] This specific combination of measures was identified as one of four feasible interim steps for dealing with the funding problem in DOE's 2001 update of the AMFM report. See U.S. Department of Energy, *Alternative Means of Financing and Managing the Civilian Radioactive Waste Management Program* (DOE/RW-0546), August 2001.

[189] Data source: U.S. Department of Energy, Office of Civilian Radioactive Waste Management Office of Business Management, "Summary of Program Financial & Budget Information" as of January 31, 2010.

[190] U.S. Department of Energy, *Alternative Means of Financing and Managing the Civilian Radioactive Waste Management Program* (DOE/RW-0546), August 2001, Fig. 3. The proposal was not accepted by the utilities because the quid pro quo was their agreement not to seek damages for delay in waste acceptance.

[191] Legal analysis performed for the BRC concluded that it may be possible to amend the Standard Contract in this way without a rulemaking proceeding to amend the rule that established the contract in the first place (10 C.F.R. § 961). See Van Ness Feldman, P.C., "Legal Analysis of Commission Recommendations for Near-Term Actions," October 11, 2011, http://www.brc.gov/sites/default/files/ documents/vnf_legal_authorities _memo_ legal_authorities_ memo_revised_20111011_final_clean_1.pdf.

[192] In proposing this approach, Secretary of Energy Peña stated that this "can be accomplished promptly within [DOE's] current authority." (See letter from Secretary of Energy Federico Peña to Alfred William Dahlberg, Chairman, President, and Chief Executive Officer, Southern Company, May 18, 1998.) Under the NWPA, the Secretary of Energy has existing authority to establish procedures for the collection and payment of the fees. In addition, the principle that fee payments can be deferred until wastes are accepted has an existing precedent in the form of the one-time fee payment imposed on utilities for spent fuel generated before the Act was passed. See Van Ness Feldman, P.C., "Legal Analysis of Commission Recommendations for Near-Term Actions," October 11, 2011, accessible at http:// www.brc.gov/sites/default/files/documents/20111011_legal_ authorities_memo_ revised_final_ clean_1.pdf.

[193] The original PAYGO requirements in the Budget Enforcement Act of 1990 have since been modified in the Statutory Pay-As-You-Go Act of 2010. The requirements apply to proposed legislation (and not administrative actions) and require that OMB maintain a "PAYGO Scorecard" of the average annual cost over a 5-year period and the annual average cost over a 10-year period of newly enacted legislation. If, at the end of the Congressional session, there is a net increase in budget costs, an across-the-board sequestration of an equal offsetting amount is triggered. Legislation that increases direct spending also is subject to points of order under the Congressional Budget Act and the rules of the House and Senate. For example, the 112th Congress adopted a Cut-As-You-Go (CUTGO) rule (part of H. Res. 5) that establishes a point of order against any legislation that increases net mandatory spending for the period of the current fiscal year, the budget year, the four fiscal years following the budget year or the nine fiscal years following the budget year. There also is a point of order against any legislation that increases mandatory budget costs in excess of $5 billion in any of the first four consecutive 10-year fiscal-year periods following the period covered by an applicable budget resolution. *It should be emphasized that PAYGO and CUTGO rules apply to legislative and not administrative actions.*

[194] Joseph S. Hezir, "Discussion of Timing of Payment of NWF Fees," presentation to the BRC Subcommittee on Transportation and Storage, January 3, 2011, Washington, D.C. If this change were a DOE-initiated proposal, rather than implementation by DOE of policy direction from the Administration, it might be subject to review under the administrative PAYGO requirements for agency proposals affecting mandatory spending established by the Bush Administration in 2005 through OMB Memorandum M-05-13 and supported by the Obama Administration in the FY 2011 budget. (The extent of its use in practice is unclear – a report by the Congressional Research Service, "OMB Controls on Agency Mandatory Spending Programs: 'Administrative PAYGO' and Related Issues for Congress," documented only a single instance where administrative PAYGO was applied.) However, the proposed renegotiation of contracts might not fall within the scope of the administrative PAYGO guidelines in any event, and even if it did, it should be subject to the provision for exceptions "...in light of extraordinary need or other compelling circumstances." In this case, the need for assured funding for the used/spent nuclear fuel management program to mitigate the magnitude of further federal budget liability to the Judgment Fund, plus the fact that reduced receipts in the near term would be offset by higher-than-projected receipts in the long term when the escrow accounts are drawn down to meet the costs of constructing and operating waste management facilities, would provide a compelling argument.

[195] Ibid.

[196] The original classification of the fee receipts as mandatory and program expenditures as discretionary was a judgment made by OMB based on general budget principles rather than on clear legislative requirements. See Hezir, op. cit.

[197] See U.S. Department of Energy, *Alternative Means of Financing and Managing the Civilian Radioactive Waste Management Program*, (DOE/RW-0546), August 2001.

[198] The Financial Report shows $19.1 billion of unpaid damages and $30 billion of unspent fees and interest that are categorized as "unearned revenue" (money received in advance of providing goods or services).

[199] The Bush Administration addressed precisely this issue in a statement by DOE to Congress explaining the Administration's proposed legislation to reclassify the revenues from the nuclear utility fees as offsetting collections "so they can be used in the way that was intended when the Nuclear Waste Policy Act was passed: to develop a repository for disposal of high-level radioactive waste and spent nuclear fuel." The specific stated objective was the same as that of our non-legislative proposal: "to ensure that Congress can focus its appropriations decisions on ensuring that the funds are used effectively and efficiently to meet the objectives of the Act, without having to worry about the impact on the funding of other programs within the Energy and Water Development appropriation." In that testimony, DOE noted: "[T]the principle supported by the proposal is specific to the highly unusual contractual arrangement required by the Nuclear Waste Policy Act, and is unlikely to be relevant to many other federal activities. Simply stated, whenever the Federal government, pursuant to an explicit statutory requirement, makes a legally binding contractual commitment specified by that statutory requirement to perform a well-defined service in exchange for payments that cover the costs of that service, it should treat those payments in a way that ensures that they are used for the statutorily-specified contracted purpose. It is hard to see how anyone could disagree with that principle. Likewise, it is hard to see how such distinctive-if not unique-statutory obligations could threaten the ability of Congress to weigh competing demands for appropriations in other, unrelated areas." Testimony by Robert G. Card, Under Secretary of Energy, before the hearing on "A Review of the Department of Energy's Yucca Mountain Project, and Proposed Legislation to Alter the Nuclear Waste Trust Fund (H.R. 3429 and H.R. 3981)," held by the Subcommittee on Energy and Air Quality of the House Committee on Energy and Commerce, March 25, 2004.

[200]. For a summary of proposals to change the Nuclear Waste Fund (NWF) funding structure from 1994 through 1999, see Figure 3 in U.S. Department of Energy, *Alternative Means of Financing and Managing the Civilian Radioactive Waste Management Program*, (DOE/RW-0546), August 2001. More recently, Senator Hagel introduced a bill in 2007 with provisions specifying that "funds from the Nuclear Waste Fund will not be subject to allocations for discretionary spending under Section 302(a) of the Congressional Budget Act or suballocations of appropriations committees under Section 302(b)." To address the issue of budget neutrality, the Hagel bill would have further required that adjustments be made "In the allocation of new budget authority to appropriate committees in amounts equal to the fees reclassified as discretionary as a result of the above provision." Legislation introduced by Senator Domenici in 2008 under the title "Strengthening Management of Advanced Recycling Technologies Act" (or SMART Act) would have established a revolving fund using $1 billion of the current NWF, as well as the annual interest on the Fund. The remaining 95 percent of the current waste Fund, as well as all future fees, would be placed in a legacy fund for the purposes of constructing a geologic repository. Expenditures from the revolving fund for the provisions of the Act could be made without further appropriations but would be subject to limitations in appropriations acts. In this way, the revolving fund could be put to use without being subject to the uncertainty of the annual appropriations process while still retaining the authority of Congress to oversee the NWF. The recent Voinovich/Upton legislation would establish two funds—an operating fund and a reserve fund—for the new waste management organization. The unexpended balance of already appropriated funds, plus accounts receivable and future revenues from NWF fees and appropriations would go to the operating fund. The corpus of the NWF would be transferred as an unfunded asset to the reserve fund (accruing interest from the NWF would go to the operating fund).

[201] This would need to take account of the current Cut-As-You-Go (CUTGO) rules that establish a point of order against (1) any legislation that increases net mandatory spending for the period of the current fiscal year, the budget year, the four fiscal years following the budget year or the nine fiscal years following the budget year, and (2) any legislation that increases mandatory budget costs in excess of $5 billion in any of the first four consecutive 10-year fiscal-year periods following the period covered by an applicable budget resolution.

[202] Testimony by Robert H. Card, Under Secretary of Energy, before the hearing on "A Review of the Department of Energy's Yucca Mountain Project, and Proposed Legislation to Alter the Nuclear Waste Trust Fund (H.R. 3429 and H.R. 3981)," held by the Subcommittee on Energy and Air Quality of the House Committee on Energy and Commerce, March 25, 2004.

[203] Section 302(b)(4) stipulates that "No high-level radioactive waste or SNF generated or owned by any department of the United States .... may be disposed of by the Secretary in any repository constructed under this Act... unless such department transfers to the Secretary, for deposit in the NWF, amounts equivalent to the fees that would be paid to the Secretary under the contracts referred to in this section if such waste or spent fuel were generated by any other person." In practice, funds for the defense wastes have been appropriated directly to the program for use each year, with no surplus to be deposited in the Fund.

[204] 52 FR 31508.

205 U.S. Department of Energy, "Fiscal Year 2007 Civilian Radioactive Waste Management Fee Adequacy Assessment Report," DOE/RW-0593, July 2008.

206 Mark Gaffigan, Managing Directorof Natural Resources and Environment, "NUCLEAR WASTE: Disposal Challenges and Lessons Learned from Yucca Mountain," statement before the Subcommittee on Environment and the Economy, Committee on Energy and Commerce, House of Representatives, June 1, 2011.

207 Information provided by DOE to the BRC. See http://brc.gov/sites/default/files/ correspon dence/blue_ribbon_request_1-6-2010_final.pdf.

208 Just as the fees paid by utilities to date are credited in determining whether they are fully "paid up" for purposes of being able to begin delivering waste for disposal, so should the defense waste appropriations to date be credited in determining when the defense share has been fully paid.

209 Recent court decisions upholding the government's obligation to accept spent fuel are backed by a long history of case law regarding the contractual obligations of the federal government, even in times of severe economic and budget crisis. In one Depression-era case involving an effort to stop payment on government-issued insurance policies, the Supreme Court concluded: "No doubt there was in March, 1933, great need of economy. In the administration of all government business economy had become urgent because of lessened revenues and the heavy obligations to be issued in the hope of relieving widespread distress. Congress was free to reduce gratuities deemed excessive. But Congress was without power to reduce expenditures by abrogating contractual obligations of the United States. To abrogate contracts, in the attempt to lessen government expenditure, would not be the practice of economy, but an act of repudiation." Lynch v. United States, 292 U.S. 571, 580 (1934).

210 The courts have ruled that "acceptance" includes the obligation to remove spent fuel from reactors. We use "accept" and "acceptance" in this broader sense.

211 Yankee Atomic Electric Co. v. United States, 536 F.3d 1268 (Fed. Cir. 2008); Pacific Gas & Electric Co. v. United States, 536 F.3d 1282 (Fed. Cir. 2008); Sacramento Municipal Utility District v. United States, Nos. 2007-5052, -5097, 2008 WL 3539880 (Fed. Cir. Aug. 7, 2008).

212 Department of Justice, "Response to Request for Information from the Blue Ribbon Commission on America's Nuclear Future", December 20, 2011. Available at http://www.brc.gov/ sites/ default/files/comments/attachments/doj_ response.12.20.11_0. pdf.

213 "Because the claims of a substantial number of utilities are not substantially affected by issues that require resolution at the appellate level, it may be possible to implement an administrative claims process with these utilities that is less expensive and more efficient than litigation and that achieves largely the same results." Testimony of Michael F. Hertz, Deputy Assistant Attorney General, Civil Division, on "Budget Implications of Closing Yucca Mountain, before the Committee on the Budget, U.S. House of Representatives, July 27, 2010.

214 Department of Justice, "Response to Request for Information from the Blue Ribbon Commission on America's Nuclear Future", December 20, 2011. Available at http://www.brc.gov/sites/default/files/comments/attachments/doj_response.12.20.11_0.pdf.

215 National Research Council, *Going the Distance: The Safe Transport of Spent Nuclear Fuel and High-Level Waste in the United States*, Washington DC: The National Academies Press, Aug. 2006.

216 Presentation of Earl Easton, U.S. Nuclear Regulatory Commission Office of Spent Fuel Storage and Transportation, to the BRC Subcommittee on Transportation and Storage, Nov. 2, 2010.

217 Examples of recommendations from the 2006 NAS report that have not been implemented include full-scale cask testing, more systematic examination of social or societal risk and risk perception, making planned shipment routes publicly available, shipping stranded spent fuel from shutdown reactor sites first, and executing technical assistance and funding under NWPA, Section 180(c).

218 Including the Pipeline and Hazardous Materials Safety Administration, the Federal Motor Carrier Safety Administration, and the Federal Railroad Administration.

219 Burnups greater than 45 gigawatt-days per metric ton (45GWd/MTU) are now common.

220 In addition, the Departments of Homeland Security and Transportation adopted regulations in 2008 to enhance the safety and security of rail shipments of hazardous materials, including spent nuclear fuel (49 CFR 172, 179, 209, 1520, 1580). The rules designated 46 High Threat Urban Areas (HTUAs) that require a chain of custody and control procedures. They also require rail route evaluations using 27 risk factors, including proximity to densely populated areas, iconic targets, and places of congregation. These rules have not been applied to large-scale spent nuclear fuel shipping campaigns; in fact, a number of observers have noted that doing so on a nationwide basis could be problematic. See presentation of Robert Halstead to the BRC

Transportation and Storage Subcommittee, Sept. 23, 2010 (available at http://www.brc. gov/sites/default /files/meetings/presentations/d_halstead_ final_sep23.pdf).

221 Ibid at p. 8.

222 BRC staff met with NRC/NSIR staff on January 11, 2011, and reviewed the classified versions of the NAS reports, as well as NRC summaries of the actions it has taken to address the issues identified. NRC staff also briefed cleared staff and Commissioners on Feb. 3, 2011.

223 Presentation of Lisa Janairo, Midwest Council of State Governments, to the BRC Transportation and Storage Subcommittee, Nov. 2, 2010 (accessible at http://www.brc. gov/index.php?q=meeting/open-meeting-3).

224 Ibid.

225 Presentation of Gary Lanthrum, Principal Engineer, RAMTASC to the Blue Ribbon Commission, Oct. 20, 2011 (available at http://brc.gov/index.php?q=meeting/ public-meeting-solicit-feedback-draft-commission-report-washington-dc).

226 Western Governors' Association, "Policy Resolution 11-5, Transportation of Radioactive Waste, Radioactive Materials, and Spent Nuclear Fuel," 2011 (available at http:// www.westgov.org/policies).

227 Ken Niles and Rick Moore, *The WIPP Transportation Program at 10 Years: Making the Case for Above-Regulatory Procedures*, Waste Management Symposium, March 2009, at p. 4 (available at http://www.brc.gov/sites/ default/files/comments/attachments/above-regulatory _transport.pdf).

228 Elizabeth Helvey, Complex Systems Group, "Overview of the Section 180(c) Program: History, Lessons Learned and Potential Next Steps," April 2011 (available at http:// www.brc.gov /sites/default/files/ documents/nwpa_ section_180c_paper_final.pdf).

229 EPA also has sole responsibility for regulating non-radiological environmental impacts.

230 10 CFR 51.23(a). The Waste Confidence decision is important because it avoids the need to resolve the issue of long-term disposition of spent fuel in each individual licensing action. See, for example, Nuclear Energy Institute press release, "Industry Applauds NRC Approval of revision of Waste Confidence Rule," Sept.15, 2010 (found at http:// www.nei.org/newsandevents/newsreleases/ industry-applaudsnrc-approval-of-revision-of-waste-confidence-rule).

231 Matthew L. Wald, "3 States Challenge Policy on Storing Nuclear Waste," New York Times, Feb.15, 2011 (available at http://www.nytimes.com/2011/02/16/ nyregion/16nuke.html?_r= 1&scp=2&sq= Nuclear &st=cse).

232 The NRC has been careful to note that despite these actions, it is not endorsing indefinite storage at reactor sites and continues to believe a mined geologic repository is necessary; in addition, the NRC has expressed "reasonable assurance" that such a repository "will be available in the foreseeable future." See Nuclear Regulatory Commission, "Staff Requirements Memorandum", Sept. 15, 2010 (found at http://www.nrc.gov /reading-rm/doc-collections/ commission/srm/meet/2010/m20100915.pdf).

233 Notably, the IAEA goes on to state that "The aim of geological disposal is not to provide a guarantee of absolute and complete containment and isolation of the waste for all time."

234 The NEA is an agency of the Organisation for Economic Cooperation and Development (OECD), which includes the world's major industrialized economies.

235 EPA's general standard also applies to WIPP and is currently in use there.

236 The change came in response to a legal challenge charging that EPA was required by law to follow the recommendation issued by the NAS in 1995 that compliance should be measured at the time of peak dose within the period of geologic stability for Yucca Mountain, which the NAS found to be on the order of 1 million years.

237 40 CFR § 197.20 (Yucca Mountain standards): "What standard must DOE meet? (a) The DOE must demonstrate, *using performance assessment,* that there is a reasonable expectation that the reasonably maximally exposed individual receives no more than the following annual committed effective dose equivalent from releases from the undisturbed Yucca Mountain disposal system: (1) 150 microsieverts (15 millirems) for 10,000 years following disposal; and (2) 1 millisievert (100 millirems) after 10,000 years, but within the period of geologic stability. (b) The DOE's performance assessment must include all potential pathways of radionuclide transport and exposure."

238. "We have recognized the strong consensus in the international radioactive waste community that dose projections extending many tens to hundreds of thousands of years into the future can best be viewed as qualitative indicators of disposal system performance, rather than as firm predictions that can be compared against strict numerical compliance criteria. In fact, international organizations have treated such numerical criteria in a more flexible way and supported their application in conjunction with other qualitative considerations in applying them to regulatory determinations over very long time frames. Further, we agree

that confidence in the way the projections were performed, and the consideration of supporting qualitative information, may be more important to an overall judgment of safety at longer times." U.S. Environmental Protection Agency (EPA), Preamble to 40 CFR Part 197, *Public Health And Environmental Radiation Protection Standards For Yucca Mountain, Nevada, Final Rule*, 73 FR 61266, October 15, 2008.

[239]. Canada's regulations, for example, call for developing a long term safety case that combines a safety assessment with complementary arguments based on appropriate selection and application of assessment strategies, (2) demonstration of system robustness, (3) the use of complementary indicators of safety, and (3) any other evidence available to provide confidence in the long term safety of the proposed system. Similarly, Finnish regulations call for a safety analysis that includes (1) a description of the disposal system and definition of barriers, an analysis of the future evolution of the system, (3) definition of performance targets for individual barriers, (4) functional description of the disposal system by means of conceptual and mathematical modeling, (5) analysis of activity releases and resulting doses from radionuclides that penetrate the barriers and enter the biosphere, (6) estimates of the probabilities of activity releases and radiation doses arising from unlikely disruptive events, (7) uncertainty and sensitivity analyses, and (8) comparison of the outcome of the safety analysis with safety requirements.

[240]. See endnote 238.

[241]. See, for example, Rodney C. Ewing, "Standards & Regulations for the Geological Disposal of Spent Nuclear Fuel and High Level Waste," prepared for the Blue Ribbon Commission on America's Nuclear Future, March 4th, 2011 and available at http://www.brc.gov/ library/commissioned_ papers/EWING% 20BRC%20white% 20paper%20FINAL.pdf.

[242]. Jukka Laaksonen (Director General, Radiation and Nuclear Safety Authority (STUK), Finland), "Regulatory Aspects of Radioactive Waste Disposal – the Finnish Approach," presented at the Conference on Geological Repositories: A Common Objective, a Variety of Paths, October 15 – 17, 2007, Berne, Switzerland.

[243] Rodney C. Ewing, "Performance Assessments: Are They Necessary or Sufficient?" in *Uncertainty Underground: Dealing with the Nation's High-Level Nuclear Waste Policy and Scientific Issues*, R. Ewing and A. Macfarlane, eds., MIT Press, 2006, p. 76.

[244] U.S. Nuclear Waste Technical Review Board, *Technical Advancements and Issues Associated with the Permanent Disposal of High-Activity Wastes: Lessons Learned from Yucca Mountain and Other Programs*, Washington DC, June 2011. Available at http://www. nwtrb.gov/reports/ technical%20lessons.pdf.

[245] U.S. Nuclear Regulatory Commission "Comments on the Blue Ribbon Commission July 2011 Draft Report to the Secretary of Energy." Available at http://www.brc.gov/ sites/default/files/comments/attachments/ nrc_comments_ on_draft brc report.pdf.

[246] 66 Federal Register 55746, November 2, 2001.

[247] Organisation for Economic Development, *Joint NEA-IAEA International Peer Review of the Yucca Mountain Site Characterisation Project's Total System Performance Assessment Supporting the Site Recommendation Process, Final Report*, December 2001.

[248] Rodney C. Ewing, "Standards & Regulations for the Geological Disposal of Spent Nuclear Fuel and High Level Waste," prepared for the Blue Ribbon Commission on America's Nuclear Future, March 4th, 2011 and available at http://www.brc.gov/library/ commissioned_papers/ EWING%20BRC%20white%20paper% 20FINAL.pdf.

[249] Nuclear Energy Institute v. Environmental Protection Agency, 373 F. 3d 1251 (D.C. Cir 2004).

[250] Notably, the statements of secondary performance standards – groundwater protection in parts 191 and 197, and human intrusion in 197 – do not include performance assessment as part of the standard, and use of performance assessment to demonstrate compliance with those standards is not absolutely required. E.g. 40 CFR § 197.30 "What standards must DOE meet? The DOE must demonstrate that there is a reasonable expectation that, for 10,000 years of undisturbed performance after disposal, releases of radionuclides from waste in the Yucca Mountain disposal system into the accessible environment will not cause the level of radioactivity in the representative volume of ground water to exceed the limits in the following Table 1:" Another example of a more flexible approach is found in EPA's regulations for uranium mill tailings, which allow the 1000-year quantitative standards to "be implemented through analysis of the physical properties of the site and the natural processes over time Computational models, theories, and prevalent expert judgment may be used to decide that a control system design will satisfy the standard." 40 CFR § 192.20.

[251] In issuing its initial repository standards, EPA stated that "unequivocal proof of compliance is neither expected nor required because of the substantial uncertainties inherent in such long-term projections."

[252] International Atomic Energy Agency and the OECD Nuclear Energy Agency, *Geological Disposal of Radioactive Waste: Safety Requirements*, WS-R-4, Vienna, 2006 (available at: http://www-pub.iaea.org/ MTCD/ publications/PDF/Pub1231_web.pdf).

[253] EPA's position on reasonable expectation was challenged as being arbitrary and capricious in the lawsuit that led to the remand of parts of 40 CFR 191 in 1987. Nevertheless, EPA's position was upheld by the Court: "Given that absolute proof of compliance is impossible to predict because of the inherent uncertainties, we find that the Agency's decision to require "reasonable expectation" of compliance is a rational one. It would be irrational for the Agency to require proof which is scientifically impossible to obtain. Any such purported absolute proof would be of questionable veracity, and thus of little value to the implementing agencies. Nor can we say that this provision is arbitrary and capricious because it will afford the implementing agencies a degree of discretion, since such imprecision is unavoidable given the current state of scientific knowledge" (Natural Resources Defense Council v. USE.P.A., 824 F.2d 1258).

[254] "As a historic matter, differences in the NRC and EPA standards are rooted in the two agencies' philosophical approach to setting limits. EPA has tended to set very aggressive goals (often based on best technology) but has been very forgiving when best efforts at compliance with the goals are made (thus: "Reasonable Expectation"). The NRC, on the other hand, has set more achievable, science-based, standards and has been very strict in enforcing the standards once set (thus: "Reasonable Assurance")." *Report of the American Nuclear Society on the EPA proposed standard for the Yucca Mountain High Level Waste Repository*, November 1999, http://www.ans.org/pi/news/sd/944200800-report. html.

[255] National Research Council, *One Step at a Time: The Staged Development of Geologic Repositories for High-Level Radioactive Waste*, Washington DC: National Academies Press, 2003, available at http://www.nap.edu/catalog/10611. html, pp. 130-131.

[256] Ibid, p. 92.

[257] Ibid, p. 91.

[258] In Appendix E of its recent "Plan for Integrating Spent Nuclear Fuel Regulatory Activities," the NRC identifies "Development of an assessment tool ("Flexible Performance Assessment"–FPA) that allows a scoping-level evaluation of the regulatory and technical aspects of various spent fuel and HLW disposition scenarios that may be identified by the Blue Ribbon Commission on America's Nuclear Future," as one of several activities to be completed by the end of FY 2010.

[259] In 1990, in the midst of ongoing debates about the EPA and NRC repository regulations, the National Research Council warned against the risks of establishing excessively rigid regulatory requirements before data on actual sites were available. National Research Council, *Rethinking High-Level Radioactive Waste Disposal*, Washington DC: National Academies Press, 1990.

[260] According to a statement submitted by Steve Frishman: "The regulatory arena associated with deep geologic disposal of highlevel radioactive waste and used nuclear fuel has been subject to an array of policy changes, changes in philosophy, and internal struggles within and between the two affected regulatory agencies – the NRC and the EPA. The interested and affected public often has been confused about the roles of the respective agencies, and the motivation, scope and meaning of the regulations proposed, while being confined in their responses to the review and comment provisions of the Administrative Procedures Act (APA), and ultimately the federal courts. Having been a participant in this process, at the affected state government level, for its entire nearly 30-year history, has been frustrating, to say the least." Summary of Statement by Steve Frishman, Consultant, Agency for Nuclear Projects, State of Nevada, presented at the public meeting focused on the issues to be addressed in development of new generic regulations for radioactive waste disposal in deep geological repositories," held by the Disposal Subcommittee of the BRC, September 1, 2010, in Washington D.C. (accessible at http://www.brc.gov/sites/default/ files/meetings/ attachments/summary_of_steve_ frishamn_to_the_disposal_ subcommittee.pdf).

[261] At a hearing in Maine concerning spent fuel stored at the shutdown Maine Yankee reactor site, an elected official described open disagreement between EPA and NRC about whether the final cleanup standard for decommissioning of the site should be 15 mrem or 25 mrem. According to this official, her constituents did not understand the technical basis for the disagreement, but the simple fact that there was a dispute between the regulatory agencies undermined public confidence in the regulatory system and the ability to safely store spent fuel at the Maine Yankee site. This ongoing dispute between the EPA and NRC was also mentioned in a paper prepared for the Commission by Dr. Rodney Ewing and described in a GAO report in 2000 (see U.S. General Accounting Office, RADIATION STANDARDS: Scientific Basis Inconclusive, and EPA and NRC Disagreement Continues, GAO/RCED-00-152, June 2000).

[262] Presentation to the BRC Subcommittee on Disposal by Robert Neill, September 1, 2010. Available at http://www. brc.gov/sites/default/files/meetings/attachments/summary_of_ bob_neill__statement_to_ the_ disposal_subcommittee.pdf.

[263] For example, the BRC's Subcommittee on Disposal has also heard a proposal that would involve forming a panel of experts from each agency and from academia or the private sector to conduct a process in accord with the Administrative Procedures Act. The aim would be to produce a report that could be used as the basis for an integrated set of disposal safety regulations to be adopted by both EPA and NRC, as was proposed by Steven Frishman at the Disposal Subcommittee meeting on September 1, 2010 (see: http://www.brc.gov/sites /default/files/meetings/ attachments/summary_of_steve_frishamn_to_the_disposal_ subcommittee.pdf). Other options such as regulatory negotiations might be possible.

[264] Current NRC regulations for geologic disposal provide requirements for physical protection (10 CFR 60.21(b)(3) and 63.21(b)(3), Part 73) and material control and accounting (10 CFR 60.78 and 63.78). The NRC is currently conducting a rulemaking to enhance these requirements.

[265] The "President's Offer" first put forward by President Lyndon Johnson, offered to place all U.S. nuclear facilities under IAEA safeguards except those of direct national security significance. This is intended to assure nonnuclear weapons states that they will not be discriminated against in having to supply information and undergo IAEA inspections and reports. For more details see Scott D. Sagan, "The International Security Implications of U.S. Domestic Nuclear Power Decisions," paper prepared for the Blue Ribbon Commission on America's Nuclear Future, April 18, 2011. Available at http://www.brc.gov/sites/ default /files/ documents/sagan_brc_ paper_final.pdf.

[266] Stoneturn Consultants: "From Three Mile Island to the Future: Improving Worker Safety and Health in the U.S. Nuclear Power Industry," March 14, 2011. In concluding that the record of occupational health and safety performance for the civilian nuclear power industry is very good (and indeed comparable to that of non-energy sectors like insurance and finance), the Stoneturn report relied on performance indicators in nuclear power plants, occupation radiation dose, and occupational injury and illness rates compared to workers in other industries.

[267] During the construction of WIPP, one construction worker was fatally injured in 1984 when he fell 1000 feet down a 6-foot diameter borehole. See: "Safety Violations Led to WIPP Worker's Death," *Albuquerque Journal*, July 4, 1984, p. D-2. Overall this was the one traumatic fatality in an estimated 17,000 person-working years needed to construct the facility. SInce WIPP opened in 2000, there have been no significant accidents involving workers. In the case of Yucca Mountain, concerns were raised about the adequacy of the industrial hygiene procedures in place to protect workers from silica exposure. A study of some 413 individuals (out of almost 3000) who worked at Yucca Mountain between 1993 and 2002 found three individuals with silicosis, however all of these individuals had previously worked in mines and two of them had been diagnosed before working at Yucca Mountain, so it was difficult to determine whether and to what extent exposures at Yucca Mountain might have contributed to their condition. The other case was a new diagnosis, but that worker also reported previous mining experience so it was not possible to attribute his disease solely to exposure at Yucca Mountain. The study was performed between 2003 and 2005 out of almost 3000 individuals who had been known to have worked in some capacity at Yucca Mountain in during the study. (See *An Investigation into the Silica Exposure of Yucca Mountain Project Workers.* Special Hearing before a Subcommittee of the Committee on Appropriations, U.S. Senate, Las Vegas, March 15. 2004. Available at http:// www.gpo.gov/fdsys/pkg/CHRG-108shrg94749/pdf/CHRG108shrg94749.pdf.) In contrast to Yucca Mountain, the WIPP facility is mined out of halite (salt) deposits. There has not been any study of whether mining halite has had any adverse health impact on workers at WIPP, even though there are significant salt dust exposures in the facility and even though exposure to salt dust is considered a risk factor for cardiovascular, gastric and kidney diseases.

[268] Note that these are not the precise legal definitions. Links or citations to the legal definitions may be found in the supplementary materials posted on the BRC website (www.brc.gov).

[269] R. H. Jones Jr., *Engineering Alternative Studies for Separations: NEPA Data Input Report*, EAS-Q-NEP-G-00001 Revision 4, June 2008, p. 3.

[270] Kennedy, J.E. 2010. "Potential Changes to the U.S. LLW Regulatory Framework," U.S. Nuclear Regulatory Commission Regulatory Information Conference 2010, March 11, 2010, Washington D.C. http://www.nrc. gov/ public-involve/conference-symposia/ric/past/ 2010/slides/ th31kennedyjpv.pdf.

[271] As Secretary Chu pointed out at the March 2010 Commission meeting, the once-through fuel cycle, as currently practiced in the United States with reactor technology that dates back to the 1970s and 1980s, uses only about 1 percent of the energy content of mined uranium. Noting the impact of future nuclear technology advances

Secretary Chu said, "Your committee should be looking at these possibilities because if you can reduce the lifetime of the waste by factors of hundreds to thousands, if you can reduce the amount of the waste by factors of tens to hundreds, that also can change things. And what you do in the interim is then an open question." See transcript of March 25, 2010 meeting at http://www.brc.gov/ sites/default/files/meetings /transcripts /0325scur.pdf.

[272] Massachusetts Institute of Technology, *The Future of the Nuclear Fuel Cycle: An Interdisciplinary MIT Study*, Cambridge, MA, 2011. For example, see (1) Bunn, et al., *The Economics of Reprocessing vs. Direct Disposal of Spent Nuclear Fuel* (Cambridge, Mass.: Project on Managing the Atom, Harvard University, 2003). http://www.publicpolicy. umd.edu/files. php/faculty/fetter/2003-Bunn-repro.pdf, (2) Dixon, et al. *Dynamic Systems Analysis Report for Nuclear Fuel Recycle*, December 2008, INL/EXT-08-15201 Rev., http://www. inl.gov/technicalpublications/Documents/4310613.pdf, (3) U.S. DOE, *Advanced Fuel Cycle Initiative Comparison Report*, FY 2003,(Updated 2004, 2006), http://www.ne.doe.gov/ pdfFiles/AFCIComp Rpt2003.pdf, (4) EPRI, *Nuclear Fuel Cycle Cost Comparison Between Once-Through and Plutonium Single-Recycling in Pressurized Water Reactors*, #1018585, 2009, http://my.epri.com/portal/server.pt?, (5) EPRI, *Advanced Nuclear Fuel Cycles – Main Challenges and Strategic Choices*, #1020307, 2010, http://my.epri.com/portal/server.pt?, (6) MIT, *The Future of Nuclear Power*, 2003 (Updated 2009), http://web.mit.edu/nuclearpower/, (7) MIT, *The Future of the Nuclear Fuel Cycle*, 2011, http://web.mit.edu/mitei/ research/studies/nuclear-fuel-cycle.shtml, (8) Shropshire, *Advanced Fuel Cycle Economic Tools, Algorithms, and Methodologies*, 2009, INL/EXT-09-15483, http://www.inl.gov/ technicalpublications/Documents/4247163.pdf, (9) Wigeland, *AFCI Options Study*, 2009, INL/EXT-10-17639, http://www. inl.gov/technicalpublications/ Documents/4480296.pdf, (10) Wilson, *Comparing Nuclear Fuel Cycle Options: Observations and Challenges*, 2011, http://brc.gov/sites/default/files/ documents/wilson.fuel _ cycle_.comparisons_final. pdf, (11) U.S. NWTRB, *Nuclear Waste Assessment System for Technical Evaluation (NUWASTE): Status and Initial Results*, 2011.

[273] Mixed oxide fuel (MOX) consists of a mix of recycled plutonium and uranium.

[274] DOE is currently planning to build a demonstration plant of this type, called the Next Generation Nuclear Plant, at the Idaho National Laboratory. The reactor would be cooled with helium gas, moderated with graphite, and use low-enriched uranium fuel. It would be capable of generating electricity as well as supplying process heat.

[275] This endnote summarizes key assumptions underlying table 4. As noted in the first paragraph in section 11.2, many of the comparisons are qualitative because the available technical literature is not comparable or consistent due to different assumptions and the current status of the technologies differs widely (LWR systems are deployed, fast reactors require more development and demonstration, and gas-cooled reactors even more).

[276] Table 4 compares four nuclear energy systems: (1) a once-through LWR system using high-burnup uranium dioxide system [at least 45 GWd/metric ton] representative of near-term technology, (2) a modified-open cycle (MOC) in which eight high-burnup uranium dioxide fuel assemblies are reprocessed to make one high-burnup MOX fuel assembly which is irradiated for one cycle and then managed as waste, (3) a high-temperature [~600 C] helium-cooled reactor (HTGR) operated on a once-through basis using veryhigh-burnup uranium dioxide fuel in a graphite matrix to produce electricity or process heat, and (4) a closed sodium-cooled fast reactor system involving sustained reprocessing and recycle of MOX fuel. The comparisons in the table are based on a hypothetical system in which all nuclear power is produced by each system, i.e., transition effects are ignored. Each of the systems is assumed to produce the same amount of electric power. The once-through LWR system is considered to be the baseline against which the other systems are compared. Noteworthy assumptions for some of the criteria that are not already stated in the table are as follows: - Uranium utilization: uranium recovered in the MOC system is assumed to be recovered and re-enriched to make fresh uranium dioxide (UOX) fuel and the MOX fuel replaces uranium dioxide fuel. In the closed cycle uranium is recovered and recycled but with ~1% losses during each recycle. - Global climate and energy security: HTGR is planned to achieve temperatures that make it possible to displace fossil fuel use in energy-intensive non-electric sectors. - Non-proliferation and counter-terrorism: Once-through systems would be sending plutonium in SNF to a repository after relatively short cooling times. MOX and closed cycle systems keep much more plutonium in reprocessing plant, MOX fabrication plant, and reactor storage. - Disposal safety: Once-through systems about the same given uncertainties in HTGR burnup and fuel composition. Reduction of TRU in MOC based on ORIGEN2 calculation. Reduction of TRU in closed cycle based on literature and staff calculations with caveats concerning duration stated. The MOC system includes disposal of intact MOX fuel after one irradiation cycle. Reduction in fuel cycle risk from MOC and closed cycles based on scaling results given in G. E. Michaels, *Impact of Actinide Recycle on Nuclear Fuel Cycle Heath Risks*, ORNL/M-1947 (June 1992) and OECD Nuclear Energy Agency, Radiological Impacts of Spent Nuclear Fuel Management Options:

A Comparative Study, (2000). - Waste volume: Volume requiring repository disposal and volume acceptable for near-surface disposal (A-B-C LLW plus uranium mill tailings plus depleted uranium) are addressed separately. Unit waste volumes are based on staff estimates using literature data. - Repository space: Assume 40%+ thermal efficiency for HTGR. Space requirements for MOC wastes (UOX HLW plus intact MOX fuel) based on integrated decay heat using ORIGEN2 results.

277 Although the safety evaluation of the once-through fuel cycle is marked as the baseline, this does not presuppose that safety is perfect. Even given consistent and approved safety design standards across fuel cycles, there is still room for improvement in this system.

278 "No existing deterministic cost study of full recycling is credible, because there has been no engineering demonstration of full recycling." Testimony received from Geoffrey Rothwell at the meeting of the Subcommittee on Reactor & Fuel Cycle Technology on August 30, 2010.

279 The table compares nuclear energy systems in the long-term which means the R&D has been successfully completed, the fuel cycle in question has been adopted, and the transition phase is over so that the U.S. is relying on just that system.

280 Assumption: Depleted uranium is deemed acceptable for near-surface disposal. If repository disposal is required the volume of repository waste increases ranging from 3 to 30 times for all but the closed fuel cycle, although decay heat and toxicity are not affected for 100,000 years. Note also that volume is less important for mined repositories than for other potential disposal options, notably boreholes.

281 President's Council of Advisors on Science and Technology (PCAST). *Report to the President on Accelerating the Pace of Change in Energy Technologies Through an Integrated Federal Energy Policy*. November 2010. Available at: http://www.whitehouse. gov/sites/default/files/microsites/ ostp/pcast-energy-tech-report.pdf.

282 The ESBWR (or Economic Simplified Boiling Water Reactor) is a reactor design marketed by GE Hitachi Nuclear Energy. It is considered a generation III+, passively safety design.

283 M. Anastasio et al., *A Sustainable Energy Future: the Essential Role of Nuclear Energy*, August 2008, page 4, http:// www.ne.doe.gov/pdffiles/rpt_sustainableenergyfuture_ aug2008.pdf.

284 World Association of Nuclear Operators (WANO), "A Post-Fukushima WANO – Applying Lessons Learned,"available at http://www.wano.info/press-release/ wano-press-release-3/.

285 World Association of Nuclear Operators (WANO), "WANO after Fukushima– Strengthening Global Nuclear Safety: A Question and Answer Document," available at http://www.wano. info/wp-content/ uploads/ 2011/10/Q-ADocument-for-Press-Pack.pdf.

286 See http://nuclearprinciples.org/the-principles/

287 Ibid.

288 Scott D. Sagan, *The International Security Implications of U.S. Domestic Nuclear Power Decisions*, 2011. Commissioned paper for the BRC, available at http://www.brc.gov/sites/ default/files/documents/sagan_ brc_paper_final.pdf.

289 "The Treaty on the Non-proliferation of Nuclear Weapons," United Nations, available at http://www.un.org/ disarmament/WMD/Nuclear/pdf/NPTEnglish_Text.pdf.

290 Scott D. Sagan, "Shared Responsibilities for Nuclear Disarmament," *Daedalus* 138:4 (Fall 2009):157-68

291 International Atomic Energy Agency, "The Safeguards System of the International Atomic Energy Agency," see http://www.iaea.org/OurWork/SV/Safeguards/.

292 See http://www.nti.org/h_learnmore/npttutorial/ chapter02_02.html.

293 International Atomic Energy Agency, "Model Protocol Additional to the Agreements Between States and the IAEA for the Application of Safeguards," INFCIRC/540, available at http://www.iaea.org/Publications /Documents/ Infcircs/1997/infcirc540c.pdf.

294 From http://www.iaea.org/OurWork/SV/ Safeguards/sg_protocol.html, accessed May 5, 2011.

295 Testimony before the BRC by Edwin Lyman on October 12, 2010.

296 The IAEA currently has 151 member states; its budget in 2011 was $447 million. The United States provided approximately 25% of that figure.

297 Regional Nuclear Fuel Cycle Centers study (1975- 1977), International Nuclear Fuel Cycle Evaluation study (1977-1980), Expert Group on International Plutonium Storage (1978-1982), IAEA Committee on Assurances of Supply (1980-1987), United Nations Conference for the Promotion of International Cooperation in The Peaceful Uses of Nuclear energy (1987).

298 World Nuclear Association - http://www.worldnuclear.org/info/reactors.html, update for Dec 1, 2011.

299 International Atomic Energy Agency, *Multilateral Approaches to the Nuclear Fuel Cycle: Expert Group Report submitted to the Director General of the International Atomic Energy Agency*, INFCIRC/640, 22 February, 2005. Available at http://www.iaea.org/Publications/ Documents/ Infcircs/2005/infcirc640.pdf.

[300] The Fund was seeded by NTI and supplemented by voluntary donations from the European Union, Kuwait, Norway, the United Arab Emirates, and the United States, according to a December 2, 2010 statement made to IAEA Board of Governors by Glyn Davies, U.S. Ambassador to the IAEA (see http://vienna.usmission.gov/101203nfs.html).

[301] Incorporated as a joint venture between Russia's Tekhsnabeksport and Kazakhstan's Kazatomprom.

[302] The fuel bank consists of two 1,000 megawatt-reactor loads of LEU.

[303] Seen as a "virtual assurance mechanism that would facilitate access to nuclear energy to avoid the huge cost and technical challenge involved in establishing a nuclear fuel cycle." Statement made at the 2010 IAEA General Conference by Charles Hendy, Minister of State for Energy and Climate Change of the United Kingdom; see http:// www.iaea.org/About/ Policy/GC/GC54/Statements/uk.pdf.

[304] Spent fuel take-away arrangements are broadly defined as negotiated agreements for governments with fuel cycle capabilities to assume liability for supplied or obligated fuel and develop permanent disposition solutions for managing used fuel in concert with countries seeking nuclear energy.

[305] U.S. Department of State – website, Principles of the Global Initiative to Combat Nuclear Terrorism, http:// www.state.gov/t/isn/c37071.htm.

[306] Official U.S. Department of State blog - Secretary Clinton, Foreign Minister Lavrov Sign Plutonium Disposition Protocol, posted April 13, 2010 http://blogs. state.gov/index. php/site/entry/clinton_lavrov_plutonium_ disposition_protocol.

[307] In 2010, The U.S. returned Russian-origin HEU from Poland, Czech Republic, Serbia, and the Ukraine. GTRI Fact Sheet, http://nnsa.energy.gov/mediaroom/ factsheets /reducing threats.

[308] Kenneth N. Luongo, "The Urgent Need for a Seoul Declaration: A Roadmap for the 2012 Nuclear Security Summit and Beyond," *Arms Control Today*, Washington, D.C., April 2012.

[309] According to the IAEA, 1,773 confirmed incidents of illegal possession, movement or attempts to illegally trade in or use nuclear material or radioactive sources occurred between January 1993 and December 2009. Information taken from the IAEA's Illicit Trafficking Database at http:// www-ns.iaea.org/security/itdb.asp.

In: United States Nuclear Waste
Editors: Gregory Hall and Steven L. Greene

ISBN: 978-1-62081-312-6

Chapter 2

STATEMENT OF CONGRESSMAN LEE HAMILTON AND GENERAL BRENT SCOWCROFT. HEARING ON "AMERICA'S NUCLEAR FUTURE"*

INTRODUCTION

Chairman Shimkus, Ranking Member Green, members of the Subcommittee, it is a pleasure to appear before you today to discuss the final recommendations of the Blue Ribbon Commission on America's Nuclear Future.

We appreciate the leadership this Subcommittee has shown in confronting some of our nation's biggest challenges, which certainly include the focus of this hearing - managing spent nuclear fuel and high level nuclear waste in the United States. Thank you for allowing us the opportunity to testify before you today.

Before we begin, we would also like to thank the 13 other members of the Commission who worked so hard in creating our final report.

As the Co-Chairmen of the Commission, we were delighted to work with such a talented and dedicated group of fellow Commissioners. We are thankful for the expertise and insights they brought to our endeavors.

Their professionalism led to our final report having unanimous approval; all of the Commissioners have agreed to our final report, a fact which we believe speaks to the strength of our recommendations.

As you aware, the Blue Ribbon Commission was formed by the Secretary of Energy at the direction of the President.

Our charge was to conduct a comprehensive review of policies for managing the back end of the nuclear fuel cycle and to recommend a new strategy. We came away from our review frustrated by decades of unmet commitments to the American people, yet confident that we can turn this record around.

* This is an edited, reformatted and augmented version of testimony given on February 1, 2012, before the House Energy and Commerce Committee, Subcommittee on Environment and the Economy United States House of Representatives, Second Session, 112th Congress.

Framing the Issue

Mr. Chairman, as we are all too well aware, America's nuclear waste management program is at an impasse. The Administration's decision to halt work on a repository at Yucca Mountain is but the latest indicator of a policy that has been troubled for decades and has now all but completely broken down. The approach laid out under the 1987 Amendments to the Nuclear Waste Policy Act has simply not worked to produce a timely solution for dealing with the nation's most hazardous radioactive materials. The United States has traveled nearly 25 years down the current path only to come to a point where continuing to rely on the same approach seems destined to bring further controversy, litigation, and protracted delay.

What we have found is that our nation's failure to come to grips with the nuclear waste issue has already proved damaging and costly. It will be even more damaging and more costly the longer it continues: damaging to prospects for maintaining a potentially important energy supply option for the future, damaging to state – federal relations and public confidence in the federal government's competence, and damaging to America's standing in the world as a source of nuclear expertise and as a leader on global issues of nuclear safety, non-proliferation, and security.

This failure is also costly to utility ratepayers who continue to pay for a nuclear waste management solution that has yet to be delivered, to communities that have become unwilling hosts of long-term waste storage facilities, and to U.S. taxpayers who face billions in liabilities as a result of the failure to meet federal waste management commitments. The national interest demands that our nuclear waste program be fixed.

The need for a new strategy is urgent, not just to address these damages and costs, but also because this generation has a fundamental ethical obligation to avoid burdening future generations with finding a safe permanent solution for managing hazardous nuclear materials they had no part in creating. At the same time, we owe it to future generations to avoid foreclosing options wherever possible so that they can make choices—about the use of nuclear energy as a low-carbon energy resource and about the management of the nuclear fuel cycle— based on emerging technologies and developments and their own best interests.

Put simply, the overall record of the U.S. nuclear waste program has been one of broken promises and unmet commitments. And yet the Commission finds reasons for confidence that we can turn this record around. To be sure, decades of failed efforts to develop a repository for spent fuel and high-level waste have produced frustration and a deep erosion of trust in the federal government. But they have also produced important insights, a clearer understanding of the technical and social issues to be resolved, and at least one significant success story – the WIPP facility in New Mexico. Moreover, many people have looked at aspects of this record and come to similar conclusions.

The Scale of the Problem

Mr. Chairman, before we discuss our recommendations it is useful to briefly review the scale of the nuclear waste problem in the U.S. As this Subcommittee is certainly aware, there are 104 commercial nuclear power reactors operating in the United States today, supplying approximately 20 percent of our nation's electricity needs. The industry as a whole generates more than 2,000 metric tons of spent nuclear fuel on an annual basis. At present, nearly all of the nation's existing inventory of approximately 65,000 metric tons of spent fuel is being stored at the reactor sites where it was generated—about three-quarters of it in shielded concrete pools and the remainder in dry casks above ground. Roughly speaking, this spent fuel would cover one football field to a depth of approximately 20 feet. This inventory also includes approximately 3,000 metric tons of what we've called "stranded" spent fuel, fuel in storage at ten sites where nuclear power reactors have been shut down and are no longer operating. In addition to the civilian spent nuclear fuel, there is a considerable inventory of DOE-managed nuclear waste – in the form of both spent nuclear fuel and of liquid high level waste. The current inventory of DOE-managed spent fuel represents a relatively small fraction of the nation's total civilian spent-fuel inventory: approximately 2,500 metric tons. Along with spent nuclear fuel, DOE manages an inventory of high level waste totaling more than 3,000 canisters of vitrified wastes and some 90 million gallons of liquids, sludges and solids from past fuel reprocessing operations for weapons production. Most of this waste is being stored at DOE's Hanford, Idaho National Laboratory, and Savannah River sites. In addition, there is a small amount of vitrified high level waste from reprocessing fuel from both commercial power reactors and government reactors at the West Valley site in New York that will also require disposal.

Our Approach

Fulfilling our charter has required the Commission to investigate a wide range of issues and listen to a broad spectrum of concerned stakeholders. It became clear to us early on that many of the problems facing our nuclear waste program have their roots in social distrust and lack of confidence in government, so we strove to make the Commission's work as inclusive, transparent, and accessible as possible. We heard from hundreds of invited witnesses, toured nuclear waste management facilities in the U.S. and abroad, and received thousands of comments at more than two dozen public meetings and through our web site. The Commission released a draft report for public comment in July of 2011. To facilitate meaningful discussion about our draft report, we arranged for a series of public meetings to be held in cooperation with regional state government groups. These meetings were held in Atlanta, Boston, Denver, Minneapolis, and Washington, DC, and were quite helpful in gaining useful insights that are reflected in our final report. In total, we received and reviewed several thousand comments on our draft report. We are indebted to the many people who have given us the benefit of their expertise, advice, and guidance. A full list of the Commission's meetings is included in a longer version of this statement that we intend to submit for the record.

Key Elements of the Blue Ribbon Commission's Final Recommendations

Mr. Chairman, the strategy we recommend in our final report has eight key elements:

1. A new, consent-based approach to siting future nuclear waste management facilities.
2. A new organization dedicated solely to implementing the waste management program and empowered with the authority and resources to succeed.

3. Access to the funds nuclear utility ratepayers are providing for the purpose of nuclear waste management.
4. Prompt efforts to develop one or more geologic disposal facilities.
5. Prompt efforts to develop one or more consolidated storage facilities.
6. Prompt efforts to prepare for the eventual large-scale transport of spent nuclear fuel and high-level waste to consolidated storage and disposal facilities when such facilities become available.
7. Support for continued U.S. innovation in nuclear energy technology and for workforce development.
8. Active U.S. leadership in international efforts to address safety, waste management, non-proliferation, and security concerns.

Although the elements of this strategy will not be new to Members and staff of this Subcommittee who have followed the U.S. nuclear waste program over the years, we are certain they are all necessary to establish a truly integrated national nuclear waste management system, to create the institutional leadership and wherewithal to get the job done, and to ensure that the United States remains at the forefront of technology developments and international responses to evolving nuclear safety, non-proliferation, and security concerns.

A few general points about the Commission's proposed strategy are worth emphasizing before our recommendations are discussed in greater detail here today. First is the issue of cost. In this time of acute concern about the federal budget deficit and high energy prices, we have been sensitive to the concern that our recommendations—particularly those that involve launching a new approach and a new organization for nuclear waste management—could add to the financial burden on the U.S. Treasury and on American taxpayers and utility ratepayers. Certainly it will cost something to implement a successful U.S. waste management program; however, trying to implement a deeply flawed program is even more costly, for all the reasons already mentioned. In fact, U.S. ratepayers are *already* paying for waste disposal (through a fee collected on each kilowatt-hour of nuclear-generated electricity)—but the program they're paying for isn't working.

Overall, we are confident that our waste management recommendations can be implemented using revenue streams *already dedicated for this purpose* – in particular the Nuclear Waste Fund and fee. Other Commission recommendations—particularly those concerning nuclear technology programs and international policies—are broadly consistent with the program plans of the relevant agencies.

Another overarching point concerns timing and implementation. All of our recommendations are interconnected and will take time to implement fully, particularly since many elements of the strategy we propose require legislative action to amend the Nuclear Waste Policy Act and other relevant laws. Nevertheless, prompt action can and should be taken in several areas, without waiting for legislative action, to get the waste management program back on track.

One of the many actions we recommend the Administration take in the near-term is to ensure that funds already being collected from nuclear utility ratepayers to cover the costs of spent fuel disposal are available to serve their intended purpose. In our report we suggest a series of actions that can be taken promptly to give the waste program the budgetary certainty that will be essential for long-term success. We also recommend steps the Department of

Energy should take to enable implementation of our consolidated storage recommendations, including efforts to provide assistance to states and regional state government groups that can be used to begin transportation planning and to support local and tribal officials in areas likely to be traversed by spent fuel shipments.

Finally, there are several questions the Commission was not chartered to address. We have not rendered an opinion on the suitability of the Yucca Mountain site or any other specific site, nor have we commented on the request to withdraw the license application for Yucca Mountain. Instead, we focused on developing a sound strategy for future storage and disposal facilities and operations that we believe *can and should be implemented regardless of what happens with Yucca Mountain*. We have also not offered a judgment about the appropriate role of nuclear power in the nation's future energy supply mix.

These are all important questions that will engage policy makers and the public in the years ahead. However, none of them alters the urgent need to change and improve our strategy for managing the high-level wastes and spent fuel that already exist and will continue to accumulate so long as nuclear reactors operate in this country. That is the focus of the Commission's work and of the specific recommendations that follow.

Further Discussion of the BRC's Recommendations

Mr. Chairman, as we mentioned previously, there are eight key elements to our strategy that are essential to the future success of the nuclear waste management program in the United States. We will now discuss those in more detail.

1. A New Consent-Based Approach to Siting

Siting storage or disposal facilities has been the most consistent and most intractable challenge for the U.S. nuclear waste management program. Of course, the first requirement in siting any facility centers on the ability to demonstrate adequate protection of public health and safety and the environment. Beyond this threshold criterion, finding sites where all affected units of government, including the host state or tribe, regional and local authorities, and the host community, are willing to support or at least accept a facility has proved exceptionally difficult. The erosion of trust in the federal government's nuclear waste management program has only made this challenge more difficult. And whenever one or more units of government are opposed, the odds of success drop greatly. The crux of the challenge derives from a federal/state/tribal/local rights dilemma that is far from unique to the nuclear waste issue—no simple formula exists for resolving it. Experience in the United States and in other nations suggests that any attempt to force a top-down, federally mandated solution over the objections of a state or community—far from being more efficient—will take longer, cost more, and have lower odds of ultimate success.

By contrast, the approach we recommend is explicitly adaptive, staged, and consent-based. Based on a review of successful siting processes in the United States and abroad—including most notably the siting of a disposal facility for transuranic radioactive waste, the Waste Isolation Pilot Plant (WIPP) in New Mexico, and recent positive outcomes in Finland, Sweden, Spain and France—we believe this type of approach can provide the flexibility and sustain the public trust and confidence needed to see controversial facilities through to completion.

In practical terms, this means encouraging communities to volunteer to be considered to host a new nuclear waste management facility while also allowing for the waste management organization to approach communities that it believes can meet the siting requirements. Siting processes for waste management facilities should include a flexible and substantial incentive program.

The approach we recommend also recognizes that successful siting decisions are most likely to result from a complex and perhaps extended set of negotiations between the implementing organization and potentially affected state, tribal, and local governments, and other entities. It would be desirable for these negotiations to result in a partnership agreement or some other form of legally enforceable agreement with the organization to ensure that commitments to and by host states, tribes, and communities are upheld. All affected levels of government must have, at a minimum, a meaningful consultative role in important decisions; additionally, both host states and tribes should retain—or where appropriate, be delegated—direct authority over aspects of regulation, permitting, and operations where oversight below the federal level can be exercised effectively and in a way that is helpful in protecting the interests and gaining the confidence of affected communities and citizens. At the same time, host state, tribal and local governments have responsibilities to work productively with the federal government to help advance the national interest.

In this context, any process that is prescribed in detail up front is unlikely to work. Transparency, flexibility, patience, responsiveness, and a heavy emphasis on consultation and cooperation will all be necessary indeed, these are attributes that should apply not just to siting but to every aspect of program implementation.

This discussion raises another issue highlighted in numerous comments to the BRC: the question of how to define "consent." The Commission takes the view that this question ultimately has to be answered by a potential host jurisdiction, using whatever means and timing it sees fit. We believe that a good gauge of consent would be the willingness of the affected units of government – the host states, tribes, and local communities – to enter into legally binding agreements with the facility operator, where these agreements enable states, tribes, or communities to have confidence that they can protect the interests of their citizens.

All siting processes take time; however, an adaptive, staged approach may seem particularly slow and open-ended. This will be frustrating to stakeholders and to members of the public who are understandably anxious to know when they can expect to see results. The Commission shares this frustration—greater certainty and a quicker resolution would have been our preference also. Experience, however, leads us to conclude that there is no short-cut, and that any attempt to short-circuit the process will most likely lead to more delay. That said, we also believe that attention to process must not come at the expense of progress and we are sympathetic to the numerous comments we received asking us to include a more detailed and specific set of milestones in our final report. Obviously there is an inherent tension between recommending an adaptive, consent-based process and setting out deadlines or progress requirements in advance. But we agree that it will be important—without imposing inflexible deadlines—to set reasonable performance goals and milestones for major phases of program development and implementation so that Congress can hold the waste management organization accountable and so that stakeholders and the public can have confidence the program is moving forward. Other countries have taken this approach, in several cases identifying target timeframes, rather than specific dates for completing stages in their process. For example the implementing organization might consider a range of, say, 15 to 20 years to accomplish site identification and characterization and to conduct the licensing process for a geologic repository. A notional timeframe for siting and developing a consolidated storage facility would presumably be shorter, perhaps on the order of 5 to 10 years.

2. A New Organization to Implement the Waste Management Program

The U.S. Department of Energy (DOE) and its predecessor agencies have had primary responsibility for implementing U.S. nuclear waste policy for more than 50 years. In that time, DOE has achieved some notable successes, as shown by the WIPP experience and recent improvements in waste cleanup performance at several DOE sites. The overall record of DOE and of the federal government as a whole, however, has not inspired widespread confidence or trust in our nation's nuclear waste management program. For this and other reasons, the Commission concludes that a new, single-purpose organization is needed to provide the stability, focus, and credibility needed to get the waste program back on track. We believe a congressionally chartered federal corporation offers the best model, but whatever the specific form of the new organization it must possess the attributes, independence, and resources to effectively carry out its mission.

The central task of the new organization would be to site, license, build, and operate facilities for the safe consolidated storage and final disposal of spent fuel and high-level nuclear waste at a reasonable cost and within a reasonable timeframe. In addition, the new organization would be responsible for arranging for the safe transport of waste and spent fuel to or between storage and disposal facilities, and for undertaking applied research, development, and demonstration (RD&D) activities directly relevant to its waste management mission (e.g., testing the long-term performance of fuel in dry casks and during subsequent transportation).

For the new organization to succeed, a substantial degree of implementing authority and assured access to funds must be paired with rigorous financial, technical, and regulatory oversight by Congress and the appropriate government agencies. We recommend that the organization be directed by a board nominated by the President, confirmed by the Senate, and selected to represent a range of expertise and perspectives. Independent scientific and technical oversight of the nuclear waste management program is essential and should continue to be provided for out of nuclear waste fee payments. In addition, the presence of clearly independent, competent regulators is essential; we recommend the existing roles of the U.S. Environmental Protection Agency in establishing standards and the Nuclear Regulatory Commission (NRC) in licensing and regulating waste management facilities be preserved but that steps be taken to ensure ongoing cooperation and coordination between these agencies.

Late in our review we heard from several states that host DOE defense waste that they agree with the proposal to establish a new organization to manage civilian wastes, but believe the government can more effectively meet its commitments if responsibility for defense waste disposal remains with DOE. Others argued strongly that the current U.S. policy of comingling defense and civilian wastes should be retained. We are not in a position to comprehensively assess the implications of any actions that might affect DOE's compliance with its cleanup agreements, and we did not have the time or the resources necessary to thoroughly evaluate the many factors that must be considered by the Administration and Congress in making such a determination. The Commission therefore urges the Administration to launch an immediate review of the implications of leaving responsibility for disposal of defense waste and other DOE-owned waste with DOE versus moving it to a new waste management organization. The implementation of other BRC recommendations, however, should not wait for the commingling issue to be resolved. Congressional and Administration efforts to implement our recommendations can and should proceed as expeditiously as possible.

3. Access to Utility Waste Disposal Fees for their Intended Purpose

The 1982 NWPA created a "polluter pays" funding mechanism to ensure that the full costs of disposing of commercial spent fuel would be paid by utilities (and their ratepayers), with no impact on taxpayers or the federal budget. Nuclear utilities are assessed a fee on every kilowatt-hour of nuclear-generated electricity as a *quid pro quo* payment in exchange for the federal government's contractual commitment to begin accepting commercial spent fuel beginning by January 31, 1998. Fee revenues go to the government's Nuclear Waste Fund, which was established for the sole purpose of covering the cost of disposing of civilian nuclear waste and ensuring that the waste program would not have to compete with other funding priorities. In contrast, costs for disposing of defense nuclear wastes are paid by taxpayers through appropriations from the Treasury.

The Fund does not work as intended. A series of Executive Branch and Congressional actions has made annual fee revenues (approximately $750 million per year) and the unspent $27 billion balance in the Fund effectively inaccessible to the waste program. Instead, the waste program must compete for federal funding each year and is therefore subject to exactly the budget constraints and uncertainties that the Fund was created to avoid. This situation must be remedied to allow the program to succeed.

In the near term, the Administration should offer to amend DOE's standard contract with nuclear utilities so that utilities remit only the portion of the annual fee that is appropriated for waste management each year and place the rest in a trust account, held by a qualified third-party institution, to be available when needed. At the same time, the Office of Management and Budget should work with the Congressional budget committees and the Congressional Budget Office to change the budgetary treatment of annual fee receipts so that these receipts can directly offset appropriations for the waste program. These actions are urgent because they enable key subsequent actions the Commission recommends. Therefore, we urge the Administration to act promptly to implement these changes (preferably in Fiscal Year 2013). For the longer term, legislation is needed to transfer the unspent balance in the Fund to the new waste management organization so that it can carry out its civilian nuclear waste obligations independent of annual appropriations (but with Congressional oversight)—similar to the budgeting authority now given to the Tennessee Valley Authority and Bonneville Power Administration.

We recognize that these actions mean no longer counting nuclear waste fee receipts against the federal budget deficit and that the result will be a modest negative impact on annual budget calculations. The point here is that the federal government is contractually bound to use these funds to manage spent fuel. The bill will come due at some point. Meanwhile, failure to correct the funding problem does the federal budget no favors in a context where taxpayers remain liable for mounting damages, compensated through the Judgment Fund, for the federal government's continued inability to deliver on its waste management obligations. These liabilities are already in the billions of dollars and could increase by hundreds of millions of dollars annually for each additional year of delay.

4. Prompt Efforts to Develop a New Geologic Disposal Facility

Deep geologic disposal capacity is an essential component of a comprehensive nuclear waste management system for the simple reason that very long-term isolation from the environment is the *only* responsible way to manage nuclear materials with a low probability of re-use, including defense and commercial reprocessing wastes and many forms of spent fuel currently in government hands. The conclusion that disposal is needed and that deep geologic disposal is the scientifically preferred approach has been reached by every expert panel that has looked at the issue and by every other country that is pursuing a nuclear waste management program.

Some commenters have urged the prompt adoption of recycling of spent fuel as a response to the waste disposal challenge, as well as a means to extend fuel supply. *It is the Commission's view that it would be premature for the United States to commit, as a matter of policy, to "closing" the nuclear fuel cycle given the large uncertainties that exist about the merits and commercial viability of different fuel cycles and technology options.* Future evaluations of potential alternative fuel cycles must account for linkages among all elements of the fuel cycle (including waste transportation, storage, and disposal) and for broader safety, security, and non-proliferation concerns. Moreover, all spent fuel reprocessing or recycle options generate waste streams that require a permanent disposal solution. In any event, we believe permanent disposal will very likely also be needed to safely manage at least some portion of the commercial spent fuel inventory even if a closed fuel cycle were adopted.

The Commission recognizes that current law establishes Yucca Mountain in Nevada as the site for the first U.S. repository for spent fuel and high level waste, provided the license application submitted by DOE meets relevant requirements. The Blue Ribbon Commission was not chartered as a siting commission. Accordingly we have not evaluated Yucca Mountain or any other location as a potential site for the storage or disposal of spent nuclear fuel and high-level waste, nor have we taken a position on the Administration's request to withdraw the license application. We simply note that regardless what happens with Yucca Mountain, the U.S. inventory of spent nuclear fuel will soon exceed the amount that can be legally emplaced at this site until a second repository is in operation. So under current law, the United States will need to find a new disposal site even if Yucca Mountain goes forward. We believe the approach set forth here provides the best strategy for assuring continued progress, regardless of the fate of Yucca Mountain.

5. Prompt Efforts to Develop One or More Consolidated Storage Facilities

Safe and secure storage is another critical element of an integrated and flexible national waste management system. Fortunately, experience shows that storage—either at or away from the sites where the waste was generated—can be implemented safely and cost-effectively. Indeed, *a longer period of time in storage offers a number of benefits because it allows the spent fuel to cool while keeping options for future actions open.*

Developing consolidated storage capacity would allow the federal government to begin the orderly transfer of spent fuel from reactor sites to safe and secure centralized facilities independent of the schedule for operating a permanent repository. The arguments in favor of consolidated storage are strongest for "stranded" spent fuel from shutdown plant sites. Stranded fuel should be first in line for transfer to a consolidated facility so that these plant sites can be completely decommissioned and put to other beneficial uses. Looking beyond the issue of today's stranded fuel, the availability of consolidated storage will provide valuable flexibility in the nuclear waste management system that could achieve meaningful cost savings for both ratepayers and taxpayers when a significant number of plants are shut down in the future, can provide emergency back-up storage in the event that spent fuel needs to be moved quickly from a reactor site, and would provide an excellent platform for ongoing R&D to better understand how the storage systems currently in use at both commercial and DOE sites perform over time.

For consolidated storage to be of greatest value to the waste management system, the current rigid legislative restriction that prevents a storage facility developed under the NWPA from operating significantly earlier than a repository should be eliminated. At the same time, efforts to develop consolidated storage must not hamper efforts to move forward with the development of disposal capacity. To allay the concerns of states and communities that a consolidated storage facility might become a *de facto* disposal site, a program to establish consolidated storage must be accompanied by a parallel disposal program that is effective, focused, and making discernible progress in the eyes of key stakeholders and the public. Progress on both fronts is needed and must be sought without further delay.

Even with timely development of consolidated storage facilities, a large quantity of spent fuel will remain at reactor sites for many decades before it can be accepted by the federal waste management program. Current at-reactor storage practices and safeguards are being scrutinized in light of the lessons that are emerging from Fukushima. In addition, the Commission recommends that the National Academy of Sciences (NAS) conduct a thorough assessment of lessons learned from Fukushima and their implications for conclusions reached in earlier NAS studies on the safety and security of current storage arrangements for spent nuclear fuel and high-level waste in the United States. This effort would complement investigations already underway by the NRC and other organizations. More broadly, it will also be vital to continue vigorous public and private research and regulatory oversight efforts in areas such as spent fuel and storage system degradation phenomena, vulnerability to sabotage and terrorism, full-scale cask testing, and others. As part of this process, it is appropriate for the NRC to examine the advantages and disadvantages of options such as "hardened" onsite storage that have been proposed to enhance security at storage sites.

6. Early Preparation for the Eventual Large-Scale Transport of Spent Nuclear Fuel and High-Level Waste to Consolidated Storage and Disposal Facilities

The current system of standards and regulations governing the transport of spent fuel and other nuclear materials appears to have functioned well, and the safety record for past shipments of these types of materials is excellent. But the current set of transport-related regulations will need to be updated to accommodate changes in fueling practices. Moreover, past performance does not guarantee that future transport operations will match the record to date, particularly as the logistics involved expand to accommodate a much larger number of shipments. Past experiences in the United States and abroad, and extensive comments to the Commission, indicate that many people fear the transportation of nuclear materials. Thus greater transport demands are likely to raise new public concerns.

As with siting fixed facilities, planning for associated transportation needs has historically drawn intense interest. Transport operations typically also have the potential to affect a far larger number of communities. The Commission believes that state, tribal and local officials should be extensively involved in transportation planning and should be given the resources necessary to discharge their roles and obligations in this arena. Accordingly, DOE should (1) finalize procedures and regulations for providing technical assistance and funds for training to local governments and tribes pursuant to Section 180(c) of the NWPA and (2) begin to provide such funding, independent from progress on facility siting. While it would be premature to fully fund a technical assistance program before knowing with some certainty where the destination sites for spent fuel are going to be, substantial benefits can be gained from a modest early investment in planning for the early transport of spent fuel from shutdown reactor sites.

Planning and providing for adequate transportation capacity while simultaneously addressing related stakeholder concerns will take time and present logistical and technical challenges. Given that transportation represents a crucial link in the overall storage and disposal system, it will be important to allow substantial lead-time to assess and resolve transportation issues well in advance of when materials would be expected to actually begin shipping to a new facility. For many years, states have been working cooperatively with DOE to plan for shipments, often through agreements with regional groupings of states and in ways that involve radiological health, law enforcement, and emergency response personnel. As has been shown with the WIPP program and other significant waste shipping campaigns, planning, training and execution involves many different parties and takes time. In addition, specialized equipment may be required that will need to be designed, fabricated and tested before being placed into service. Historically, some programs have treated transportation planning as an afterthought. No successful programs have done so.

7. Support for Advances in Nuclear Energy Technology and for Workforce Development

Advances in nuclear energy technology have the potential to deliver an array of benefits across a wide range of energy policy goals. The Commission believes these benefits—in light of the environmental and energy security challenges the United States and the world will confront this century—justify sustained public- and private-sector support for RD&D on advanced reactor and fuel cycle technologies. In the near term, opportunities exist to improve the safety and performance of existing light-water reactors and spent fuel and high-level waste storage, transport, and disposal systems. Longer term, the possibility exists to advance "game- changing" innovations that offer potentially large advantages over current technologies and systems.

The Commission believes the general direction of the current DOE research and development (R&D) program is appropriate, although we also urge DOE to take advantage of the Quadrennial Energy Review process to refine its nuclear R&D "roadmap." We are not making a specific recommendation concerning future DOE funding for nuclear energy RD&D; in light of the extraordinary fiscal pressures the federal government will confront in coming years, we believe that budget decisions must be made in the context of a broader discussion about priorities and funding for energy RD&D more generally.

One area where the Commission recommends increased effort involves ongoing work by the NRC to develop a regulatory framework for advanced nuclear energy systems. Such a framework can help guide the design of new systems and lower barriers to commercial investment by increasing confidence that new systems can be successfully licensed.

Specifically, the Commission recommends that adequate federal funding be provided to the NRC to support a robust effort in this area. We also support the NRC's risk-informed, performance-based approach to developing regulations for advanced nuclear energy systems, including NRC's ongoing review of the current waste classification system (changes to the existing system may eventually require a change in law).

Another area where further investment is needed is nuclear workforce development. Specifically, the Commission recommends expanded federal, joint labor-management and university-based support for advanced science, technology, engineering, and mathematics training to develop the skilled workforce needed to support an effective waste management program as well as a viable domestic nuclear industry. At the same time, DOE and the nuclear energy industry should work to ensure that valuable existing capabilities and assets, including critical infrastructure and human expertise, are maintained. Finally, the jurisdictions of safety and health agencies should be clarified and aligned. New site-independent safety standards should be developed by the safety and health agencies responsible for protecting nuclear workers through a coordinated joint process that actively engages and solicits input from all relevant constituencies. Efforts to support uniform levels of safety and health in the nuclear industry should be undertaken with federal, industry, and joint labor–management leadership. Safety and health practices in the nuclear construction industry should provide a model for other activities in the nuclear industry.

8. Active U.S. Leadership in International Efforts to Address Safety, Non-Proliferation and Security Concerns

As more nations consider pursuing nuclear energy or expanding their nuclear programs, U.S. leadership is urgently needed on issues of safety, non-proliferation, and security/counter-terrorism. Many countries, especially those just embarking on commercial nuclear power development, have relatively small programs and may lack the regulatory and oversight resources available to countries with more established programs. International assistance may be required to ensure they do not create disproportionate safety, physical security, and proliferation risks. In many cases, mitigating these risks will depend less on technological interventions than on the ability to strengthen international institutions and safeguards while promoting multilateral cooperation and coordination. From the U.S. perspective, two further points are particularly important: First, with so many players in the international nuclear technology and policy arena, the United States will increasingly have to lead by engagement and by example. Second, the United States cannot exercise effective leadership on issues related to the back end of the nuclear fuel cycle so long as its own program is in disarray; effective domestic policies are needed to support America's international agenda.

The Fukushima accident has focused new attention on nuclear safety worldwide. Globally, some 60 new reactors are under construction and more than 60 countries that do not have nuclear power plants have expressed interest in acquiring them. These nations will have to operate their facilities safely and plan for safe storage and disposition of spent nuclear fuel. The United States should help launch a concerted international safety initiative—encompassing organizations like the International Atomic Energy Agency (IAEA) as well as regulators, vendors, operators, and technical support organizations—to assure the safe use of nuclear energy and the safe management of nuclear waste in all countries that pursue nuclear technology.

Nuclear weapons proliferation has been a central concern of U.S. nuclear policy from the earliest days of the nuclear era. These concerns are still prominent, especially where the deployment of uranium enrichment, reprocessing, and recycled fuel fabrication technology is being contemplated. As countries with relatively less nuclear experience acquire nuclear energy systems, the United States should work with the IAEA, nuclear power states, private industry, and others in the international community to ensure that all spent fuel remains under effective and transparent control and does not become "orphaned" anywhere in the world with inadequate safeguards and security.

Longer term, the United States should support the use of multi-national fuel-cycle facilities, under comprehensive IAEA safeguards, as a way to give more countries reliable access to the benefits of nuclear power while simultaneously reducing proliferation risks. U.S. sponsorship of the recently-created IAEA global nuclear fuel bank is an important step toward establishing such access while reducing a driver for some states to engage in uranium enrichment. But more is needed. The U.S. government should propose that the IAEA lead a new initiative, with active U.S. participation, to explore the creation of one or more multi-national spent fuel storage or disposal facilities.

In addition, the United States should support the evolution of spent fuel "take-away" arrangements as a way to allow some countries, particularly those with relatively small national programs, to avoid the costly and politically difficult step of providing for spent fuel disposal on their soil and to reduce associated safety and security risks. An existing program to accept highly-enriched uranium fuel from research reactors abroad for storage in the United States has provided a demonstration—albeit a limited one—of the national security value of such arrangements. The capability to accept limited quantities of spent fuel from foreign commercial reactors could be similarly valuable from a national security perspective. As the United States moves forward with developing its own consolidated storage and disposal capacity, it should work with the IAEA and with existing and emerging nuclear nations to establish conditions under which one or more nations, including the United States, can offer to take foreign spent fuel for ultimate disposition.

The susceptibility of nuclear materials or facilities to intentional acts of theft or sabotage for terrorist purposes is a relatively newer concern but one that has received considerable attention since 9/11. The United States should continue to work with countries of the former Soviet Union and other nations through initiatives such as the Nunn-Lugar Cooperative Threat Reduction Program and the Global Initiative to Combat Nuclear Terrorism to prevent, detect, and respond to nuclear terrorism threats. Domestically, evolving terrorism threats and security risks must be closely monitored by the NRC, the Department of Homeland Security, and other responsible agencies to ensure that any additional security measures needed to counter those threats are identified and promptly implemented. The recent events at Fukushima have – as they should – prompted the NRC and the industry to re-examine the adequacy of "mitigative strategies" for coping with large-scale events (like an explosion or fire) or catastrophic system failures (like a sudden loss of power or cooling); as noted previously, we also recommend that Congress charter the National Academy of Sciences to assess lessons learned from Fukushima with respect to the storage of spent fuel.

Tying It Together

In conclusion, the problem of nuclear waste may be unique in the sense that there is wide agreement about the outlines of the solution. Simply put, we know what we have to do, we know we have to do it, and we even know how to do it. Experience in the United States and abroad has shown that suitable sites for deep geologic repositories for nuclear waste can be identified and developed. The knowledge and experience we need are in hand and the necessary funds have been and are being collected. Rather the core difficulty remains what it has always been: finding a way to site these inherently controversial facilities and to conduct the waste management program in a manner that allows all stakeholders, but most especially host communities, states, and tribes, to conclude that their interests have been adequately protected and their well-being enhanced—not merely sacrificed or overridden by the interests of the country as a whole.

This is by no means a small difficulty, but we have witnessed other countries make significant progress with a flexible approach to siting that puts a high degree of emphasis on transparency, accountability, and meaningful consultation. Indeed, our friends in Spain have just succeeded in selecting a site for a consolidated storage facility by using the kind of consent-based process we recommend. Here at home, we have had more than a decade of successful operation of WIPP. And most recently, the Fukushima accident in Japan has reminded Americans that we have little physical capacity at present to do anything with spent nuclear fuel other than to leave it where it is. Against this backdrop, the conditions for progress are arguably more promising than they have been in some time. But we will only know if we start, which is what we urge the Administration and Congress to do, without further delay.

Thank you for having us here today, and we look forward to your questions.

In: United States Nuclear Waste
Editors: Gregory Hall and Steven L. Greene
ISBN: 978-1-62081-312-6

Chapter 3

TESTIMONY OF LAKE H. BARRETT. HEARING ON "AMERICA'S NUCLEAR FUTURE"*

Thank you, Chairman Shimkus and Ranking Member Green, for the opportunity to provide my personal views on the Blue Ribbon Commission (BRC) Report on America's Nuclear Future recommendations. I have spent several decades of my professional life working to implement the Nuclear Waste Policy Act that established, by law; a scientific, regulatory, and administrative process to safely dispose of our Nation's spent nuclear fuel and high level radioactive wastes in an environmentally protective manner. Unfortunately this Administration has succumbed to politics and has reversed much of the progress that has been made over the past 25 years. Although, this Commission was created partially as a cover for the Administration's actions, it has produced a very thoughtful report with excellent useful recommendations that should be incorporated into our statutes and programs.

It is not the fault of the BRC that it was not allowed to examine the obvious and lawful Yucca Mountain high level radioactive waste deep geologic repository disposal solution concluded under the Nuclear Waste Policy Act process. Despite such a politically imposed shortcoming, this highly qualified and dedicated group produced a thoughtful and well-reasoned report with excellent recommendations. I strongly urge the Administration and Congress to act promptly to incorporate the BRC recommendations into our remaining national program to get our nation's high level nuclear waste program moving forward again.

I strongly agree with the BRC's finding that a solution to the nation's spent fuel management needs is urgently needed and that the cost of inaction is many billions of dollars and mounting every day. The BRC corroborates what every objective study has found since the National Academies of Sciences looked at spent fuel management in 1957 – namely, that a national geologic repository is the foundation of any national spent fuel management program. Under existing U.S. law, a repository site complying with regulatory standards could be available immediately for construction of a disposal facility at Yucca Mountain - if the political opposition from Nevada's elected officials were overcome.

The BRC report also basically affirms the key components of the Nuclear Waste Policy Act. It confirms that there is no technological "silver bullet" that can replace the need for a geologic repository, such as Yucca Mountain.

* This is an edited, reformatted and augmented version of testimony given on February 1, before the House Energy and Commerce.

However, there is a timing issue in the BRC report that I would like to point out. The BRC report properly explains the urgency of developing a repository; however, the only repository development process they recommend will take well over 20 years to achieve. As you may know, the NWPA statute contemplates two geologic repositories, thus this issue could be addressed by utilizing a fresh start BRC consensus process for the NWPA second repository and modifying the existing Yucca Mountain program based on many of the BRC recommendations. The BRC partnership consensus approach for the second repository should be started immediately, independent of whatever happens with Yucca Mountain.

Although they did not address Yucca Mountain, I believe most of their recommendations are applicable to Yucca Mountain as well. Clearly everyone wishes that a consensus agreement could have been established between the DOE and State of Nevada along the lines of the BRC report consensus recommendation and as somewhat originally contemplated in the NWPA Section 117, "Consultation and Cooperation". Clearly that was my personal goal when I was in the Program. Unfortunately, the political situation, in those times, prevented any meaningful negotiations to find a common ground that could resolve Nevada's issues. One major step in the right direction now would be to complete the Yucca Mountain licensing process in an open and transparent manner to independently resolve all of the Nevada safety and environmental concerns.

After spending well over 8 billion dollars on scientific study and engineering evaluations, I am personally very confident that the scientific evidence supports the conclusion that the site will be safe and protective of the environment indefinitely. However, that is not the message that one hears in the Nevada public media. Completing the nearly finished NRC licensing process would hopefully make safety evident to all Nevadans, such that politically driven fear mongering sound bites would be seen for what they are and meaningful negotiations begin so that the Nevada fairness, equity, and benefits concerns can be equitably addressed by the federal government. Such a negotiation could lead to the changes, assurances and the substantial benefits that Nevada deserves from the federal government for acting in the national interest. Such a binding agreement would be of great value and be of mutual benefit to all Nevadans and the rest of the nation.

As the BRC correctly points out, the Carlsbad community and State of New Mexico have made cooperation in the national, state and local interest work well together for the WIPP facility. Nye County, Nevada also has been an effective model with its strong science program and focus on citizen protection as well as understanding benefits.

The BRC report properly highlights the need for federal action to remove the spent nuclear fuel stranded at isolated shutdown reactors. This action would eliminate all nuclear risks at those sites, reduce costs of all (including the federal taxpayers), keep promises, and the sites could be returned to useful societal purposes. To achieve this important goal in a timely, effective manner the BRC correctly recommends a partnership; consensus based consolidated interim storage facility. In my view, this process should begin immediately. This is not a technical problem: it is a problem of our collective failure to act in our mutual national interest with respect to the host state.

DOE has the authority under existing law and the capability with its commercial contractors to act now on many of the BRC recommendations, such as working to develop consensus hosting partnerships. Indeed, the recently passed FY 2012 budget provides $3M for "development of models for potential partnerships." It is also a fact that the DOE's commercial contractors made significant progress in developing over 10 state and local hosting expressions of interest for the past Global Nuclear Energy Partnership (GNEP) facilities. Although the GNEP program no longer exists, the volunteer hosting partnership concept used in that program should be employed to put into practice the BRC recommendation and FY 2012 budget guidance. Hopefully the DOE will integrate these into a prompt effective voluntary host exploration program. The DOE should finance their commercial nuclear contractor support teams, which are already in place under existing support contracts, to engage interested communities and states. They can start this process now to explore potential willing state and local community hosting partnership arrangements.

This simple first step would begin the process of developing what I envision as an integrated phased used nuclear fuel management R&D center(s). Such a facility would initially serve as a consolidated interim storage facility and could eventually expand into other high technology missions if the host community and host state so desire. There are multiple other synergistic Nuclear Energy, Environmental Management, NNSA and Science missions that could complement used nuclear fuel management. These could be negotiated as a package voluntary agreement that could be beneficial to all local, state, federal, and commercial interests. The exact nature of such missions would of course depend upon the actual site and host community and state interests, but there are many opportunities for mutually supportive creative planning to meet everyone's needs. It is possible that such a negotiation could even result in a volunteer host for a second geologic repository.

Working in the DOE Office of Civilian Radioactive Waste Management for over a decade, I would like to take this opportunity to commend the thousands of scientists, engineers, and trades people that completed the most complex scientific study of an underground complex that has ever been accomplished. This achievement was very difficult as the program was politically attacked and interfered with constantly. All of the people who worked on this project tried to fulfill their duties to implement the Nuclear Waste Policy Act law within the constraints externally imposed by the political structure. Very regrettably this political interference has now led to its current unlawful termination and has severely impacted the lives of many dedicated people.

The report is also crystal clear that the DOE statutory and contractual commitments to utility and electricity consumer stakeholders have not been met, that spent fuel policy should be augmented by a new organization dedicated solely to carry-out implementation, and that funds collected for waste management should be made available for their intended purpose by taking the nuclear waste fund off-budget. Hopefully progress can be made on all of these superb recommendations.

I would like to point out that many of the complex policy challenges, such as state veto authority, organizational independence, financing, and contract obligations were studied and debated extensively during the past decades. As the complex NWPA was implemented over these decades, there were many more lessons learned about Yucca and the former second repository program than were included in the BRC report.

The BRC recommendations are an excellent start, but they are only a beginning. Please beware of over simplification, wishful thinking, and be prepared for hard work ahead to create a new durable sustainable policy infrastructure. A realistic and objective evaluation of the factors that worked and did not work with the NWPA must be compared with a realistic evaluation of alternative paths forward. Improvements and enhancements can and should be made. Many are exactly what the BRC recommends, but there are also many old devils hiding within the details of the BRC recommendations that will have to be practically resolved.

Our nation now stands at a critical ethical crossroad on nuclear waste management. We owe our children and grandchildren a protective disposal solution for used nuclear fuel and high level radioactive waste that our generation has made. It is irresponsible to just continue kicking the problem down the road to the next generation whenever there is political pressure from a few during a primary campaign. We must reestablish a will to act in the national interest. The issue has been exploited long enough for the local political gains for a few at great cost to many. Solutions are at hand, and the Blue Ribbon Commission, despite its politically imposed restrictions, has given the federal government useful actionable recommendations to greatly enhance what has already been achieved. Let us not waste this opportunity. Thank you.

In: United States Nuclear Waste
Editors: Gregory Hall and Steven L. Greene

ISBN: 978-1-62081-312-6

Chapter 4

TESTIMONY OF D. WARNER NORTH. HEARING ON "AMERICA'S NUCLEAR FUTURE"*

I welcome the opportunity to provide my views in this Hearing regarding the Final Report to the Secretary of Energy by the Blue Ribbon Commission on America's Nuclear Future (BRC). Thank you, Chairman Shimkus and Ranking Member Green.

I provide these comments as a private citizen concerned about America's nuclear future, based on my experience. I served from 1989 to 1994 on the Nuclear Waste Technical Review Board (NWTRB), and then for the NAS as a member of the Board on Radioactive Waste Management and as chair of the NAS committee that produced the 2001 report referenced in Endnotes 18 and 125 in the BRC Report, and as a member of the NAS committee that produced the 2008 report referenced in Endnote 136. I have a background in physics, but my career has been in decision analysis and risk analysis. I have served as president of the Society for Risk Analysis, and I am a national associate of the National Research Council. (This honor enables me to use the Members' Library in the National Academy of Sciences Building.)[1]

I strongly endorse the BRC Final Report and its recommendations. The Commission and its staff have produced an excellent document, well researched and well-written, at a level appropriate for a broad audience.

The scope of the BRC report does not include discussion of any facilities at Yucca Mountain in America's nuclear future. Characterizing the Yucca Mountain site as the sole candidate for our nation's first geological repository has been the principal focus of the federal program over the past several decades. While I can accept language such as in the first paragraph of the Executive Summary, that national policy has "been troubled for decades and has now all but broken down." I would have preferred more clarity at the outset in this report as to where responsibility for this shortfall lies. I believe that the staff of the federal agencies (DOE, NRC, NWTRB), their contractors, and consultants have worked diligently to implement the Nuclear Waste Policy Act passed by Congress in 1982 and amended in 1987, and that their accomplishments have been commendable. I would have liked to see more explicit recognition of these technical achievements. The achievements are reflected in the many Endnotes in the BRC Report citing the work of these organizations. My reading of the

* This is an edited, reformatted and augmented version of testimony given on February 1, 2012, Before the House Energy & Commerce Committee, Subcommittee on Energy and Environment Blue Ribbon Commission Hearing.

full text of the Report is that the BRC recognizes that the shortfall in the nation's program lies with the law established by Congress, the deficiency and inconsistency in national leadership in implementing this law (for example, the cancellation of the second repository), and the strong and ongoing opposition by state political leaders, especially those from Nevada.

A major finding in the 2001 NAS report, from a distinguished committee of international experts that I had had the honor to chair, was that "the biggest challenges ... are societal" (meaning policy or political) and not technical. Our first principal recommendation was that national governments should provide the leadership and support for solving the problems. In the United States there has not been consistent Congressional and senior Administration leadership adequate to meet the challenges. As a result, there is now a situation badly in need of corrective action. The BRC has, within its scope, provided excellent guidance on this corrective action.

Much of this guidance is consistent with findings and recommendations of earlier reports. There are not any major breakthroughs in understanding or from the emergence of new technologies. The nation needs a program to move spent nuclear fuel and defense waste from where these materials are now located into geological disposal, where these dangerous materials will be safe over a "long-term" determined by the time for radioactivity to diminish, ten thousand to one million years. As the BRC report describes, there is general international consensus on how to accomplish this goal, and many nations are making good progress. In this country, we gone taken 25 years since the law was last revised, spent over ten billion dollars, and we have a policy that BRC judges, "has all but broken down." Our country has a liability of nearly $50 billion dollars: $30 billion in the Nuclear Waste Fund[2], money (including interest) the federal government has received from electricity ratepayers but not yet spent in providing disposal services, plus at least $20 billion[3] in legal penalties for failure to take possession of spent fuel beginning in 1998, as mandated under law and existing contracts. BRC states, in large type at the outset of Chapter 4, "The central flaw of the U.S. nuclear waste program to date has been its failure to develop permanent disposal capability."

BRC's first recommendation is a "new consent-based approach" to siting nuclear waste management facilities. This is not a new idea, but one that has been around for decades. It would be new as an approach for the U.S. federal government, replacing law that at first required a technical evaluation of candidate sites, and then designated only a single site to be characterized. The 2001 NAS report states on page 132, "When an agency seeks permission to investigate the technical feasibility of a proposed site, it is usually necessary first to obtain the assent of the local community and its various councils and representatives that the site has been selected fairly and the project is provisionally acceptable." The context for this statement is a report dealing with this problem internationally. Many other countries have the need for disposal of nuclear waste. The word "usually" in the quote above indicates an exception. When our report was written the U.S. was proceeding with a program established by U.S. law, in which such "assent" from the state of Nevada had not been obtained in advance. There was a provision in the law for state disapproval, but such disapproval could be, and was, overridden by Congress.

Many U.S. state governments opposed a nuclear waste disposal facility in their state. Nevada established under state law an organization to oppose such a facility, the Nevada Nuclear Waste Project Office (NWPO) in 1985, when the Yucca Mountain site was under active consideration. This date preceded by two years the 1987 Amendments Act designating Yucca Mountain as the sole site to be characterized. As noted on page 22 of the BRC report, this 1987 legislation has been characterized as the "Screw Nevada" bill. The Nevada NWPO and the state's elected leaders have for 27 years carried out a strong campaign opposing development of the Yucca Mountain site. I met with the NWPO Board, including former Nevada Governor Grant Sawyer, early in my service on the NWTRB. The Board expressed to me quite clearly that their mission was to oppose. I was equally clear that the NWTRB mission established in the 1987 Amendments Act was oversight to improve the technical aspects of the federal nuclear waste management program. Accessing the NWPO website as I write this testimony on January 29, I find the NWPO mission remains the same. It is not improved scientific understanding and support for wise decision making, but opposition. I note that the allegations of dangers posed by a Yucca Mountain disposal facility are on a web page first posted in 1998 and unchanged since. (Attachment A).

In contrast, local government entities near Yucca Mountain, such as Nye County, have cooperated with DOE and have carried out research in support of understanding better the safety of the site. Nye County's Q&A webpage (Attachment B) provides a strong contrast to the NWPO page in attachment A. As noted in the BRC report, page 23, Nye and other local counties have expressed support for the facility, or at least, for allowing the license approval process to go forward.

Recommendations 3 and 6 in the BRC report describe other urgently needed remedial action to prepare for the time when waste management facilities become available, at Yucca Mountain or at another location. During my initial time on NWTRB more that 20 years ago, DOE had a program in place to develop the system planning for packaging and transporting spent fuel from its location at many reactor sites to either interim storage or a final disposal site. NWTRB reviewed and encouraged this planning. But Congress cut the appropriations, and DOE had to reduce its activities in this area. The limited funds made available to DOE were spent almost entirely on site characterization activities supporting the license application for Yucca Mountain, called for under the law. The other aspects were forced to be deferred. The BRC added recommendation 6, on preparing for transport, to the seven recommendations in its draft report, because the Commission heard from many parties on the need to do this planning and preparation. A lead time on the order of a decade is needed before waste transport begins. Shorting the needed appropriation of funds for this purpose, funds already being paid by ratepayers, were the result of decisions by Congress, and not a failure by DOE, by the technical community, or by state and local government. In contrast, for WIPP, timely and effective advance planning and preparation for the transport of waste was done by DOE and its contractors in cooperation with state and local agencies, as described in the BRC report.

Recommendations 4 and 5 have an important interaction with recommendation 1. The time scale to achieve success, which BRC defines as a legally binding agreement (Box, page 57) is essentially the time to reach what was described in the Republican

Presidential candidates' debate in Las Vegas last October 18, as "a pretty good deal." (See Attachment C, my letter to the BRC of October 31, 2011). Other states - such as New Mexico, which negotiated a "pretty good deal" with the federal government on WIPP - might agree to a new facility or expansion of existing facilities (i.e., WIPP) to take additional nuclear waste. The "new approach" for the U.S. nuclear waste management program is negotiation to achieve consent, not overcoming the opposition based on a federate mandate in existing law.

Steve Frishman's statement to BRC is cited in Endnote 260. Steve has for more than two decades worked for the Nevada NWPO. He and I came to know each other well when I served on the NWTRB. I agree with his point that "the interested public has often been confused about the roles of the respective agencies, and the motivation, scope, and meaning of the regulation proposed ..." and I will add, the existing regulation. Performance assessment for these facilities can become mind-numbingly complex, even for those of us who are specialists. The goals and compliance process need to be explained in simple language to the interested and affected members of the public.

Let me illustrate such risk communication by putting the summary of U.S. disposal facility regulations, a box on Page 91 of the BRC report, into terms more readily understood. Members of the Subcommittee who represent States far from Washington, D.C. travel often by plane, as I have done from my home in California to this Hearing. Modern commercial jet aircraft fly above much of the atmosphere, so radiation we receive from cosmic rays is not attenuated as it is when we are at ground level. Figure 7 on page 15 of the BRC report lists 40 microSieverts of radiation exposure for a one-way flight from New York to Los Angeles, roughly the same distance as my flight from San Francisco to Washington-Dulles. Two round trips is 4 x 40 = 160 microSieverts, slightly above 150 microSieverts, listed on page 91 in the Box as the limit for the first 10,000 years. The goal set forth in the regulation is that there should be a reasonable expectation that no members of the public will receive more than this dose of radiation annually: two cross-country round trip flights worth. The standard for ten thousand to one million years is less stringent. It corresponds to about one round tip flight across our country per month. (For many years I flew this much, one round trip across the country per month. I expect many Members of the Subcommittee fly even more.) Most members of the public do not understand what microSieverts and rems signify. But if they come to understand that the requirement for nuclear waste disposal facilities is for limiting radiation exposure to all members of the public to such low levels, comparable to exposures most of us accept without concern, they might come to understand that the regulations should assure acceptable safety, now and far into the distant future.

Implementation of these regulations to assure a reasonable expectation of compliance, across an extremely long time period with all its attendant uncertainties, is, indeed, complex. But if we focus on sharing our understanding in good-faith negotiation, I am confident success on public acceptance can be achieved, as has been done at WIPP and in the nuclear waste management programs of many other countries.

I hope the BRC Report and this Hearing will motivate Members of this Subcommittee and more generally, both the House and the Senate, to proceed quickly toward bipartisan consensus to implement the recommendations in the BRC Report. Enhanced national leadership was recommendation #1 in the 2001 NAS report. It did not happen then; I hope it can happen now. Our country has a liability of $50 billion. We need to take the steps to deliver on our commitments and work this number down, not let it grow as we have done over the past decade.

Yucca Mountain remains in my judgment a viable siting option. The BRC Report stresses not precluding options, but engaging in adaptive management. The 2001 NAS report and its successor, the "One Step at a Time" NAS report of 2003, are cited as references on adaptive management. As urged in a letter (Attachment D) to BRC of September 29, signed by me and five other distinguished scientists, two of whom played key leadership roles in the success for WIPP, the NRC evaluation of the license application for a disposal facility at Yucca

Mountain should be completed. The public should have full access to the expert evaluation by NRC professional staff, of allegations by the Nevada opposition and the plans and analysis submitted by the DOE staff and its contractors.

Release of a positive evaluation by NRC of the license application for a Yucca Mountain waste repository does not commit the nation to go forward with construction of this facility at Yucca Mountain. Congress must first appropriate the money, and that will require a positive vote by the Senate. Five years might pass before that could happen. But it will take 15-20 years, perhaps longer, before a disposal facility for commercial spent nuclear fuel at another site can be brought to the point of approval on a construction license application, even when consent has been achieved. The type of site characterization and technical analysis carried out at Yucca Mountain takes time and resources; no other nation has been able to do it more quickly than on the order of 20 years. Our nation needs two sites for disposal facilities, as were called for 30 years ago in the 1982 Nuclear Waste Policy Act. If there may be a 15-20 year wait before the first disposal site is ready for facility construction, the nation will need an interim consolidated storage facility, strong motivation for BRC recommendation 5. It will take the order of ten years to do the negotiating, planning, design, licensing, training and other preparation for transport and storage of the waste materials. This transportation planning and preparation should be funded and restarted, via BRC recommendations 3 and 6. In about 20 years, even more serious penalties may occur if the federal government remains unable to accept spent fuel. For example, the U.S. Navy may be unable, under existing agreements, to continue to ship the spent fuel from its nuclear-powered ships to Idaho (BRC Report, page 28).

The steps listed in chapter 13, *Near-Term Actions*, should be undertaken by DOE and the current Administration. Congress should move quickly to enact new legislation to enable the BRC recommendations. To support discussion and debate on these changes, the DOE should designate a senior official, as BRC has recommended at the beginning of its chapter 13. The BRC has ended its work and has no further budget. This senior official should be supported by staff and consultants at the same level of excellence as the staff and consultants who participated in the preparation of the BRC Final Report.

Thank you for this opportunity to present my views.

ATTACHMENTS

http://www.state
http://www.nyecounty.com/YMIC/YMICQandA.pdf
http://brc.gov/sites/default/files/comments/attachments/commentstobrcbydwn10- 31-2011.pdf
http://www.sustainablefuelcycle.com/resources11.pdf

In: United States Nuclear Waste
Editors: Gregory Hall and Steven L. Greene

ISBN: 978-1-62081-312-6

Chapter 5

TESTIMONY OF MARTIN G. MALSCH. HEARING ON "AMERICA'S NUCLEAR FUTURE"*

Mr. Chairman and members of the Subcommittee, I appreciate the opportunity to provide testimony today regarding the recommendations of the Blue Ribbon Commission on America's Nuclear Future (BRC), released late last week. My name is Martin G. Malsch. I am a partner in the law firm Egan, Fitzpatrick, Malsch & Lawrence, PLLC. I have practiced law in the nuclear energy and nuclear waste fields for over forty years, including many years as the Deputy General Counsel or Acting General Counsel for the U.S. Nuclear Regulatory Commission. In private practice I have advised both public and private entities regarding nuclear issues.

I believe I am one of the few practitioners in the nuclear energy field who has represented both proponents and opponents of nuclear facilities. I currently represent the State of Nevada on matters related to the Yucca Mountain repository, but I am not testifying today in a representative capacity. My testimony today presents my personal opinions regarding the BRC's recommendations based on my experience and expertise in the nuclear energy and nuclear waste fields; it does not necessarily represent the views of Nevada or anyone else.

BACKGROUND

The BRC report includes a brief summary of the U.S. experience in developing geologic repositories and draws some conclusions based on this experience (report at pp. 19-24). I would like to add a few details about this experience, focusing on Lyons, Kansas and Yucca Mountain, Nevada, because I believe this will add substantial additional context and support for the key BRC conclusion that "any attempt to force a top-down, federally mandated solution over the objections of a state or community – far from being more efficient – will take longer, cost more, and have lower odds of ultimate success" (report at pg. ix). I also believe that consideration of some additional historic details will support the conclusion that a premature commitment to a site before sufficient supporting scientific evidence is available

* This is an edited, reformatted and augmented version of testimony given on February 1, 2012, before the U.S. House of Representatives, Committee on Energy and Commerce Subcommittee on Environment and the Economy.

also creates a high risk of program failure and erodes credibility. The BRC mentions this second point as well (report at pg. 23) but I think it warrants additional emphasis.

Lyons, Kansas

In the 1960s a clamor arose over the potential that liquid high-level radioactive wastes would leak from Atomic Energy Commission (AEC) storage facilities located at the National Reactor Testing Station in Idaho, the Savannah River Site in South Carolina, and the Hanford Site in Washington. As a result, the AEC promised Idaho Senator Church that the Idaho wastes would be transferred out of Idaho to a permanent geologic repository by the end of the 1970s. The AEC pinned all its hopes on an abandoned salt mine in Lyons, Kansas. However, rather than taking the time to complete necessary scientific investigations, the AEC offered disputable safety conclusions and pressed ahead. The AEC believed state and local support was essential, but it lost that support when it failed to give any credence to the legitimate concerns of Kansas experts. Ultimately, the Lyons, Kansas site proved to be unsuitable.

Yucca Mountain

After the failure of Lyons, Kansas, the AEC's successor agencies continued to investigate other possible repository sites and the Congress enacted the Nuclear Waste Policy Act of 1982 (NWPA). In accordance with the NWPA, DOE selected five sites for more detailed study (characterization): salt deposits in Mississippi, Texas, and Utah; basalt formations in Hanford, Washington; and volcanic tuff rock in Nevada. In perhaps a hint of what was to come, potential sites in Louisiana were excluded based on a political side agreement between Louisiana Senator Johnston and the Secretary of DOE. The NWPA then called upon DOE to narrow the choices to three, all three of which were to be fully characterized (studied) so that any one failure would not prematurely destroy the whole repository program. In 1986, the DOE Secretary announced that the final three choices were the ones in Deaf Smith County, Texas; Yucca Mountain, Nevada; and Hanford, Washington. The designation prompted angry protests from all three areas, whose representatives believed that the scientific investigations were not completed, and the protests became part of a nationwide movement when DOE cancelled the search for an eastern site, notwithstanding a clear informal agreement among NWPA supporters that the second site called for by the NWPA would be located in an eastern State.

The program was now in shambles and Congress reacted by enacting the NWPA Amendments Act of 1987. That Act directed DOE to limit its future site characterization and selection efforts to a single site in Yucca Mountain, Nevada, notwithstanding the advice from NRC (and others) that the scientific information was insufficient to make an informed safety conclusion about the suitability of the site. In fact, the selection of the Yucca Mountain site was based on DOE's so-called "Multiattribute Utility Analysis of Sites," which depended in important part on the assumption that little groundwater would move downward from the mountain top and seep into the tunnels where the waste would be disposed of.

This assumption later proved to be false.

The NWPA Amendments Act of 1987 attempted to place the entire high-level waste disposal burden on one western state with no nuclear power plants or other high-level waste generating facilities.

The Act's supporters ignored the incompleteness of the scientific information and ignored the objections of the host State, which believed (with good reason) that Nevada had been singled out simply because it was "the small kid on the block." Lessons that should have been learned from the history of Lyons, Kansas were ignored.

In February 2002, DOE Secretary Abraham formally recommended the Yucca Mountain site to President Bush, notwithstanding the Nuclear Waste Technical Review Board's conclusion that DOE "has yet to make a convincing case that nuclear waste can safely be buried at Yucca Mountain," and President Bush recommended the site to the Congress. Citing numerous scientific flaws, Nevada Governor Guinn formally disapproved of the site, using the state veto procedure set forth in the NWPA. Congress then formally overrode Nevada's veto by enacting H.J. Res. 87.

The designation of Yucca Mountain as a repository site became effective on July 23, 2002, when President Bush signed S.J. Res. 34 into law.

The NWPA required DOE to file its license application within 90 days after the President's site recommendation became effective, or by October 21, 2002. October 21, 2002 came and receded into history without any application being filed. This was not a surprising development, given the scientific and engineering challenges DOE still faced when Nevada's veto was overridden. Obviously, DOE's recommendation of the Yucca Mountain site to the President was another example of a premature commitment to the site, continuing the trend set in 1987. The application was not filed and docketed by the NRC until September 8, 2008, almost six years after the statutory deadline expired. Final repository safety regulations were not even in place until 2009. DOE moved to withdraw its license application on March 3, 2010 and the presiding Licensing Board denied DOE's motion on June 29, 2010. On September 9, 2011, the Commission announced that it could neither affirm nor reverse the Licensing Board's decision because it was evenly divided on the matter. However, because of budgetary limitations, the Commission also directed the Licensing Board to take steps that would facilitate an orderly suspension of the licensing proceeding. The Licensing Board suspended the licensing proceeding on September 30, 2011.

Thus, the Yucca Mountain license application proceeding before the NRC is now suspended indefinitely because of budgetary limitations. All four participating NRC Commissioners agreed with this result (one NRC Commissioner previously recused himself), including the two Commissioners who believed DOE lacked the authority to withdraw the license application. No party to the proceeding has asked the NRC to reconsider the suspension. Congress "zeroed out" Yucca Mountain in the final FY 2012 Appropriations Act and, as a result, there are not enough funds to come anywhere close to completing the proceeding.

Past Mistakes and Lessons Learned

Two things stand out in this brief history of the two most significant attempted U.S. geologic repository projects. First, in both cases project proponents made a premature commitment to the site in the face of incomplete scientific information. In the case of Yucca Mountain, the BRC correctly observed that the "short-circuiting of the initial site selection process ... had the effect of tainting all subsequent state-federal interactions over the project" (report at pg. 48). Indeed the process as a whole "created a widespread perception that the repository location was being determined on the basis of primarily political, rather than technical and scientific, considerations" (report at pg. 23).

Second, in both cases, the site proponents ignored the legitimate objections of the host State. The BRC observed correctly (report at pg. 58) that "determined opposition at any level of government can at a minimum significantly complicate and delay, and in many cases defeat, best efforts to site a facility."

While there may be some residual hope in some quarters that Yucca Mountain might somehow be revived, I believe most informed observers would agree with the BRC that tying the entire fate of the U.S. high-level waste program to Yucca Mountain "has not worked to produce a timely solution for dealing with the nation's most hazardous radioactive materials" (report at pg. vi).

BRC Recommendations

I agree with the BRC's recommendations. The BRC studied the lessons of history and its recommendations are well supported.

We owe a large debt of gratitude to the BRC members and the BRC staff for their willingness to serve, their dedication to the task, their openness to divergent ideas and opinions, and their careful analysis of problems and possible solutions to the nuclear waste management issues confronting Americans today.

I would like to focus my testimony today on three key and closely connected BRC recommendations:

(1) that there should be prompt efforts to develop one or more geologic disposal facilities;
(2) that there should be a new single-purpose organization dedicated to implementing the waste management program; and
(3) that a new consent-based approach to site selection should be adopted.

Developing a New Geologic Disposal Facility

The BRC recommended that there should be "prompt efforts to develop one or more geologic disposal facilities" (report at pg. vii).

I agree there should be prompt efforts to develop one or more geologic disposal facilities, although not in the sense that we need to select and license a repository in the near term.

We are not facing any disposal crisis because vitrified high-level waste and spent nuclear fuel can be stored safely for a long time and we therefore have plenty of time to get things right. But amending the Nuclear Waste Policy Act, organizing and staffing a new waste management organization, and implementing a new consent-based site selection process, as

BRC recommends, will take considerable time. We should start the process promptly, especially the necessary changes to the NWPA.

The BRC recommended an "adaptive, staged facility siting and development process" whereby "[p]roject managers are able and willing to reevaluate earlier decisions and redesign or change course when new information warrants" (report at pg. 54). This recommendation addresses the key lesson from the past that premature commitments to one site should be avoided.

The development of a geological disposal facility can take decades, and a step-by-step, iterative process is required. The amount of time and effort required creates a grave danger that project momentum will overcome common sense and sound science. There should be multiple opportunities to assess the quality of the technical program and the safety case supporting the decision-making process and no reluctance to "pull the plug" when warranted.

A New Organization

I support the BRC's recommendation that there should be "a new organization dedicated solely to implementing the waste management program" (report at pg. vii). This means that the responsibility for the geologic repository program should be taken away from DOE and assigned to a new single-purpose organization, perhaps a government-chartered entity like TVA. DOE has not performed well here and a new single organizational approach is clearly needed. In fact, I made a similar recommendation several years ago at an NRC Regulatory Information Conference.

The new organization should be dedicated solely to implementing the waste management program and empowered with the authority and resources to succeed. It will also be important that the new organization be subject to independent licensing and regulation of its waste management activities (including transportation) in the same way that any other private entity would be.

Consent Based Site Selection

The BRC recommended "a new consent-based approach to siting future waste management facilities" (report at pg. vii). I believe a consent-based site selection approach is not just good government – it is a frank concession to reality and, as I indicated above, one of the two key lessons that must be learned from history. We should not assume that the objections of a host state or local government or tribe will melt away and that they will be "ready to deal" if the NRC grants the construction authorization. Nor should we assume that the pre-emptive powers of the federal government are so great, and that state and local rights and preferences are so undeserving of respect, that a site can always be thrust upon an unwilling host state, local government, or tribe. Even a site located on Federal land is subject to numerous state and local laws and regulations that can be used to vindicate states' rights absent draconian and dubious Federal legislation preempting state law. The BRC put it well when it concluded that "determined opposition at any level of government can at a minimum significantly complicate and delay, and in many cases defeat, best efforts to site a facility" (report at pg. 58).

Therefore means must be found to enlist the cooperation, or at least the acquiescence, of the host state, local government, and tribe. Generous financial or other incentives (a so-called benefits package) can be provided to the affected governmental entities, but this approach can easily operate or be construed as a kind of unacceptable bribe in return for ignoring the safety of current and future generations of citizens.

Therefore an affected governmental entity should never be asked to agree completely with, and withhold its objections to, a site while site investigations and safety analyses are still underway. This asks for too much. Instead, as the BRC recommended, there should be a step-wise and iterative process that avoids premature commitments on all sides.

I suggested how such a step-wise and iterative process might be constructed in written comments I provided the BRC, dated November 10, 2010. But I agree with the BRC that a successful site selection decision will most likely result from "a complex and perhaps extended set of negotiations between the implementing organization and the potentially affected state, tribal, and local governments, and other entities," and that it would be desirable for these negotiations to result in "a partnership agreement or some other form of legally enforceable agreement" (report at pg. ix).

I also recognize that a state, local or tribal government's ability to "veto" a repository project cannot last indefinitely; otherwise the uncertainty over whether the project could ever be successful would be so great that any significant investment in it would be imprudent. Ending the "veto" can be a matter subject to negotiations between the waste management organization and the governmental entity.

Some Reservations

I agree with the BRC that independent regulation is "an essential element of a safe, secure, environmentally responsible and ultimately effective nuclear waste management strategy" (report at pg. 88) and also that site-specific regulations like those that apply now to Yucca Mountain "undermine confidence" (report at pg. 23). I also agree with the BRC that there is no need to readjust the assignment of generic repository regulatory authority as between NRC and EPA.

However I have some reservations about certain parts of the BRC's discussion of regulatory standards issues (report at pp. 89-92). I suspect the BRC would agree with me that these particular topics are best addressed by the NRC in a future standards rule making and I see no need to address them in any depth today. Suffice it to say here that there is some tension between (1) concluding, as most experts have, that very long-lived radioactive wastes can be safely disposed of if a suitable site is selected, but (2) cautioning that it may be unrealistic to have a very long requirement for demonstrating compliance with a traditional safety regulation and suggesting that a different standard of proof should be applied. If we are overly aggressive in limiting compliance time frames and relaxing the burden of proof we will undercut the first conclusion and detract from the credibility of the licensing process.

CONCLUSION

Almost everything that could go wrong with a geologic repository program in the United States has now gone wrong. It would be unfortunate if the nuclear power program in this country foundered because of poorly chosen policies for managing spent fuel and high-level radioactive waste, and the citizens living near DOE nuclear legacy sites deserve a better program than the one they got. The BRC has now offered a path forward. We have ample time to consider the BRC's recommendations and get things right. But history suggests that amending the Nuclear Waste Policy Act, organizing and staffing a new waste management organization, and implementing a new consent-based site selection process, as the BC recommends, will take considerable time. Therefore we should start the process now – especially the legislative process.

Moreover, the BRC has also recommended some significant useful steps that can be taken without new legislation, especially regarding the equitable treatment of Nuclear Waste Fund (NWF) fees (report at pp. 75-78). These also should be considered promptly.

Thank you for your consideration of my testimony. I would be pleased to answer any questions you may have and to assist the Subcommittee in its future efforts to develop a sound nuclear waste management policy.

In: United States Nuclear Waste
Editors: Gregory Hall and Steven L. Greene
ISBN: 978-1-62081-312-6

Chapter 6

TESTIMONY OF DR. EDWIN LYMAN, SENIOR SCIENTIST, GLOBAL SECURITY PROGRAM. HEARING ON "AMERICA'S NUCLEAR FUTURE"*

SUMMARY OF UCS TESTIMONY

- The Union of Concerned Scientists (UCS) commends the commissioners and staff of the Blue Ribbon Commission (BRC) for doing an excellent job in addressing an extremely challenging set of issues.
- UCS agrees with most of the eight recommendations in the BRC report.
- UCS strongly concurs with the BRC's conclusion that "no currently available or reasonably foreseeable reactor and fuel cycle technology developments --- including advances in reprocessing and recycling technologies --- have the potential to fundamentally alter the waste management challenge over at least the next several decades, if not longer." UCS believes that if the BRC had endorsed reprocessing, it would have sent the wrong message to the rest of the world, undermining efforts to control the growth of weapon-usable material stockpiles.
- UCS supports the consent-based siting approach and the creation of a new waste management organization that is independent of DOE, provided that its operations are limited to transport, storage and direct disposal of spent fuel and high-level waste.
- UCS is not persuaded that new legislation to facilitate the siting and development of consolidated interim storage facilities is necessary, either for spent fuel from operating reactors or from shutdown reactors.
- UCS believes that spent fuel can be managed safely at reactor sites provided that the Nuclear Regulatory Commission appropriately upgrades its requirements to minimize the safety and security risks associated with long-term (up to 100 years) storage at reactors.

* This is an edited, reformatted and augmented version of testimony given on February 1, 2012, before the Subcommittee on Environment and the Economy, Committee on Energy and Commerce, U.S. House of Representatives.

- UCS supports limited taxpayer-funded nuclear energy R&D on improving safety, security and efficiency of existing nuclear plants and the once-through fuel cycle.

Good morning. On behalf of the Union of Concerned Scientists, I would like to thank Chairman Shimkus, Ranking Member Green, and the other distinguished members of the Subcommittee for the opportunity to provide our views on the recommendations of the Blue Ribbon Commission on America's Nuclear Future (BRC).

The Union of Concerned Scientists (UCS) is neither pro nor anti-nuclear power, but has served as a nuclear power safety and security watchdog for over 40 years. UCS is also deeply concerned about global climate change and has not ruled out an expansion of nuclear power as an option to help reduce greenhouse gas emissions—provided that it is affordable relative to other low-carbon options and that it meets high standards of safety and security. However, the Fukushima Daiichi crisis has revealed significant vulnerabilities in nuclear safety and has shaken public confidence in nuclear power. Regulators around the world must seriously address these vulnerabilities in order to reduce the risk of another Fukushima in the future. Otherwise, the viability of nuclear power as a reliable electricity option will be in doubt.

Before proceeding, I would like to mention that although UCS supports the development of one or more geologic repositories for the direct disposal of spent fuel, UCS does not have a position on the suitability of the Yucca Mountain site, or for that matter, any other potential site in the United States. The UCS Global Security Program does not have the geological expertise necessary to evaluate site suitability. However, we concur with the BRC's assessment that the process by which Yucca Mountain was selected was flawed and contributed to the program's ultimate failure. UCS supports the BRC's call for a new, consent-based repository siting approach that will be more likely to lead to selection of sites that are both technically suitable and broadly acceptable to the public.

UCS commends the BRC commissioners and staff for doing an excellent job in addressing an extremely challenging set of issues. The BRC's report is clear, well-written and compelling, and provides a comprehensive roadmap for moving toward achieving a national consensus on this highly controversial issue. And the BRC's exhaustive effort to conduct its business in a transparent way and to solicit and seriously consider public input was apparent. UCS staff had the opportunity to testify three times before the BRC and also to participate in more informal BRC-sponsored forums.

UCS has reviewed the eight recommendations in the final report and agrees with most of them. However, perhaps our greatest area of agreement concerns the *absence* of a recommendation. The BRC, after careful consideration, did <u>not</u> recommend that the United States reverse a 35- year precedent and proceed immediately with development of facilities for spent fuel reprocessing and plutonium fuel production and use. UCS strongly concurs with the BRC's conclusion that "no currently available or reasonably foreseeable reactor and fuel cycle technology developments --- including advances in reprocessing and recycling technologies --- have the potential to fundamentally alter the waste management challenge over at least the next several decades, if not longer."

UCS has long opposed reprocessing because it produces plutonium and other materials that could be used in nuclear weapons, greatly increasing the risks of nuclear terrorism and proliferation, yet provides no benefits for radioactive waste management. In contrast, reprocessing actually worsens the radioactive waste disposal problem. For instance, the Energy Department calculated in a 2008 draft environmental impact statement that a 50-year reprocessing program would only reduce the volume of high-level waste by 15,000 cubic meters compared to the once-through cycle, while generating an additional 400,000 cubic meters of greater-than-class C low level waste, a category of waste that itself will likely require deep geologic disposal.

UCS believes that if the BRC had endorsed reprocessing, it would have sent the wrong message to the rest of the world, undermining efforts to control the growth of weapon-usable material stockpiles. For instance, Japan is on the verge of restarting its reprocessing plant at Rokkasho-mura, a troubled $20 billion project that has been buffeted by technical problems, massive cost escalation and, since the Fukushima accident, renewed concerns about its vulnerability to accidents and severe natural phenomena. Japan has already accumulated 45 metric tons of plutonium from overseas and domestic reprocessing operations, of which 10 metric tons— enough for more than one thousand Nagasaki-type nuclear weapons—is on Japanese territory. Japan will be unable to use any of this plutonium in its nuclear reactors for the foreseeable future because of public doubts about reactor safety in the wake of Fukushima, and the technical failure of the Monju experimental fast breeder reactor program. Thus Japan does not need to add to its stockpile of separated plutonium by resuming reprocessing. UCS appreciates that the BRC report will give no support to advocates of a reprocessing restart in Japan.

Concerning BRC recommendation 2, creation of a new congressionally chartered federal corporation for managing the disposal of spent fuel and high-level waste, UCS supports creation of a new entity that is independent of DOE, fully transparent in its deliberations and decision-making, and free of undue influence from any of the multiple stakeholders that it must serve. Most importantly, however, the entity's operation should be strictly limited to the activities recommended by the BRC: transport, storage and direct disposal of spent fuel and high-level waste, with limited research and development as needed to support the safety and security of those activities. The entity should not be given any authority to use the Nuclear Waste Fund or any other funds to conduct research, development or deployment of reprocessing plants or any other fuel cycle technology or facility not needed for direct disposal of spent fuel and high-level waste. In any event, the huge additional cost of such activities would require a significant increase in the waste fee assessment that would be unpopular among ratepayers.

One area where UCS disagrees with the BRC recommendations concerns its strong endorsement of prompt efforts to develop centralized interim storage facilities (Recommendation 5). UCS is not persuaded that new legislation and other actions to facilitate the siting and development of consolidated interim storage facilities are necessary, either for spent fuel from operating reactors or from shutdown reactors. The argument for consolidating spent fuel from shutdown reactors is more compelling than for fuel from operating reactors, but UCS has yet to see an analysis clearly demonstrating that the benefits of interim storage outweigh the additional costs and risks associated with siting and licensing new storage facilities and the additional transportation that would be required—even for spent fuel from shutdown reactors. An alternative that might be more desirable would be to arrange to ship spent fuel from each shutdown reactor to the nearest operating reactor that has the space to accommodate it, thus eliminating the need to license greenfield facilities, capitalizing on existing infrastructure and reducing transport distances.

It is not apparent that siting a consolidated interim storage facility would be any easier politically to achieve than siting a geologic repository. Prospective host communities for new centralized storage sites would likely demand significant incentives, such as new research and development facilities, in exchange for their acceptance. Such costly incentives would best be reserved for potential repository host communities, as there is unlikely to be enough funding to support multiple endeavors. Also, efforts to site interim storage facilities could distract from or even derail the far more important goal of finding a repository site. There was a good reason why the 1987 Nuclear Waste Policy Act amendments linked construction of a monitored retrievable storage facility to progress on licensing a repository; UCS does not support the BRC's proposal to sever that link. UCS also does not agree that the "flexibility" a retrievable interim storage facility could provide is necessarily a desirable property, should that flexibility facilitate reprocessing of spent fuel in the future. We believe that the principle of intergenerational equity requires that action must be taken today to preclude easy access in the future to the plutonium in spent fuel, which will become more vulnerable over time as the spent fuel radiation barrier provided by cesium-137 decays away. This can best be accomplished by direct geologic disposal of spent fuel as soon as practicable.

UCS believes that spent fuel can be managed safely at reactor sites provided that the Nuclear Regulatory Commission appropriately upgrades its requirements to minimize the safety and security risks associated with long-term (up to 100 years) storage at reactors. To this end, we support the BRC's call for a new review by the National Academy of Sciences (NAS) of the safety and security issues associated with spent fuel storage, both in wet pools and in dry casks. This review should consider all that the NRC has—or hasn't—done since the 2006 NAS study on spent fuel security to address the risk of a zirconium fire and widespread fuel damage at densely packed spent fuel pools. As was the case at the time of the 2006 study, much of the information associated with this issue is classified. Now, however, there should be additional efforts to declassify the information necessary to fully inform Americans of the risks they face from overstuffed spent fuel pools in the event of a terrorist attack or severe accident. Any lessons learned from Fukushima, where the spent fuel pools were not nearly as full as those at U.S. plants, will have to be interpreted appropriately for the U.S. case.

Although a new NAS study would be useful for a number of reasons, we do not believe that more study is needed to support a new requirement by the NRC to thin out densely packed spent fuel pools by accelerating transfer to dry cask storage. With regard to addressing the potential risk of a zirconium fire in a spent fuel pool, NRC Chairman Gregory Jaczko commented at an October 2011 meeting that it "should be an issue we should have a handle on today, there really is no excuse for that. This came up in 9/11, we've done experiments so, I think if we do this the way we've always done things we will not get these things done in a reasonable period of time ..." In other words, NRC appears to already have sufficient information. All it needs now is the political will to follow through and do what is necessary to protect the public.

The NRC must also comprehensively address the potential sabotage threat to dry storage casks and transport casks. It must consider a wide variety of plausible attack modes that could lead to significant radiological releases when setting its requirements for physical protection systems for dry cask storage facilities and spent fuel shipments.

With regard to BRC Recommendation 7, UCS supports a limited program for nuclear energy research and development. However, it does not support BRC's endorsement of a major DOE research and development program on spent fuel reprocessing and related technologies. UCS maintains that the proliferation, nuclear terrorism and environmental risks posed by reprocessing-based fuel cycles are so intractable that continuing to spend scarce taxpayer dollars on studying these systems is a clear case of throwing good money after bad. Instead, we believe that taxpayer funded R&D needs to focus on enhancing the safety, security and effectiveness of nuclear plants and the once-through fuel cycle, and the safe interim storage, handling, transportation and direct geologic disposal of spent fuel.

In the current and foreseeable fiscal climate, DOE should not continue to spend money on failed technologies, such as actinide-burning fast reactors, that cannot meet basic waste management objectives even if the systems were to perform perfectly. For instance, the BRC points out that "many decades to a couple of centuries" would be needed to decrease required repository space by 75% in a fast-reactor based closed fuel cycle, and that this is fundamentally due to the low rates of consumption of plutonium and other long-lived elements in fast reactors. We believe that this fact illustrates the futility of such approaches, as well as their incompatibility with the principle of intergenerational equity. DOE has already spent decades and many millions of dollars studying these systems even though their limitations were widely known. DOE also continues to research advanced reprocessing technologies that it calls "proliferation-resistant," even though the U.S. nuclear weapons labs have concluded there is little value to such approaches. Yet the BRC apparently ignores this history, specifically citing "fast-spectrum reactors ... capable of continuous actinide recycling" as a good example of potential "game-changing" technologies worthy of further R&D.

For this reason, an external, independent peer review process for DOE fuel cycle R&D should be established by an entity such as the NAS. Simply relying on a quadrennial internal review, as the report recommends, is not sufficient. The review should be based on clear and quantitative objectives and milestones, and should reject technologies without a realistic chance of achieving program goals, such as actinide "recycling."

Finally, UCS agrees with Recommendation 8 that U.S. leadership is an important factor in promoting safety, nonproliferation and security, and believes that the best approach is for the United States to lead by example. With regard to the nuclear fuel cycle, the most valuable signal the United States could send to the rest of the world is the demonstration that direct disposal of spent fuel in a nation with a very large nuclear power program is both politically and technically feasible. In addition, this would show the rest of the world that reprocessing spent fuel as a waste management strategy is neither necessary nor desirable.

UCS supports the concept of multi-national fuel cycle facilities with regard to those facilities needed for the once-through fuel cycle, such as uranium conversion, uranium enrichment and uranium fuel fabrication. However, UCS does not believe that a multi-national model could mitigate the profound proliferation and nuclear terrorism risks associated with spent fuel reprocessing plants, MOX fuel fabrication plants and other facilities that produce or process separated weapon-usable materials.

In particular, the threats of sub-national diversion or theft would not be effectively addressed merely by adopting a multi-national approach, because they would be as challenging to control at a multi-national facility as they would at a national facility. Such arrangements would also involve the international transport of weapon-usable materials such as MOX fuel, presenting additional opportunities for theft.

Thank you for your attention. I would be happy to answer your questions.

In: United States Nuclear Waste
Editors: Gregory Hall and Steven L. Greene

ISBN: 978-1-62081-312-6

Chapter 7

TESTIMONY OF THOMAS A. SCHATZ, PRESIDENT, CITIZENS AGAINST GOVERNMENT WASTE. HEARING ON "AMERICA'S NUCLEAR FUTURE"*

My name is Thomas A. Schatz and I am president of Citizens Against Government Waste (CAGW). CAGW was founded in 1984 by the late industrialist J. Peter Grace and nationally-syndicated columnist Jack Anderson to build support for implementation of President Ronald Reagan's Grace Commission recommendations and other waste-cutting proposals. Since its inception, CAGW has been at the forefront of the fight for efficiency, economy, and accountability in government. CAGW has more than one million members and supporters nationwide, and, over the past 28 years, has helped save taxpayers $1.2 trillion through the implementation of Grace Commission findings and other recommendations.

CAGW's mission reflects the interests of taxpayers. All citizens benefit when government programs work cost-effectively, when deficit spending is reduced and government is held accountable. Not only will representative government benefit from the pursuit of these interests, but the country will prosper economically because government mismanagement, fiscal profligacy, and chronic deficits soak up private savings and crowd out the private investment necessary for long-term growth.

Since the 1970s, the U.S. has been searching for a long-term site to dispose of its nuclear waste. As far back as 1957, and in as many as 50 separate official reports and scholarly essays, including several by the National Academy of Sciences, disposing of nuclear waste in deep geologic repositories has been recognized as the best solution. This method is also the state-of-the-art method of nuclear waste disposal internationally. It is the preferred method of disposal for long-lived radioactive waste in countries such as Australia, Belgium, France, the Netherlands, Russia, Spain, Sweden, and Switzerland.

In 1983, President Ronald Reagan signed the Nuclear Waste Policy Act of 1982 (NWPA), which acknowledged that disposing of spent nuclear fuel was a national priority. The bill authorized the Department of Energy (DOE) to begin searching for an appropriate deep geologic repository. Nine sites were reviewed and in 1987, Congress amended the NWPA and directed DOE to focus only on Yucca Mountain, Nevada, a site which is 100 miles north of Las Vegas on the Nevada Test Site, where more than 800 nuclear weapons

* This is an edited, reformatted and augmented version of testimony given on February 1, 2012, before the House Subcommittee on Environment and the Economy.

tests were conducted during the Cold War. Yucca Mountain is virtually surrounded by Nellis Air Force Base. It was chosen for its extreme isolation, arid conditions, and the prospect of being able to sequester the spent fuel beneath 1,000 feet of solid rock, yet 1,000 feet above the water table.

The 1987 amendment also directed the DOE to begin entering into contracts with commercial nuclear reactor operators to take custody of their spent nuclear fuel for disposal at the repository beginning in January 1998. Former Energy Secretary Spencer Abraham certified the suitability of Yucca Mountain in February, 2002 and that recommendation was forwarded to Congress for a final decision. Congress approved the site in July 2002.

As directed in the NWPA, nuclear reactor operators and their millions of customers have been paying fees into the Nuclear Waste Fund (NWF) since 1983, in exchange for the government's guarantee that the revenue collected would be used to create a safe, long-term repository for the spent nuclear fuel. Through assessments in their utility bills, ratepayers have contributed between $750 and $780 million each year since 1983 into the NWF. DOE has spent $15 billion to evaluate various possible sites, to develop Yucca Mountain, and to submit the licensing application.

Yet, today, the national inventory of spent nuclear fuel stands at 65,000 metric tons and not one spent fuel rod has been moved to the Yucca Mountain facility. The spent fuel languishes at 75 sites in 33 states, stored either in cooling pools or, when the pools have reached capacity, in expensive dry cask storage facilities adjacent to operational reactor sites.

As a presidential candidate, Barack Obama pledged to voters (and Nevada voters in particular) that, if elected, he would do all that he could to make sure the Yucca Mountain project never saw the light of day. In a May 20, 2007 letter to the editor of the *Nevada Review Journal*, candidate Obama stated, "I believe all spending on Yucca Mountain should be redirected to other uses...All Nevadans should know that as president, I will bring to this issue not just independent judgment and careful deliberation, but a personal appreciation that comes from my own experience of living in the back yard of hazardous nuclear materials."

It became official on January 29, 2010, when the White House's top energy adviser, Carol Browner declared that "The debate over Yucca Mountain is over as the president has made clear...We're done with Yucca. We need to be looking at other alternatives." The DOE under the Obama administration moved to terminate the Office of Civilian Radioactive Waste Management and the Yucca Mountain repository project. The president's FY 2011 budget contained no further funding for the project. DOE Secretary Stephen Chu, in his statement explaining the decision to terminate the facility, did not mention safety or environmental issues as the rationale for shuttering the operations. Instead, the administration simply stated that the decision was predicated upon "a proposed change of department policy for managing spent nuclear fuel." That is quintessential Washington double talk, meaning "log rolling and politics as usual."

The DOE's announcement to terminate Yucca came with a prototypical kick-the-can-down the-road tactic of calling for more taxpayer-funded studies and the establishment of the Blue Ribbon Commission (BRC) on America's Nuclear Future. Although the BRC was led by two distinguished veterans of public policy and public service, it was predisposed to spin its wheels, since it was prohibited by the administration from reviewing the administration's decision on Yucca or to revisit the suitability of the site.

Consequently, the BRC spent another more time and additional scarce federal resources to reiterate the obvious; that national nuclear waste storage and disposal policies have deadlocked over the pressures of political favoritism rather than following science and technology, even as demand for electricity rises, even as stockpiles of spent fuel become more mountainous, even as lawsuits and tens of billions of dollars worth of judgments have sapped the U.S. Treasury, and even as utility ratepayers are still being forced to pay more than $750 million in annual payments into a fund that has a $26.7 billion balance and earns annual interest in excess of $1 billion.

Ironically, the commission itself cost taxpayers $5 million, and in the two years it took to deliver its litany of tepid recommendations, the judgment fund paid out another $2 billion in liability payments.

The panel's final report, which came out on January 26, 2012, made eight recommendations, but offered little that is new. The report admits that it was "not chartered as a siting commission. Accordingly, we have not evaluated Yucca Mountain or any other location as a potential site for the storage or disposal of spent nuclear fuel and high-level waste, nor have we taken a position on the Administration's request to withdraw the license application."

One of the commission's top recommendations is that, next time around, lawmakers and DOE officials should adopt a "new consent-based approach for siting nuclear waste facilities."

While it is good public policy and completely appropriate to ensure that DOE officials and lawmakers fully engage with state and local officials, it is worth remembering that, years ago, Yucca Mountain had support in the state of Nevada. In fact, in 1975, the Nevada state legislature strongly urged federal officials to choose the Nevada Test Site for the storage and processing of nuclear material. The Yucca Mountain project owes its ultimate demise to years of delays, manipulation, and obstructionism by Senate Majority Leader Harry Reid and the exigencies of election-year politics. Nuclear waste policy decisions are complex and should be driven, first and foremost, by scientific and technological rigor.

The BRC report offers the usual platitudes about the urgency of the nuclear waste disposal issue, supports more public funding for research and development, continued preparation for large-scale transportation challenges, and, as always, more taxpayer money for research and development.

Congress bears an enormous amount of responsibility for the Yucca Mountain program's current status. While members of Congress meddled, apparently more interested in using the NWF fees to spend on other projects or to make the deficit numbers look better, the spent nuclear fuel piled up at operational reactors across the country, forcing utility companies to construct temporary storage facilities to house the material until the Yucca facility was ready. The BRC commissioners implicitly acknowledge that Congress and the administration can no longer be counted upon to handle the issue without rank politicization and mismanagement, and call for the establishment of a federally-chartered corporation to run the nuclear waste effort.

The BRC states that it believes "a congressionally chartered federal corporation offers the best model, but whatever the specific form of the new organization it must possess the attributes, independence, and resources to effectively carry out its mission. The central task of the new organization would be to site, license, build, and operate facilities for the safe consolidated storage and final disposal of spent fuel and high-level nuclear waste at a reasonable cost and within a reasonable timeframe."

Termination of the Yucca project poses serious challenges to the funding of any future disposal site. According to the Government Accountability Office's (GAO) April 2011 report, "If DOE were to pursue an alternate repository – assuming an alternate repository would have costs similar to the Yucca Mountain repository – it is not certain that the fund will have built up a sufficient surplus to site, license, construct, and operate it. DOE makes an annual assessment of the adequacy of the nuclear waste fund to ensure that full costs of a disposal program will be fully recovered.

"In November 2010, the Secretary determined that the fund was adequate, even though an attachment stated that DOE had no alternative to the Yucca Mountain repository, and that the Yucca Mountain repository provided the closest 'proxy' – in terms of cost – to an alternative. If the nuclear waste fund does not have a sufficient surplus for an alternate repository, additional funding would have to be found... Moreover, since the taxpayers have paid a proportion of the costs to establish a repository for DOE-managed high-level waste and spent nuclear fuel, the taxpayers may also end up paying more for an alternate repository. In addition, the proposed termination has prompted calls from industry for DOE to suspend collection of payments into the Nuclear Waste Fund. Industry has argued that their customers should not pay for a repository effort that has been shut down, with no work being done on an alternative. Suspending payments into the Nuclear Waste Fund could reduce the funds set aside for a repository."

Although CAGW agrees with the BRC that the funds in the NWF should no longer be made to go through the appropriations process, where the money has subjected to years of inappropriate diversion by Congress, the BRC recommendations do not have the force of law. The NWPA is still the law; Yucca Mountain is still the only repository available; and any changes to the siting procedures and funding mechanisms would require congressional action.

Questions remain about whether there should be an immediate suspension of payments by utilities and ratepayers into the NWF until such time as the courts decide whether or not Yucca will be resurrected or permanently moribund and whether Congress should prohibit the use of the Judgment Fund to pay any liabilities of the DOE resulting from litigation. Taxpayers are now left with deep and abiding questions of whether the administration has the legal authority to terminate Yucca and understandable suspicions about whether, given the failures of the current administration to abide by current law and Congress's predilection for injecting politics into every nook and cranny of public policymaking, the next go-around won't be just another exorbitantly expensive sequel to the first one.

The stockpile of spent nuclear fuel increases by about 2,000 metric tons each year, and according to the GAO, the volume of commercial spent nuclear fuel is projected to more than double by the year 2055. The NWF fund balance stands at $26.7 billion as of September 30, 2011 and ratepayers continue to be forced to pay into the fund. But those expenditures only scratch the surface of the total costs associated with the development and subsequent rejection of Yucca Mountain, not to mention the cost of starting all over.

Originally, utility companies and their consumers were promised that the Yucca Mountain site would begin accepting spent fuel in 1998. Almost from the start, Congress began siphoning and interfering with the project's funding. For example, Edward Sproat, then-Director of the Office of Civilian Radioactive Waste Management, testified on October 4, 2007 before the House Budget Committee that "The projected budget authority needed through repository construction is well above current and historic levels, and the current funding levels are insufficient to build the repository and the transportation system. If the

Program is funded at its current levels without fixing the current funding mechanism, the shortfall in the funding needed would be between $1.0 billion and $1.5 billion per year. This funding shortfall will not allow the placement of the design and construction contracts for the repository or the transportation systems. In short, DOE will not be able to execute its responsibilities under the Nuclear Waste Policy Act and will not be able to set a date for meeting its contractual obligations. Government liability will continue to grow with no apparent limit... So basically, unless we fix this issue and are able to achieve these cash flows to build this repository, the liability of the Federal Government and the U.S. taxpayers will continue to grow unabated."

Director Sproat warned the Budget Committee that, even at that point, had the Yucca Mountain facility begun accepting spent fuel in 2017, nineteen years beyond its statutory deadline, the DOE's liabilities for the delays would have lasted far beyond the date on which the facility became operational.

Director Sproat also pointed out that Congress had been starving the Yucca Mountain project of necessary funds, which triggered a cascade of delays and deferrals that invited costly lawsuits from utility companies. He pointed out that the budget rules required NWF revenue fees to be classified as mandatory receipts, while the program funding would be classified as discretionary. Sproat noted that "the fees are not dedicated to offset the appropriations on the program, and therefore, both the fees and the interest generated on the fund are used as offsets against the total federal deficit. Essentially, we have $20.5 billion in the Nuclear Waste Fund, but we are not able to use it for its intended purpose to actually build the repository."

The first lawsuit was filed in 1998. The U.S. Court of Federal Claims found that the DOE had failed to begin accepting spent nuclear fuel from nuclear power plants when Yucca Mountain was supposed to be operational. That suit was the first of 74 separate lawsuits to date against the agency for its abject failure to comply with the provisions of the NWPA. In each of those cases, the courts have ordered the DOE to compensate the utilities for failure to begin moving the spent fuel to Yucca, as mandated by law, and to reimburse them for the cost of storing the fuel at the reactor sites. Those fees come straight out the taxpayers' pockets.

If that were not bad enough, GAO estimates that the Department of Justice spent $188 million by the end of 2011 just to defend the DOE in court, unsuccessfully, against the suits. That excludes the cost of DOJ and DOE staff. The DOJ anticipates as many as six more lawsuits could move forward in 2012.

According to a December 22, 2009 Congressional Research Service (CRS) report, "...contract damages will continue to build as there seems to be no prospect for a completed facility capable of storing [spent nuclear fuel] anywhere on the horizon."

Although Yucca Mountain has been terminated, the fiscal fallout of this convoluted and costly tale of governmental and congressional ineptitude will hang over taxpayers and eat away at the federal budget for decades. The DOE's latest estimates for taxpayer liability for the agency's failure to comply with the NWPA are stunning.

To date, the Treasury Department's Judgment fund has paid $1.7 billion to settle cases for those utilities which have incurred damages as a result of the DOE's delays. The DOE's FY 2011 Audit Report stated that "Additional payments under these settled and adjudicated cases may be made if the utilities incur additional costs before the Department permanently disposes of the spent nuclear fuel. The Department believes its assumptions and methodology provide a reasonable basis for the contingent liability estimate."

Utility industry estimates for the ultimate liability costs to taxpayers, including what has already been paid out, are in the range of $50 billion. DOE officials dismiss those predictions as highly inflated and peg the losses at closer to $20.7 billion. However, the department's estimates are predicated upon the promise that Yucca Mountain would have begun accepting spent fuel for disposal in 2020. Now that the administration has shuttered Yucca and terminated the program, the 2020 assumption is null and void and it is clear that the taxpayer liability will be exponentially higher than the DOE's estimates; in fact, it is certainly arguable now that even the industry estimates of $50 billion might be too low.

In a June 10, 2011 report accompanying the FY 2012 Energy and Water Appropriations bill, the House Appropriations Committee castigated the administration's plans to shut down Yucca, noting that "The Department of Energy now estimates that taxpayers will have to pay nearly $16.2 billion in damages by 2020, and an additional $500 million for each year after 2020 that the Department does not fulfill its legal obligations." According to a March 16, 2011 article in the *Orange County Register*, "Payouts to nuclear plant operators – to essentially cover their costs for storing the spent nuclear fuel that the government was supposed to handle – could total as much as $50 billion."

Adding to the tens of billions of dollars in liability costs for nuclear power facilities and their ratepayers associated with the abandonment of the Yucca Mountain, the ill-advised decision to close the facility has also left the U.S. Navy with a serious conundrum.

The DOE has an agreement with five states, including Idaho, and the Navy has a separate agreement with Idaho, regarding the storage and disposal of nuclear waste at DOE sites. The DOE promised to remove the waste from Colorado and the DOE and the Navy promised to remove the waste from Idaho by January 1, 2035. The two states have penalties, respectively, or $15,000 per day and $60,000 per day for each day the waste remains in the state after that date.

According to a May 5, 2011 GAO report, the penalties for failure to comply with the contractual obligations in Idaho and Colorado could amount to about $27.4 million annually. The Navy has no other location for its waste disposal, and the GAO quotes Navy officials as being most concerned about the fact that the failure to open Yucca Mountain in a timely manner could "interfere with the Navy's ability to refuel its nuclear warships...it would likely extend on-site storage and increase storage costs, which could be substantial."

There are additional costs for extended storage at the sites in those states. A DOE Office of Environmental Management analysis estimates that it would need another $918 million in appropriations to extend storage, again, assuming a 20-year delay in a repository's opening. Since there is now no known alternative to Yucca Mountain and the administration intends to begin this process from scratch, the total additional storage costs stemming from terminating the repository will certainly be higher than that.

Not only is the spent nuclear fuel piling up around the country, so are the massive costs associated with this blatantly politicized decision. What confidence can the states have that DOE will make good on its commitment to them when the administration and the Congress have so casually disregarded their other legal obligations?

The saga of Yucca Mountain has been one of gross disregard for the demands of nuclear-generated electricity, which constitutes 20 percent of the electricity generated in the country; for the ratepayers, who have contributed and continue to contribute billions to what was supposed to be a lockbox to build a spent nuclear fuel repository; and to the taxpayers, who will pay for the debacle far into the future. Congress and the DOE have squandered not only the money in the trust fund, but also the trust of the ratepayers and the taxpayers.

Notwithstanding the politics of the issue, the United States must, at some point, construct a deep, geologic disposal repository and the nation cannot afford to begin at square one. The GAO has signaled that the death of the Yucca Mountain project could delay the opening of a new waste disposal site by more than 20 years. The Waste Isolation Pilot Plant (WIPP) just outside Carlsbad, New Mexico, is the only operating deep geological waste repository in the United States, and it is not suitable as a long-term solution for all nuclear waste. That site was chosen by the U.S. Atomic Energy Commission, now the Department of Energy (DOE), in 1974. In 1979, DOE estimated that the total cost for the WIPP, both for construction and to operate the facility for 25 to 25 years, would be $1.1 billion. By the time the facility opened and took its first shipment in 1999, construction alone had cost approximately $3 billion, three times the original estimate.

Furthermore, when it comes to Yucca Mountain and nuclear power, the Obama administration has transmitted a series of decidedly mixed messages. A few weeks after the announcement to ax funds for Yucca and pull the plug on its licensing application, President Obama announced his support for $54 billion in federal loan guarantees to build new nuclear power plants. The President, who has been described as a "resolute advocate" for nuclear power, claims he has never wavered in his support for nuclear power, and that he supports accelerating the nuclear licensing process. This policy is patently incoherent and irresponsible, given the administration's do-nothing attitude toward managing the spent fuel at the back end.

In addition, there appears to be no discernible plan in sight for the disposition of spent nuclear fuel currently located at 121 sites in 39 states around the country. This includes the Savannah River nuclear complex in South Carolina, which currently manages 36 million gallons of high-level liquid radioactive waste, and the Hanford site in Washington, which has 56 million gallons of high-level radioactive tank waste, as well as nuclear fuel. All of that waste was lined up to go to Yucca.

There are countless problems associated with the politicized and short-sighted decision to abandon Yucca Mountain. First, and perhaps most importantly, it contravenes the law. The NWPA, as amended, clearly designates Yucca Mountain as the only permanent nuclear waste storage facility and any alteration of that location will require a legislative change to the NWPA. Neither Congress nor the Obama administration has moved to amend the Act.

In its April, 2011 report, the GAO suggested that in any future attempts to address the nuclear waste disposal issue Congress must consider infusing the entire process with more transparency, predictability, and independence. With GAO's recommendation in mind, CAGW respectfully requests that the House Committee on Science, Space, and Technology release an unredacted copy of the Safety Evaluation Report (SER), Volume 3, which it currently has in its possession. If they will not release it to the public, the committee should publicize the legal rationale it is relying upon for withholding the report.

Taxpayers and ratepayers have paid tens of billions of dollars over the last 25 years and will pay tens of billions more in the future for a national nuclear waste repository. They have been dismissively informed that Yucca Mountain is simply no longer "workable." They deserve to know now exactly what the DOE discovered and what the SER has to say about its comprehensive review of the Yucca Mountain application in terms of the site's suitability and its "workability."

This troubled saga has destroyed the public's confidence in the government's ability to provide a safe and timely solution for the disposal of spent nuclear fuel. Whether Yucca Mountain is reopened or the taxpayers are forced to initiate another costly and time-consuming search, as long as the process and the funds that attend that process are accessible for manipulation by politicians, there is little reason to believe that the taxpayers and ratepayers won't be forced to buy into yet another boondoggle that will cost more than projected and arrive later than promised, if ever. The entire process has set an appalling legal precedent; DOE has repeatedly been found guilty in court of breaching its contractual obligations to millions of ratepayers and their utility companies, yet the department has been allowed to impose the cost of those verdicts back onto the backs of taxpayers.

There are already too many examples of the federal government breaching, sidestepping, and blatantly ignoring its contractual obligations, even thumbing its nose at the law, as in the case of Yucca Mountain. Even more insidiously, the Yucca Mortmain debacle graphically illustrates a deepening cultural flaw which is becoming almost organic in the federal government. Lawmakers and executive branch officials appear to be increasingly ill-equipped and unwilling, often for shallow political reasons, to address larger national issues. Year in and year out, lawmakers and government officials are permitted to shy away from their constitutionally-mandated obligations to grapple with momentous national issues until they have become even more damaging and destructive to our country's bottom line and to the rule of law.

Legislative action on vital public policy issues is increasingly pushed off to bloated commissions, indolent study committees, portentous but impotent boards, and unaccountable regulators who patch together short-term quick fixes, all paid for with taxpayer dollars. More often, nothing at all is done and the problems fester. It would be difficult to come up with a more pertinent example of this kind of reckless Washington behavior, and certainly not one as costly and egregious as Yucca Mountain.

The BRC report included the following statement: "This generation has a fundamental, ethical obligation to avoid burdening future generations with the entire task of finding a safe, permanent solution for managing hazardous nuclear materials they had no part in creating." That statement is just as true about the staggering financial liabilities related to this epic failure as it is regarding the spent nuclear fuel. Without a rational plan for dealing with nuclear waste disposal, which had already been decided and enacted into law in 1982, the viability of future energy policy, along with the fiscal well-being of our children and grandchildren, will remain tenuous.

In: United States Nuclear Waste
Editors: Gregory Hall and Steven L. Greene

ISBN: 978-1-62081-312-6

Chapter 8

TESTIMONY OF THE HONORABLE DAVID A. WRIGHT, PRESIDENT, NATIONAL ASSOCIATION OF REGULATORY UTILITY COMMISSIONERS. HEARING ON "AMERICA'S NUCLEAR FUTURE"*

Good Morning, Chairman Shimkus, Ranking Member Green, and Subcommittee Members. Thank you for the opportunity to appear before you today.

My name is David Wright. I am a commissioner with the South Carolina Public Service Commission and I serve as president of the National Association of Regulatory Utility Commissioners (NARUC), on whose behalf I am speaking this morning. I appreciate the opportunity to present NARUC's views on the subject of disposition of spent or used nuclear fuel from commercial nuclear power plants.

NARUC is a quasi-governmental, non-profit organization founded in 1889. Our membership includes the public utility commissions serving all States and territories. NARUC's mission is to serve the public interest by improving the quality and effectiveness of public utility regulation. Our members regulate the retail rates and services of electric, gas, water, and telephone utilities. We are obligated under the laws of our respective States to assure the establishment and maintenance of such utility services as may be required by the public convenience and necessity and to assure that such services are provided under rates and subject to terms and conditions of service that are just, reasonable, and non-discriminatory.

NARUC and State utility commissions in forty States served by nuclear-generated electricity have been involved in the troubled history of nuclear waste disposal since 1983. That is when the utilities, which own the used fuel, were required by the Nuclear Waste Policy Act to enter into contracts with DOE. Those contracts called for payments of fees for nuclear-generated electricity to the Treasury for deposit into the Nuclear Waste Fund to pay for the cost of disposal of the used fuel beginning in 1998. As you know, that disposal has not happened, but the fee payments continue to be made. Or, as a former Florida utility commissioner summarized the status in 1991, "The government has our money—we have their waste." It is now 20-plus years later and we still have the government's waste. Utility commissioners care because the utilities pass the cost of the fees to their customers through

* This is an edited, reformatted and augmented version of testimony given on February 1, 2012, before the United States, House of Representatives, Committee on Energy and Commerce, Subcommittee on Energy.

their electric bill. In addition, and because of the government's failure to open Yucca, customers, through their rates, have had to pay additional amounts to cover the cost of re-racking of the utility spent fuel pools to accommodate more spent fuel, and have had to pay for on-site dry cask storage as well as the increased security required there. Moreover, all taxpayers, through the Judgment Fund, have had to pay damages for the lawsuits brought to date as well as those to come.

We followed the slow progress of the civilian radioactive waste management program as it met a variety of setbacks and advances, exacerbated by chronic budget cuts even as the illusion of a multi-billion dollar corpus grew in the Nuclear Waste Fund. A significant milestone was met in 2002 when Congress passed the joint resolution approving Yucca Mountain as the site for the geologic repository, subject to the Department of Energy obtaining a construction license from the Nuclear Regulatory Commission. The next setback was the court remand to the Environmental Protection Agency to revise the regulation setting the radiation standard for the facility. Finally, DOE submitted the license application in June 2008. The NRC began its review of the 8,000-page application for the first-of-a-kind facility which was expected to take three to four years.

In 2009, the Administration pronounced Yucca Mountain not to be a "workable option" and that it intended to terminate the repository development there. In March 2010, DOE asked the NRC's Atomic Safety Licensing Board for permission to withdraw the application with prejudice. In June, the ASLB rejected the request, ruling that once a valid license application was submitted under the NWPA, the NRC was required to review and act upon the application. The decision was appealed to the NRC. While the NRC was disposing of the license matter, the President directed that the Secretary of Energy appoint the Blue Ribbon Commission on America's Nuclear Future (BRC) to consider and recommend a new strategy; a strategy that soon became evident would be a "post-Yucca" strategy. In 2010, NARUC, and several other parties, petitioned the Court of Appeals under the NWPA to challenge DOE's authority to withdraw the Yucca Mountain license application, but the case was dismissed because there had been no final agency action by the NRC on the appeal of the Board's decision rejecting DOE's request. After lengthy and unnecessary delays, the NRC Chairman ultimately released a decision. The NWPA mandates that once the Yucca Mountain license was submitted the NRC only had three years to complete the review proceedings. Those three years have expired. Currently, the NRC faces a mandamus action to force it to complete the required review in the United States Court of Appeals for the District of Columbia Circuit. NARUC is one of several petitioners in that suit. Our reply briefs were just filed last Friday. Notwithstanding our position on Yucca, NARUC was closely involved in the work of the BRC. We wrote letters, gave testimony, provided comments on the Subcommittee, and attended most of the public meetings. We were impressed with the distinguished members, their approach to the task, the talented professional staff, and the sincere interest in public input. We have asked DOE to preserve and maintain access to the Commission website. As for the recommendations, while we welcome them all, we have the following points:

1. Reform of the Nuclear Waste Fund is essential for most of the others to occur.
2. Regardless of Yucca Mountain, we need another repository. The lessons of Yucca and the better lessons of Finland, Sweden and WIPP suggest the "consent-based" siting approach may get better results, but will require patience.

3. We have long favored consolidated interim storage, but find the Report vague as to quantity, duration, and cost. We are not sure what the effect will be on the fee if the Nuclear Waste Fund is to be used to pay for storage.
4. We agree with the concept and benefits of a new federal corporation that can focus solely on the waste management mission, hopefully with a fresh partnership attitude for encouraging the consent-based approach. We look forward to refining the concept in enabling legislation.

Transportation planning and coordination with States and others cannot begin soon enough.

We would add that the time is not right to commit to a reprocessing strategy, although R&D should continue, as the BRC recommends. Also, we encourage DOE to take steps to seek volunteer host communities to step forward in storage siting without waiting to form the new management organization.

There are two areas where we disagree with the Commission Report:

1. The Report says: "Overall, we are confident that our waste management recommendations can be implemented using revenue streams already dedicated for this purpose." There are no cost estimates to substantiate that belief, which likely also assumes the $26.7 billion in the Nuclear Waste Fund is assured.
2. The Report further says: "We know what we have to do; we know we have to do it, and we even know how to do it." While we may wish that were true, our assessment is that there are too many people who are content to pass the problem along to future generations and "leave the waste where it is." It is fitting for the Commission to call for prompt action developing both consolidated interim storage and beginning the search for a new repository, but we may need public education and outreach to help persuade some who seem to favor the "no action" alternative. Continuing to "kick the dry cask down the road" should not be an option.

So, yet another study calls for prompt action, yet despite (on paper) a financing plan, implementation relies on leadership from the Administration and Congress. NARUC stands ready to assist on behalf of the ratepayers who may not realize that they are paying for safe waste disposition.

INDEX

#

A

B

C

D

E

F

G

H

I

N

O

P

Q

R

S

T